Intermittency in Transitional Shear Flows

Intermittency in Transitional Shear Flows

Editor

Yohann Duguet

MDPI • Basel • Beijing • Wuhan • Barcelona • Belgrade • Manchester • Tokyo • Cluj • Tianjin

Editor
Yohann Duguet
LIMSI-CNRS
Université Paris Saclay
France

Editorial Office
MDPI
St. Alban-Anlage 66
4052 Basel, Switzerland

This is a reprint of articles from the Special Issue published online in the open access journal *Entropy* (ISSN 1099-4300) (available at: https://www.mdpi.com/journal/entropy/special_issues/ Intermittency_shear_flow).

For citation purposes, cite each article independently as indicated on the article page online and as indicated below:

LastName, A.A.; LastName, B.B.; LastName, C.C. Article Title. *Journal Name* **Year**, *Volume Number*, Page Range.

ISBN 978-3-0365-0942-6 (Hbk)
ISBN 978-3-0365-0943-3 (PDF)

Contents

About the Editor . . . . **vii**

Yohann Duguet
Intermittency in Transitional Shear Flows
Reprinted from: *Entropy* **2021**, *23*, 280, doi:10.3390/e23030280 . . . . **1**

Kerstin Avila and Björn Hof
Second-Order Phase Transition in Counter-Rotating Taylor–Couette Flow Experiment
Reprinted from: *Entropy* **2021**, *23*, 58, doi:10.3390/e23010058 . . . . **3**

Pavan V. Kashyap, Yohann Duguet and Olivier Dauchot
Flow Statistics in the Transitional Regime of Plane Channel Flow
Reprinted from: *Entropy* **2020**, *22*, 1001, doi:10.3390/e22091001 . . . . **15**

Rishav Agrawal, Henry C.-H. Ng, Ethan A. Davies, Jae Sung Park, Michael D. Graham, David J. C. Dennis and Robert J. Poole
Low- and High-Drag Intermittencies in Turbulent Channel Flows
Reprinted from: *Entropy* **2020**, *22*, 1126, doi:10.3390/e22101126 . . . . **31**

Jinsheng Liu, Yue Xiao, Mogeng Li, Jianjun Tao and Shengjin Xu
Intermittency, Moments, and Friction Coefficient during the Subcritical Transition of Channel Flow
Reprinted from: *Entropy* **2020**, *22*, 1399, doi:10.3390/e22121399 . . . . **59**

Xiangkai Xiao and Baofang Song
Kinematics and Dynamics of Turbulent Bands at Low Reynolds Numbers in Channel Flow
Reprinted from: *Entropy* **2020**, *22*, 1167, doi:10.3390/e22101167 . . . . **75**

Kazuki Takeda, Yohann Duguet, and Takahiro Tsukahara
Intermittency and Critical Scaling in Annular Couette Flow
Reprinted from: *Entropy* **2020**, *22*, 988, doi:10.3390/e22090988 . . . . **93**

Hirotaka Morimatsu and Takahiro Tsukahara
Laminar–Turbulent Intermittency in Annular Couette–Poiseuille Flow: Whether a Puff Splits or Not
Reprinted from: *Entropy* **2020**, *22*, 1353, doi:10.3390/e22121353 . . . . **109**

Paul Manneville and Masaki Shimizu
Transitional Channel Flow: A Minimal Stochastic Model
Reprinted from: *Entropy* **2020**, *22*, 1348, doi:10.3390/e22121348 . . . . **127**

Daniel Feldmann, Daniel Morón and Marc Avila
Spatiotemporal Intermittency in Pulsatile Pipe Flow
Reprinted from: *Entropy* **2021**, *23*, 46, doi:10.3390/e23010046 . . . . **153**

About the Editor

Yohann Duguet (Dr), was born on t 21st April 1977 (43 yrs old) and is of French nationality. He graduated from Ecole Centrale de Lyon France in 1999 and achieved his Doctorate in Mechanics at Ecole Centrale de Lyon, France, in 2004. He has been a researcher in hydrodynamics at the CNRS-Université Paris-Saclay (LISN) since 2009.

Editorial

Intermittency in Transitional Shear Flows

Yohann Duguet

LIMSI-CNRS, University Paris-Saclay, F-91400 Orsay, France; duguet@limsi.fr

Received: 9 February 2021; Accepted: 18 February 2021; Published: 26 February 2021

The study of the transition from a laminar to a turbulent flow is as old as the study of turbulence itself. Since the seminal pipe flow experiments of O. Reynolds at the end of the XIXth century, it is understood that turbulent velocity fluctuations do not emerge in a regular way. Instead they appear via intermittent bursts of activity in an otherwise laminar environment [1]. A series of experiments carried out in the last century have demonstrated, in most incompressible fluid flows occurring near solid walls, the existence of a transitional range at the onset of the turbulent regime. This specific parameter range corresponding to low velocities has been labelled by hydraulic engineers the "uncertainty zone" because of the difficulty to perform either deterministic nor statistical prediction of the flow. Only around the end of the XXth century did researchers begin to understand that laminar-turbulent intermittency features a higher degree of organization, in the statistical sense, than previously thought. Yet, technical limitations as well as finite-size effects have made rigorous investigation notoriously difficult because of the different length scales and timescales involved.

The last decade has witnessed a quickly growing number of decisive contributions, made possible by the huge progress in computational power, in experimental measurements and in visualization techniques. Theoretical progress, notably due to an exciting analogy with the thermodynamical formalism of phase transitions, has motivated most of these recent advancements [2]. It is now well established that the transitional range, parameterized by the so-called Reynolds number proportional to the fluid velocity, features a regime of laminar-turbulent patterning. It has been advanced for several decades that the lower transitional range features a continuous transition belonging to the universality class of directed percolation. The hydrodynamical and statistical organization of these coherent structures considered individually remain however not well understood. Finally, there are open issues about to how universal these results are, given the variety of different fluid flow cases.

The goal of the present special issue is to give an up-to-date overview of this cross-disciplinary topic. It contains nine original research articles written by specialists from the most active research teams in the field.

No less than three detailed experimental investigations of the transitional regime of Taylor-Couette and plane channel flow are part of this special issue. The study by K. Avila and B. Hof [3] establishes with minimal finite-size effects that the turbulent fraction evolves continuously with the Reynolds number, rather than discontinuously as often believed. The two experimental investigations of channel flow by J. Liu et al. [4] and by M. Agrawal et al. [5] contain a rich and complementary database on friction fluctuations in channel flow.

A series of careful direct numerical studies explore the dynamics of individual coherent structures at the onset of turbulence, in possible connection with the directed percolation regime expected theoretically. Morimatsu and Tsukahara [6] focus on the mechanisms leading localized turbulent structures in annular Couette-Poiseuille flow to split into two. Takeda et al. [7] verify the existence of a critical range of annular Couette flow using artificial extensions of numerical domains. X. Xiao and B. Song [8] focus on the dynamics of localized turbulent bands at the onset of turbulence in plane channel flow.

The upper transitional range of channel flow features clear oblique patterns, as investigated numerically by P. Kashyap et al. [9]. They demonstrate there an unusual link across the transitional and full-fledged turbulent regime via high-order statistics of the wall shear stress. Low-order modelling covering all these intermittent sub-regimes of plane channel flows is also considered in the contribution by P. Manneville and M. Shimizu, based on the simple concept of cellular automata [10].

Eventually, the special issue includes an original extension of the intermittency concepts to pulsatile flows by D. Feldmann et al. [11]. Possible applications to cardiovascular diseases up a new line of research in connection to biological applications.

This special issue is meant to represent a snapshot of the field at the beginning of this new decade. It aims at fostering interaction and debates, and not at all to close possible debates. Overall, the associated articles represent a timely perspective of the current research in hydrodynamics as well as, more generically, in complexity science. They suggest that the field of intermittent hydrodynamics has now reached the age of maturity.

Funding: This work received no particular funding.

Institutional Review Board Statement: Not applicable.

Informed Consent Statement: Not applicable.

Acknowledgments: We express our thanks to the authors of the above contributions and to the journal Entropy and MDPI for their support during the preparation of the special issue.

Conflicts of Interest: The authors declare no conflict of interest.

References

1. Reynolds, O. An experimental investigation of the circumstances which determine whether the motion of water shall be direct or sinuous, and of the law of resistance in parallel channels. *Philos. Trans. R. Soc. Lond.* **1883**, *174*, 935–982.
2. Manneville, P. Laminar-Turbulent Patterning in Transitional Flows. *Entropy* **2017**, *19*, 316. [CrossRef]
3. Avila, K.; Hof, B. Second-Order Phase Transition in Counter-Rotating Taylor–Couette Flow Experiment. *Entropy* **2020**, *23*, 58. [CrossRef] [PubMed]
4. Liu, J.; Xiao, Y.; Li, M.; Tao, J.; Xu, S. Intermittency, Moments, and Friction Coefficient during the Subcritical Transition of Channel Flow. *Entropy* **2020**, *22*, 1399. [CrossRef] [PubMed]
5. Agrawal, R.; Ng, H.C.-H.; Davis, E.A.; Park, J.S.; Graham, M.D.; Dennis, D.J.; Poole, R.J. Low- and High-Drag Intermittencies in Turbulent Channel Flows. *Entropy* **2020**, *22*, 1126. [CrossRef] [PubMed]
6. Morimatsu, H.; Tsukahara, T. Laminar–Turbulent Intermittency in Annular Couette–Poiseuille Flow: Whether a Puff Splits or Not. *Entropy* **2020**, *22*, 1353. [CrossRef] [PubMed]
7. Takeda, K.; Duguet, Y.; Tsukahara, T. Intermittency and Critical Scaling in Annular Couette Flow. *Entropy* **2020**, *22*, 988. [CrossRef] [PubMed]
8. Xiao, X.; Song, B. Kinematics and Dynamics of Turbulent Bands at Low Reynolds Numbers in Channel Flow. *Entropy* **2020**, *22*, 1167. [CrossRef]
9. Kashyap, P.V.; Duguet, Y.; Dauchot, O. Flow Statistics in the Transitional Regime of Plane Channel Flow. *Entropy* **2020**, *22*, 1001. [CrossRef]
10. Manneville, P.; Shimizu, M. Transitional Channel Flow: A Minimal Stochastic Model. *Entropy* **2020**, *22*, 1348. [CrossRef]
11. Feldmann, D.; Morón, D.; Avila, M. Spatiotemporal Intermittency in Pulsatile Pipe Flow. *Entropy* **2020**, *23*, 46. [CrossRef]

Publisher's Note: MDPI stays neutral with regard to jurisdictional claims in published maps and institutional affiliations.

Article

Second-Order Phase Transition in Counter-Rotating Taylor–Couette Flow Experiment

Kerstin Avila [1,2,3,*] **and Björn Hof** [3]

1   Faculty of Production Engineering, University of Bremen, Badgasteiner Strasse 1, 28359 Bremen, Germany
2   Leibniz Institute for Materials Engineering IWT, Badgasteiner Strasse 3, 28359 Bremen, Germany
3   Institute of Science and Technology Austria, Am Campus 1, 3400 Klosterneuburg, Austria; bhof@ist.ac.at
*   Correspondence: kavila@uni-bremen.de

Received: 4 December 2020; Accepted: 28 December 2020; Published: 31 December 2020

Abstract: In many basic shear flows, such as pipe, Couette, and channel flow, turbulence does not arise from an instability of the laminar state, and both dynamical states co-exist. With decreasing flow speed (i.e., decreasing Reynolds number) the fraction of fluid in laminar motion increases while turbulence recedes and eventually the entire flow relaminarizes. The first step towards understanding the nature of this transition is to determine if the phase change is of either first or second order. In the former case, the turbulent fraction would drop discontinuously to zero as the Reynolds number decreases while in the latter the process would be continuous. For Couette flow, the flow between two parallel plates, earlier studies suggest a discontinuous scenario. In the present study we realize a Couette flow between two concentric cylinders which allows studies to be carried out in large aspect ratios and for extensive observation times. The presented measurements show that the transition in this circular Couette geometry is continuous suggesting that former studies were limited by finite size effects. A further characterization of this transition, in particular its relation to the directed percolation universality class, requires even larger system sizes than presently available.

Keywords: phase transition; Couette flow; lifetimes

---

1. Introduction

In shear flows, turbulence tends to first appear in spatially localized patches that are interspersed by quiescent, laminar regions, a phenomenon commonly referred to as spatio-temporal intermittency. The resulting flow pattern chaotically changes in time and unless the entire flow relaminarises, it never settles to a steady state. One of the earliest reports of laminar turbulent intermittency dates back to Osborne Reynolds and his study of pipe flow [1]. The corresponding turbulent "flashes" or "puffs" are quasi-one-dimensional, meaning that they tend to fill out the radial-azimuthal pipe cross-section, whilst being localized in the streamwise direction [2]. Puffs have a well defined mean length; however, their spacing and hence the size of the laminar gaps is irregular and continuously changes. The resulting overall flow pattern can be accurately modeled as one dimensional [3]. In flows that are extended in two spatial dimensions, but strongly confined in the third (such as channel and Couette flows), turbulence forms elongated stripes [4–7]. Here turbulence fills the wall normal gap and is localized in the extended streamwise and spanwise directions. The resulting laminar-turbulent intermittent stripe pattern can be regarded as quasi-two-dimensional.

In quasi-one- and two-dimensional cases alike, individual patches of turbulence have finite lifetimes and eventually decay. Early propositions that individual turbulent patches (or turbulence in small domains) become sustained at a critical point [8–11] turned out to be incorrect. Despite their often long lifetimes individual patches remain transient and eventually decay following a memoryless process [12–16]. In line with other contact processes such as directed percolation [17] and coupled map lattices [18,19] and as pointed out in the context of shear flows [20,21] spatial proliferation of active sites can give rise to a phase transition to sustained turbulence. Specifically it has been demonstrated for pipe flow [22,23] that turbulence becomes sustained via a contact process where individual localized patches remain transient but can seed new patches before they decay. Also puff splitting has been found to be a memoryless process, a circumstance that allowed to determine the critical point for pipe flow as the balance point between lifetimes and splitting rates [23].

A key remaining question regarding the onset of turbulence, for both one-dimensional and two-dimensional cases alike, is whether the transition is of first or second order (in the context of contact processes and phase transitions in statistical physics, see [24]). In a second-order phase transition, the turbulent fraction decreases continuously to zero as the Reynolds number is decreased toward the critical point, whereas in a first-order phase transition the turbulent fraction jumps from a finite value to zero at the critical point. Hence first-order transitions are referred to as discontinuous and second-order transitions as continuous. In both cases, however, the laminar flow is linearly stable and because of the hysteresis the flow must be initialized with turbulence to measure the transition. While for pipe flow the transition is presumed continuous [3], this so far could not be shown explicitly due to the excessive time scales that prohibit to reach a statistical steady state sufficiently close to the critical point [25]. In a circular Couette experiment of large azimuthal and small axial aspect ratio, where flow patterns like in pipe flow can only evolve in one spatial dimension, the transition has been shown to be continuous [26] and to fall into the directed percolation (DP) universality class.

In an earlier study Bottin and Chatté [8] characterized the transition to turbulence in an experimental study of planar Couette flow in a moderately large aspect-ratio ($190d \times 35d$ in the streamwise and spanwise direction, where d is the gap). In this two dimensional setting, the turbulent fraction was about 30% close to the onset of sustained turbulence and dropped dramatically to zero (laminar flow) as the Reynolds number was reduced. The authors suggested that the onset of turbulence in plane Couette flow corresponds to first-order phase transition. Duguet et al. [27] did direct numerical simulations of a larger system ($400d \times 178d$), but with substantially shorter observation times (2×10^4 advective time units), and reported similar results. More recently, Chantry et al. [28] examined numerically the onset of turbulence in Waleffe flow. In contrast to Couette flow, in this case stress-free boundary conditions are applied at the walls and the flow is driven by a sinusoidal body force. The choice of boundary conditions greatly reduces computational cost and allowed direct numerical simulations of a very large aspect-ratio system ($1280d \times 1280d$) for very long observation times (exceeding 2×10^6 advective time units). Their simulations compellingly show that transition in this simple model system falls in the universality class of two-dimensional directed percolation. While suggestive, it nevertheless remains unclear if for quasi-two-dimensional Couette type flows the transition is either of first or second order. For a recent review of the flow patterns and dynamics of wall-bounded flows extended in two directions, see Tuckerman et al. [7].

In Taylor–Couette flow between two counter-rotating cylinders, the flow dynamics is qualitatively similar to plane Couette flows [4,5,16,29,30] provided that the laminar velocity profile is linearly stable. Indeed, in the narrow-gap limit $\eta = r_i/r_o \to 1$, where r_i and r_o are the radii of the inner and outer cylinders, Taylor–Couette flow turns into rotating plane Couette flow [31]. For fully turbulent flows, the dynamics of Taylor–Couette flow converges to that of rotating plane Couette flow already for moderately small gaps $\eta \geq 0.9$ [32]. By contrast, the dynamics of transition for exactly counter-rotating cylinders is alike that of plane Couette flow only for very narrow gaps $\eta \geq 0.993$ [33]; for larger gaps the linear centrifugal (Rayleigh) instability occurs at lower Reynolds number than the subcritical transition. We note that a new linear instability of counter-rotating Taylor–Couette flow was recently discovered [34], however this instability occurs for extremely high Reynolds numbers (for $\eta > 0.9$, $|Re_o| > 10^8$, where Re_o is the Reynolds number of the outer cylinder) and disappears in the narrow gap limit. This instability is far away in parameter space of the experiments performed here, with $Re_o = O(10^3)$. In Figure 1 we show a regime diagram of counter-rotating Taylor–Couette flow of radius ratio of $\eta = 0.98$. In the infinite-cylinders case, the onset of Taylor vortices is at $Re_i = 292$ when the outer cylinder does not rotate ($Re_o = 0$). For increasing counter-rotation of the outer cylinder, the linear stability threshold rises to higher Re_i and the stability boundary previously measured with our experimental setup [35] is in excellent quantitative agreement with the linear stability analysis of the infinite-cylinder case (solid line in Figure 1), and to a lesser extent also with the experimental measurements of Prigent and Dauchot [36]. For moderately strong counter-rotation ($Re_o < -800$), turbulence can be triggered via finite amplitude perturbations well below the linear instability. Such perturbations occurred naturally in the experimental setup of Prigent and Dauchot [36], whereas in our setup a progressively growing band of hysteresis between the onset of linear instability and the decay of sub-critical turbulence can be observed.

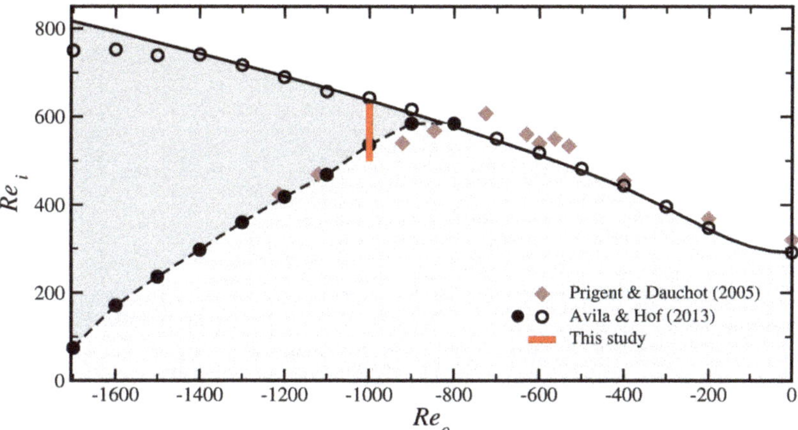

Figure 1. Stability diagram of counter-rotating Taylor–Couette flow with radius ratio $\eta = 0.98$ and stationary lids, $Re_{\text{lids}} = 0$. The solid line shows the linear stability boundary in the infinite–cylinder case. As the the Reynolds number of the outer cylinder (Re_o) is decreased, the linear instability of the laminar, circular Couette flow is shifted to higher Reynolds number of the inner cylinder (Re_i). The empty symbols denote our experimental measurements of the onset of instability, obtained by increasing Re_i at fixed Re_o, which we reported previously in [35]. Subcritical turbulence in the form of turbulent stripes and spots is found in the shaded region starting at $Re_o \approx -800$; the full symbols mark the relaminarization of subcritical turbulence and were obtained by decreasing Re_i at fixed Re_o, in order to detect hysteresis. In this paper, the subcritical transition at $Re_o = -1000$ is analyzed in more detail (statistically) to shed light on the nature of this phase transition to turbulence. For comparison the data of Prigent et al. and coworkers (diamonds) [5,36,37] for a similar radius ratio η are shown, indicating the sensitivity of the flow to finite amplitude perturbations.

In Avila and Hof [35], the critical Reynolds number for self-sustained turbulence was measured by quasi-statically decreasing Re_i in steps of 1 min. This measurement procedure is suited to obtain a rough estimate of the transition border, but does not take into account the stochastic nature of turbulence decay. Measurements of the lifetimes statistics are required here, as previously performed in a small aspect ratio Taylor–Couette flow ($55d \times 34d$) [16]. Compared to all previous quasi-two dimensional Couette or Taylor–Couette experiments, our system's streamwise-spanwise area is at least 12 times larger ($311d \times 263d$), see Table 1. This allows us to study the nature of the turbulence transition with a reduced influence of finite-size effects. We show that lifetimes are exponentially distributed below the critical point and that the increase of the turbulent fraction beyond the critical point is continuous and therefore of second order.

Table 1. Summary of experiments (first four rows) and direct numerical simulations (last four rows) of plane Couette and Taylor–Couette flows in the sub-critical regime. Only published works in which lifetimes were determined statistically and/or the turbulent fraction close to onset was measured are listed. The systems investigated by Lemoult et al. [26] and Shi et al. [38] are quasi-one-dimensional (strongly confined in the spanwise direction). In their experiments and DNS the minimum measurable turbulent fraction is constrained by the streamwise length (instead of the area).

Reference	System	Streamwise Length	Spanwise Length	Area
Bottin and Chatté [8]	pCf ($\eta = 1$)	$190d$	$35d$	$6650d^2$
Borrero et al. [16]	TCf ($\eta = 0.9$)	$55d$	$34d$	$1870d^2$
Lemoult et al. [26]	TCf ($\eta = 0.998$)	$2750d$	$8d$	———
This work	TCf ($\eta = 0.98$)	$311d$	$263d$	$81{,}793d^2$
Duguet et al. [27]	pCf ($\eta = 1$)	$400d$	$178d$	$71{,}200d^2$
Shi et al. [26]	TCf ($\eta = 0.993$)	$480d$	$5d$	———
Lemoult et al. [26]	TCf ($\eta = 0.993$)	$960d$	$5d$	———
Chantry et al. [28]	Waleffe flow	$1280d$	$1280d$	$1{,}638{,}400d^2$

2. Experimental Methods

The Taylor–Couette experiment used in this study consists of two concentric cylinders with radii $r_i = (110.25 \pm 0.025)$ mm and $r_o = (112.53 \pm 0.05)$ mm leading to a radius ratio $\eta = 0.98$ and an azimuthal length of 311 gap width $d = r_o - r_i = 2.28$ mm. The Reynolds number of the inner (outer) cylinder with angular velocity ω_i (ω_o) is defined as $Re_i = \omega_i r_i d / \nu$ ($Re_o = \omega_o r_o d / \nu$), where ν is the kinematic viscosity of the working fluid. The azimuthal direction is in our system the streamwise direction and is naturally periodic (in contrast to Couette flow experiments); this eliminates end effects in the streamwise direction. The axial (spanwise) direction is bounded by the axial lids and has a length of $263d$. The lids can be rotated independently of the cylinders. Their Reynolds number is based on the radius of the outer cylinder ($Re_{\text{lid}} = \omega_{\text{lid}} r_o d / \nu$). In many Taylor–Couette experiments, the lids are attached to the outer cylinder to reduce the Ekman pumping, see, e.g., [16,39–41]. For example, spiral patterns are less influenced by the axial lids, when the lids co-rotate with the outer cylinder, than when they are stationary [42]. The effect of axial boundary conditions was investigated systematically in experiments [43] and in simulations [44], that showed that rotating the lids at angular speeds between the inner and outer cylinder leads to laminar flows closest to circular Couette flow. For our setup and selected parameter regime, $Re_{\text{lids}} = -800$ minimized end effects, but the spatio-temporal dynamics was identical for lids attached to the outer cylinder $Re_{\text{lids}} = -1000$, and for stationary lids $Re_{\text{lids}} = 0$, because of the large height-to-gap aspect ratio. Rotating the lids merely led to a slight stabilization of the laminar flow and hence to a small shift of the onset of turbulence to slightly higher Re_i.

The viscosity of the working fluid silicone oil was determined by measuring the onset of Taylor vortices for stationary outer cylinder as the inner cylinder rotation was increased. Specifically, the value of the viscosity was selected to match the critical inner Reynolds number obtained with a linear stability analysis of laminar, circular Couette flow between infinite cylinders ($Re_{i,c} = 292$ at $Re_o = 0$). The accuracy of this method and of our experiment is verified in the excellent agreement obtained with the linear stability results throughout the counter-rotating regime. In particular, the discrepancy is less than 1% in Re_i when comparing the experimentally measured and the theoretical stability curves. For the visualization of the flow the working fluid silicone oil was seeded with aluminium platelets.

The turbulent fraction was determined by analyzing the images from a high speed camera used to monitor the flow. The flow was seeded with highly reflective aluminum platelets (Eckart, Effect Pigments, STAPA WM Chromal V/80 Aluminum) in a concentration below 1% in weight (and volume). In turbulent flows these tracers are randomly oriented and reflect light efficiently. Turbulent flow patches appear therefore brighter than laminar regions. In our image processing code we use this difference in the light intensity to distinguish laminar from turbulent regions by thresholding. The turbulent fraction is calculated at each instant of time in the spatio-temporal diagrams (see, e.g., Figure 2) as axial length covered by turbulent flow in comparison to the axial length of the field of view. Further details of the image analysis are provided in [35]. Videos were typically recorded with 80 Hz and the resolution in the axial direction was 1920 pixels and in the azimuthal direction between 5 and 1080 pixels, from which only 3 were used for the generation of the spatio-temporal diagrams and hence the quantitative analysis. The measurements shown in this paper consist of three independent sets of experiments with slightly different viscosities and different field of views of the camera, each of them optimized for the specific analysis. For the measurements shown in Figures 1 and 3, the working fluid silicon oil has a viscosity of $\nu = (4.65 \pm 0.02)cSt$. The field of view of the camera in Figure 3 was $(50d \times 80d)$, corresponding to about 10% of the total area and was located $46d$ above the lower lid. For the measurements in Figures 4 and 5 the viscosity was $\nu = (4.55 \pm 0.02)cSt$ and the field of view consisted of a line of 3 pixel width and an axial length of $245d$, which started $5d$ above the lower lid. For the measurements in Figures 6 and 7 the viscosity was $\nu = (4.41 \pm 0.02)cSt$ and the field of view was $(5d \times 170d)$ and started $25d$ above the lower lid. More details of the setup and the image analysis and processing that are omitted here can be found in [35].

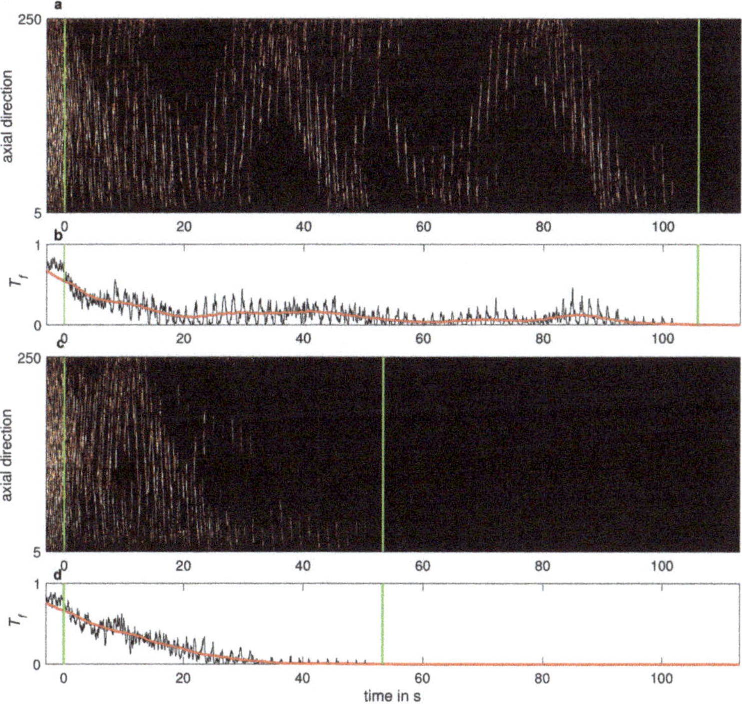

Figure 2. (**a,c**) Spati-temporal dynamics of two selected lifetime measurements at $Re_i = 530$, $Re_{\text{lids}} = -800$. (**b,d**) Corresponding instantaneous (black solid line) and averaged (red thick line) turbulent fraction. The average turbulent fraction is calculated in windows of about 9 s (moving-average technique) to illustrate the long-time dynamics and is used to detect the relaminarization of the flow. The left green line marks the time of the reduction in Re_i and the right green line the decay of turbulence. The determined lifetime corresponds to the time interval between the two green lines.

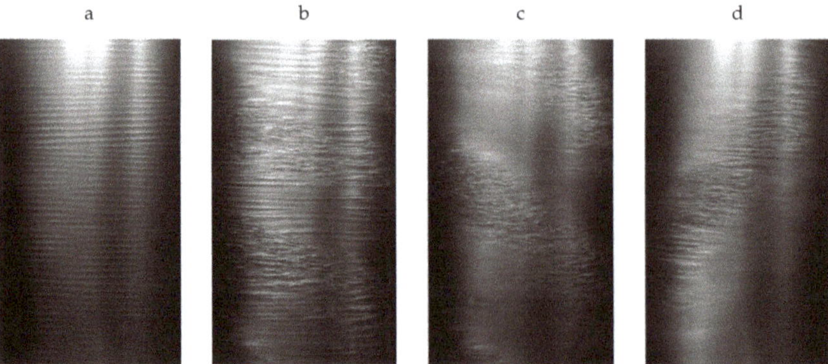

Figure 3. Snapshots of typical flow patterns in counter-rotating Taylor–Couette flow. (**a**) The linear instability arises in the counter-rotating regime in the form of laminar spirals (snapshot taken at $Re_i = 560$, $Re_o = -700$). (**b**) Laminar spirals can coexist with turbulent spots frequently decaying and arising, or they can aligne into stripes, as shown here ($Re_i = 700$, $Re_o = -700$). (**c**) Laminar-turbulent intermittency in the form of subcritical turbulent stripes ($Re_i = 600$, $Re_o = -1000$). (**d**) For decreasing Re_i the regions of laminar flow around the turbulent stripes increase in area ($Re_i = 540$, $Re_o = -1000$). The field of view corresponds here to about 10% of the total system size area. The axial lids are stationary in all snapshots ($Re_{\text{lids}} = 0$). All snapshots were taken in the statistically steady regime.

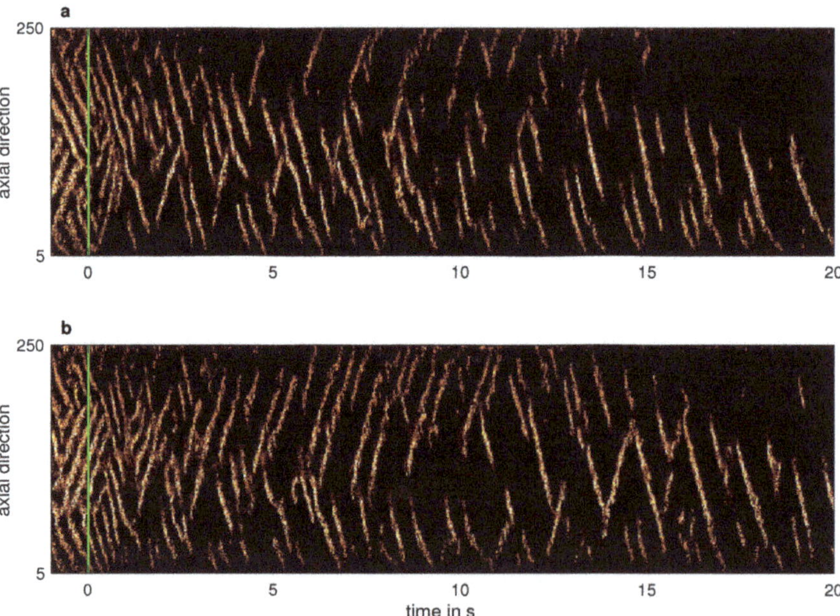

Figure 4. Spatio-temporal dynamics of subcritical turbulence $Re_{\text{lids}} = -800$ (**a**) and $Re_o = -1000$ (**b**) following a reduction in Re_i. Turbulent stripes dominate the dynamics at $Re_i = 630$, prior to an abrupt reduction to $Re_i = 530$ (green line). The turbulent fraction decreases immediately after the reduction in Re_i, but it takes about 20 s for the flow to adjust into a (metastable) statistically steady state. The long-time dynamics of these two cases are is displayed in Figure 2a,b, respectively. The axial direction is in dimensionless units (i.e., normalized with the gap width d).

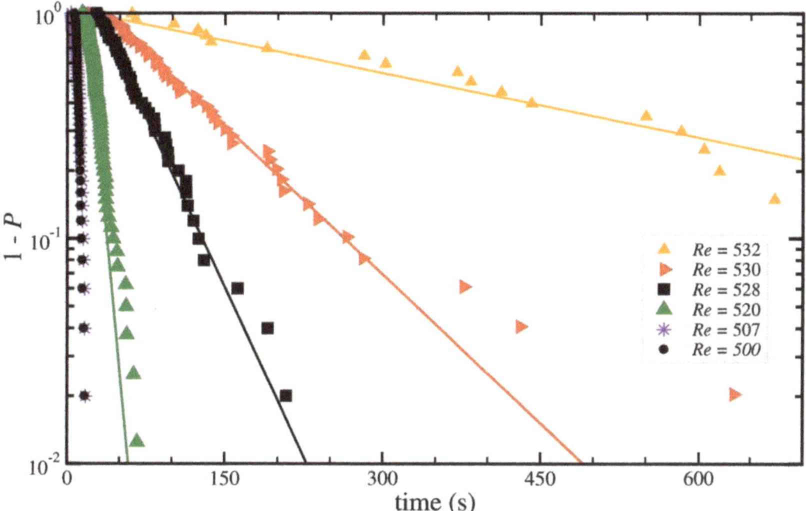

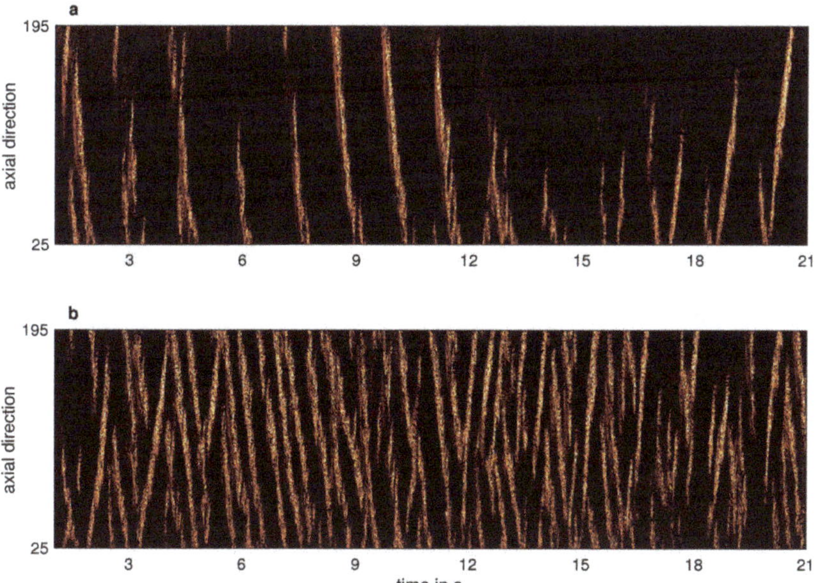

Figure 5. Lifetime statistics at $Re_o = -1000$ and $Re_{\mathrm{lids}} = -800$. Shown is the survival probability of turbulence (in a logarithmic scale) as a function of time for several Re_i, as indicated in the legend. The symbols denote individual measurements, which are sorted in increasing survival time to construct the survival probability function. In all cases, the initial condition was a turbulent flow at $Re_i = 630$ and the rotation of the cylinder was suddenly changed to the desired Re_i.

Figure 6. Excerpt of the spatio-temporal dynamics of turbulent stripes ($Re_o = -1000$ and $Re_{\mathrm{lids}} = 0$) above the critical point for the onset of sustained turbulence. (**a**) $Re_i = 525$, (**b**) $Re_i = 532$.

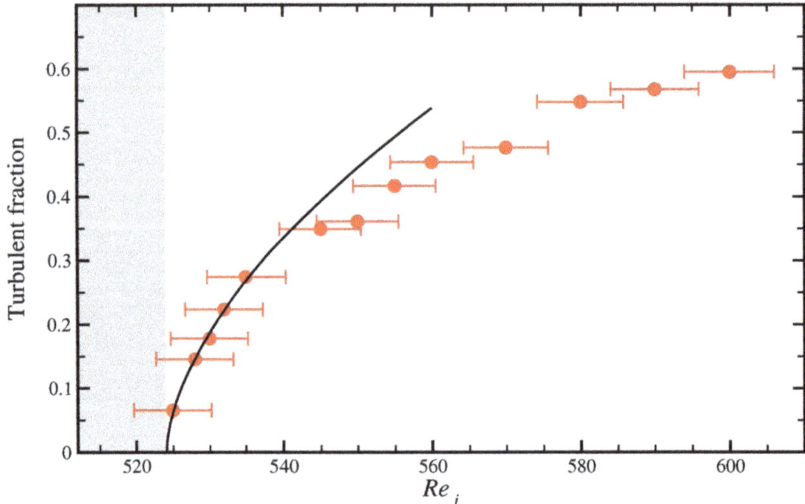

Figure 7. Second-order phase transition in counter-rotating Taylor–Couette flow ($Re_o = -1000$ and $Re_{lids} = 0$). The turbulent fraction increases smoothly from a minimum of 7% at $Re_i = 525$ up to 60% at $Re_i = 600$. The error bars indicate a 1% deviation in Re_i, which estimated from the discrepancy between linear stability analysis and experiment in Figure 1. The black line is a fit of the form $T_f = a\left(Re_i - Re_{i,c}\right)^{\beta_{DP}}$, with $\beta_{DP} = 0.583$, to the data points in the vicinity of the critical point ($524 < Re_i < 540$). The fit parameters are $a = 0.0667$ and $Re_{i,c} = 524.1$. Below the critical point (grey region), turbulence is transient.

3. Results

The experiments reported in this work were performed at $Re_o = -1000$ as indicated by the red line in Figure 1. The dynamics obtained at this selected Re_o is representative for the subcritical regime and hence also for other Re_o.

3.1. Lifetimes of Turbulent Stripes and Spots

For the lifetimes measurements, the speed of the lids was held constant at $Re_{lids} = -800$. The system was perturbed by rapidly accelerating the speed of the inner cylinder to $Re_i = 630$. This excited at first a linear instability in the form of laminar spirals (see Figure 3a), which quickly evolved into an intermittent pattern of laminar-turbulent stripes (see Figure 3b). The flow was then given sufficient time to reach a statistical steady state pattern. The camera started to record the flow pattern 20 s before Re_i was abruptly reduced to one of the six values indicated in the legend of Figure 5. The flow was continuously recorded until it relaminarised. Rather independently of the Re_i, the turbulent fraction typically dropped monotonically within the first 20 s, as the flow adapted to the new Re_i. Two examples of the corresponding spatio-temporal dynamics are shown in Figure 4, where the green line indicates the change of the Re_i in time. Despite the apparently similar dynamics, the long time behavior of these two cases is very different, leading to different lifetimes, see Figure 2. The complete decay of turbulence was systematically detected by determining the time at which the moving average of the turbulent fraction dropped permanently below a threshold.

During most of the runtime, the axial extent of the turbulent stripes was shorter than the cylinder length. The stripes moved in the axial and azimuthal direction exhibited a rich dynamics, including growth, shrinkage, splitting, merging and decay. Interactions with the axial lids occurred frequently. Specifically, the decay occurred often close to the lids. We thus believe that end effects are likely to influence the turbulent dynamics despite the large axial aspect ratio of our setup.

The probability of survival of turbulence as a function of time is shown in Figure 5. For the two lowest Re_i investigated the lifetimes are all shorter than 20 s, which corresponds to the time in which the (averaged) turbulent fraction continuously decreases without developing intrinsic dynamics. Therefore it is unclear wether the corresponding lifetimes are exponentially distributed or not in these two cases. A similar behavior was observed for the decay of puffs in pipe flow at low Re [14] in which the distribution deviated from an exponential one. However, in our measurements the distributions still seems to be exponential and for $Re_i > 507$ the probability follows $P(t) = 1 - \exp[(t - t_0)/\tau$, with the equilibration time $t_0 \approx 20$ s and τ the characteristic lifetime. This indicates that the decay of turbulence in this regime is a memoryless process, as reported for spatially extended plane Couette flow [8], quasi-one-dimensional [26,38] and moderate aspect-ratio [16] Taylor–Couette flows, and also for pipe flow [14] and quasi-one-dimensional channel flow [45].

3.2. Second-Order Phase Transition

In this section we present measurements of the turbulent fraction above the critical point. The measurement procedure was as in the previous section, with the exception that the recording started after a few minutes in order to ensure that the flow reached steady state conditions at the Re_i of interest. Since turbulence was sustained in these measurements, the recording time was set from 90 s at the largest Re_i to 15 min at $Re_i = 525$ (corresponding to 1.4×10^6 advective units), which was the lowest Re_i at which turbulence was sustained. In general, the observation time was increased, as the critical point was approached (in order to account for the expected critical slowing down). Note that the lids were stationary in these experiments (as for the results shown in Figure 1), which slightly stabilized turbulence when compared to the lifetime measurements with rotating lids discussed in the previous section; with stationary lids turbulence was sustained for $Re_i \geq 525$, whereas with rotating lids transient turbulence was found up to $Re_i = 532$.

As shown in Figure 6a, the spatio-temporal dynamics of turbulent patterns at $Re_i = 525$ is very rich. Oftentimes a single turbulent stripe spanning the whole system in the axial direction was observed. This then receded and eventually split into two or more arms, one of which would survive and extend to fill the system axially again. Only a slight increase of Re_i to 532 was sufficient to almost triple the turbulent fraction, which is reflected by the persistence of more than two turbulent spiral arms (in average) as shown in Figure 6b.

The retrieved turbulent fractions from all measurements are plotted in Figure 7. The minimum measured turbulent fraction is about five times smaller than in previous plane Couette experiments [8], and the maximum observation time in advective units is about 30% longer. The turbulent fraction increases continuously with increasing Re_i from its minimum value of about 7% ($Re_i = 525$) to more than 50%, suggesting a second-order phase transition. The scaling of the turbulent fraction in the vicinity of the critical point is consistent with that expected from directed percolation in two dimensions, $T_f = a(Re_i - Re_{i,c})^\beta$, where $\beta = 0.583$, $Re_{i,c}$ is the critical Reynolds number and a is a proportionality constant. A least-square fit of this function to the data close to the critical point ($524 < Re_i < 540$) yields $a = 0.0667$ and $Re_{i,c} = 524.1$ and approximates very well the data (see the black line in Figure 7). However, measurements closer to the critical point (including a direct determination of the critical point itself) would be necessary to test the robustness and accuracy of this fit. For example, if the function above is fitted with a free exponent, then $a = 0.0493$, $Re_{i,c} = 523.5$ and $\beta = 0.703$ is obtained. Finally, we stress that our system is too small to accurately determine critical exponents. Studies of quasi-one-dimensional Couette flow [26] and of quasi-two-dimensional Waleffe flow [28] show that determining the critical exponents requires a considerably larger system size. Indeed the observed interactions of the stripes with the axially bounding lids demonstrate that the the axial aspect ratio may be insufficient to probe the question of whether transition to turbulence in quasi-two-dimensional Couette flow falls into the directed percolation universality class.

4. Discussion

We investigated transient turbulence and the transition to sustained turbulence in a high-radius-ratio Taylor–Couette experiment. The presented lifetime measurements confirm the transient nature of turbulent stripes and show that their decay is memoryless in agreement with the study by Borrero et al. [16] for a smaller Taylor–Couette setup and more generally with transitional turbulence in other shear flows. At lower Reynolds numbers the lifetimes are shorter than the equilibrium time of the flow to adapt to the reduction in Reynolds number, but distributions remain exponential unlike in pipe flow where at low Reynolds numbers the tails deviated from exponential [14]. Our system area is more than 10 times larger than previous Couette and Taylor–Couette experiments, which enables us to approach the critical point much closer without suffering from finite size effects. Whereas such studies in smaller aspect ratio Couette flow had suggested a discontinuous drop form considerably larger turbulent fractions in our case the scaling is continuous, consistent with a second-order phase transition. Our observation of a continuous phase transition is also in line with recent studies of Waleffe flow [28] and of channel flow (see Figure 9a of [46]). An even closer approach to the critical point also leads to a sudden drop in turbulent fraction in the present case. As the critical point is approached length scales diverge and once typical laminar gap sizes exceed the system size the flow relaminarizes. Finite size effects can therefore be mistaken for a discontinuous transition. To resolve this question and to potentially obtain critical exponents, would require an even larger system size which sets a challenge for future experiments. Because of the long laminar gaps separating stripes in the vicinity of the critical point, and of the results of simulations and experiments of quasi-one-dimensional Couette flow [26,38] and Waleffe flow [28], we estimate that order of 1000 gap width are needed in the azimuthal and axial directions to probe for scale invariant flow patterns sufficiently close to the critical point. Such a study would however require cylinders manufactured to considerably higher precision than the already very precise ones used in the present study.

Author Contributions: The experiments and analyses was performed by K.A. and the work was conceptualized by K.A. and B.H. Both authors wrote the manuscript. All authors have read and agreed to the published version of the manuscript.

Funding: This research was funded by the Central Research Development Fund of the University of Bremen grant number ZF04B /2019/FB04 Avila_Kerstin ("Independent Project for Postdocs").

Institutional Review Board Statement: Not applicable.

Informed Consent Statement: Not applicable.

Data Availability Statement: Data sharing not applicable.

Acknowledgments: Shreyas Jalikop is acknowledged for recording some of the lifetime measurements.

Conflicts of Interest: The authors declare no conflict of interest. The funders had no role in the design of the study; in the collection, analyses, or interpretation of data; in the writing of the manuscript, or in the decision to publish the results.

References

1. Reynolds, O. An experimental investigation of the circumstances which determine whether the motion of water shall be direct or sinuous, and of the law of resistance in parallel channels. *Proc. R. Soc. Lond.* **1883**, *35*, 84–99.

2. Wygnanski, I.J.; Sokolov, M.; Friedman, D. On transition in a pipe. Part 2. The equilibrium puff. *J. Fluid Mech.* **1975**, *69*, 283–304. [CrossRef]

3. Barkley, D. Simplifying the complexity of pipe flow. *Phys. Rev. E* **2011**, *84*, 016309. [CrossRef] [PubMed]

4. Coles, D. Transition in circular Couette flow. *J. Fluid Mech.* **1965**, *21*, 385–425. [CrossRef]

5. Prigent, A.; Grégoire, G.; Chaté, H.; Dauchot, O.; van Saarloos, W. Large-scale finite-wavelength modulation within turbulent shear flows. *Phys. Rev. Lett.* **2002**, *89*, 14501. [CrossRef] [PubMed]

6. Ishida, T.; Duguet, Y.; Tsukahara, T. Transitional structures in annular Poiseuille flow depending on radius ratio. *J. Fluid Mech.* **2016**, *794*. [CrossRef]

7. Tuckerman, L.S.; Chantry, M.; Barkley, D. Patterns in Wall-Bounded Shear Flows. *Annu. Rev. Fluid Mech.* **2020**, *52*, 343–367. [CrossRef]
8. Bottin, S.; Chaté, H. Statistical analysis of the transition to turbulence in plane Couette flow. *Eur. Phys. J. B* **1998**, *6*, 143–155. [CrossRef]
9. Faisst, H.; Eckhardt, B. Sensitive dependence on initial conditions in transition to turbulence in pipe flow. *J. Fluid Mech.* **2004**, *504*, 343–352. [CrossRef]
10. Peixinho, J.; Mullin, T. Decay of turbulence in pipe flow. *Phys. Rev. Lett.* **2006**, *96*, 094501. [CrossRef]
11. Willis, A.P.; Kerswell, R.R. Critical behavior in the relaminarization of localized turbulence in pipe flow. *Phys. Rev. Lett.* **2007**, *98*, 014501. [CrossRef] [PubMed]
12. Hof, B.; Westerweel, J.; Schneider, T.M.; Eckhardt, B. Finite lifetime of turbulence in shear flows. *Nature* **2006**, *443*, 59–62. [CrossRef] [PubMed]
13. Hof, B.; De Lozar, A.; Kuik, D.J.; Westerweel, J. Repeller or attractor? Selecting the dynamical model for the onset of turbulence in pipe flow. *Phys. Rev. Lett.* **2008**, *101*, 214501. [CrossRef] [PubMed]
14. Avila, M.; Willis, A.P.; Hof, B. On the transient nature of localized pipe flow turbulence. *J. Fluid Mech.* **2010**, *646*, 127. [CrossRef]
15. Kuik, D.J.; Poelma, C.; Westerweel, J. Quantitative measurement of the lifetime of localized turbulence in pipe flow. *J. Fluid Mech.* **2010**, *645*, 529–539. [CrossRef]
16. Borrero-Echeverry, D.; Schatz, M.F.; Tagg, R. Transient turbulence in Taylor-Couette flow. *Phys. Rev. E* **2010**, *81*, 025301. [CrossRef]
17. Pomeau, Y. Front motion, metastability and subcritical bifurcations in hydrodynamics. *Physica D* **1986**, *23*, 3–11. [CrossRef]
18. Kaneko, K. Spatiotemporal intermittency in coupled map lattices. *Prog. Theor. Phys.* **1985**, *74*, 1033–1044. [CrossRef]
19. Kaneko, K. Supertransients, spatiotemporal intermittency and stability of fully developed spatiotemporal chaos. *Phys. Lett. A* **1990**, *149*, 105–112. [CrossRef]
20. Chaté, H.; Manneville, P. Transition to turbulence via spatio-temporal intermittency. *Phys. Rev. Lett.* **1987**, *58*, 112–115. [CrossRef]
21. Manneville, P. Spatiotemporal perspective on the decay of turbulence in wall-bounded flows. *Phys. Rev. E* **2009**, *79*, 25301. [CrossRef] [PubMed]
22. Moxey, D.; Barkley, D. Distinct large-scale turbulent-laminar states in transitional pipe flow. *Proc. Natl. Acad. Sci. USA* **2010**, *107*, 8091. [CrossRef] [PubMed]
23. Avila, K.; Moxey, D.; de Lozar, A.; Avila, M.; Barkley, D.; Hof, B. The onset of turbulence in pipe flow. *Science* **2011**, *333*, 192–196. [CrossRef] [PubMed]
24. Hinrichsen, H. Non-equilibrium critical phenomena and phase transitions into absorbing states. *Adv. Phys.* **2000**, *49*, 815–958. [CrossRef]
25. Mukund, V.; Hof, B. The critical point of the transition to turbulence in pipe flow. *J. Fluid Mech.* **2018**, *839*, 76–94. [CrossRef]
26. Lemoult, G.; Shi, L.; Avila, K.; Jalikop, S.V.; Avila, M.; Hof, B. Directed percolation phase transition to sustained turbulence in Couette flow. *Nat. Phys.* **2016**, *12*, 254–258. [CrossRef]
27. Duguet, Y.; Schlatter, P.; Henningson, D.S. Formation of turbulent patterns near the onset of transition in plane Couette flow. *J. Fluid Mech.* **2010**, *650*, 119. [CrossRef]
28. Chantry, M.; Tuckerman, L.S.; Barkley, D. Universal continuous transition to turbulence in a planar shear flow. *JFM* **2017**, *824*, R1. [CrossRef]
29. Andereck, C.D.; Liu, S.S.; Swinney, H.L. Flow regimes in a circular Couette system with independently rotating cylinders. *J. Fluid Mech.* **1986**, *164*, 155–183. [CrossRef]
30. Meseguer, A.; Mellibovsky, F.; Avila, M.; Marques, F. Instability mechanisms and transition scenarios of spiral turbulence in Taylor-Couette flow. *Phys. Rev. E* **2009**, *80*, 046315. [CrossRef]
31. Drazin, P.G.; Reid, W.H. *Hydrodynamic Stability*; Cambridge University Press: Cambridge, UK, 2004.
32. Brauckmann, H.J.; Salewski, M.; Eckhardt, B. Momentum transport in Taylor–Couette flow with vanishing curvature. *J. Fluid Mech.* **2016**, *790*, 419–452. [CrossRef]
33. Faisst, H.; Eckhardt, B. Transition from the Couette-Taylor system to the plane Couette system. *Phys. Rev. E* **2000**, *61*, 7227. [CrossRef] [PubMed]

34. Deguchi, K. Linear instability in Rayleigh-stable Taylor-Couette flow. *Phys. Rev. E* **2017**, *95*, 021102. [CrossRef] [PubMed]

35. Avila, K.; Hof, B. High-precision Taylor-Couette experiment to study subcritical transitions and the role of boundary conditions and size effects. *Rev. Sci. Instrum.* **2013**, *84*, 065106. [CrossRef]

36. Prigent, A.; Dauchot, O. Transition to versus from turbulence in subcritical Couette flows. In *IUTAM Symposium on Laminar-Turbulent Transition and Finite Amplitude Solutions*; Springer: Dordrecht, The Netherlands, 2005; pp. 195–219.

37. Prigent, A.; Grégoire, G.; Chaté, H.; Dauchot, O. Long-wavelength modulation of turbulent shear flows. *Physica D* **2003**, *174*, 100–113. [CrossRef]

38. Shi, L.; Avila, M.; Hof, B. Scale invariance at the onset of turbulence in Couette flow. *Phys. Rev. Lett.* **2013**, *110*, 204502. [CrossRef]

39. Ravelet, F.; Delfos, R.; Westerweel, J. Influence of global rotation and Reynolds number on the large-scale features of a turbulent Taylor–Couette flow. *Phys. Fluids* **2010**, *22*, 055103. [CrossRef]

40. Paoletti, M.S.; Lathrop, D.P. Measurement of angular momentum transport in turbulent flow between independently rotating cylinders. *Phys. Rev. Lett.* **2011**, *106*, 024501. [CrossRef]

41. Van Gils, D.P.M.; Huisman, S.G.; Bruggert, G.W.; Sun, C.; Lohse, D. Torque scaling in turbulent Taylor-Couette flow with co-and counterrotating cylinders. *Phys. Rev. Lett.* **2011**, *106*, 24502. [CrossRef]

42. Heise, M.; Hochstrate, K.; Abshagen, J.; Pfister, G. Spirals vortices in Taylor-Couette flow with rotating endwalls. *Phys. Rev. E* **2009**, *80*, 045301. [CrossRef]

43. Schartman, E.; Ji, H.; Burin, M.J. Development of a Couette–Taylor flow device with active minimization of secondary circulation. *Rev. Sci. Instrum.* **2009**, *80*, 024501. [CrossRef] [PubMed]

44. Lopez, J.M.; Avila, M. Boundary-layer turbulence in experiments on quasi-Keplerian flows. *J. Fluid Mech.* **2017**, *817*, 21–34. [CrossRef]

45. Gomé, S.; Tuckerman, L.S.; Barkley, D. Statistical transition to turbulence in plane channel flow. *Phys. Rev. Fluids* **2020**, *5*, 083905. [CrossRef]

46. Shimizu, M.; Manneville, P. Bifurcations to turbulence in transitional channel flow. *Phys. Rev. Fluids* **2019**, *4*, 113903. [CrossRef]

Publisher's Note: MDPI stays neutral with regard to jurisdictional claims in published maps and institutional affiliations.

Article

Flow Statistics in the Transitional Regime of Plane Channel Flow

Pavan V. Kashyap [1,*], Yohann Duguet [1] and Olivier Dauchot [2]

[1] LIMSI-CNRS, UPR 3251, Université Paris-Saclay, 91405 Orsay, France; yohann.duguet@limsi.fr
[2] Gulliver, ESPCI-CNRS, 10 Rue Vauquelin, 75005 Paris, France; olivier.dauchot@espci.fr
* Correspondence: pavan.kashyap@limsi.fr

Received: 16 August 2020; Accepted: 4 September 2020; Published: 8 September 2020

Abstract: The transitional regime of plane channel flow is investigated above the transitional point below which turbulence is not sustained, using direct numerical simulation in large domains. Statistics of laminar-turbulent spatio-temporal intermittency are reported. The geometry of the pattern is first characterized, including statistics for the angles of the laminar-turbulent stripes observed in this regime, with a comparison to experiments. High-order statistics of the local and instantaneous bulk velocity, wall shear stress and turbulent kinetic energy are then provided. The distributions of the two former quantities have non-trivial shapes, characterized by a large kurtosis and/or skewness. Interestingly, we observe a strong linear correlation between their kurtosis and their skewness squared, which is usually reported at much higher Reynolds number in the fully turbulent regime.

Keywords: transition to turbulence; spatio-temporal intermittency; channel flow

---

1. Introduction

Laminar and turbulent flows are two different regimes encountered sometimes at the same parameters for a given geometry. In many flows they are in competition from the point of view of the state space. Shear flows next to solid walls however show this surprisingly robust property that both laminar and turbulent regions coexist spatially on very long time scales, when the laminar state is locally stable. This phenomenon, called 'laminar-turbulent intermittency' is well known in circular pipe flow since the days of O. Reynolds [1] and has lead recently to a burst of interest, a review of which is provided in Reference [2]. Such laminar-turbulent flows have been identified and partly characterized in Taylor-Couette flow [3,4] and in plane Couette flow [4–6]. They also have been identified in other set-ups involving curvature [7–9] or stabilizing effects [10]. The transitional regimes of plane Poiseuille flow, the flow between two fixed parallel plates driven by a fixed pressure gradient, have not received as much attention although this flow is the archetype of wall-bounded turbulent flows. Although this flow is frequently cited as an example of flow developing a linear instability (under the form of Tollmien–Schlichtling waves) [11], coherent structures typical of laminar-turbulent coexistence have been frequently reported in channel flow well below the linear instability threshold and a series of experimental and cutting-edge numerical studies in the 1980s and 1990s have focused on the development of spots [12–16]. Sustained intermittent regimes have not been identified as such before the mid-2000s, when Tsukahara [17] reported large-scale coherent structures from numerics in larger numerical domains. Like their counterpart in Couette flows, these structures display obliqueness with respect to the mean flow direction and a complicated long-time dynamics. The dynamics at onset in particular have remained mysterious [18] and, although this is currently debated, could follow a scenario different from the directed percolation one proposed for Couette flow. [9,19,20]. In recent years, the so-called transitional regime of plane channel flow has attracted renewed attention after new experimental studies. Although the works in Refs [21–23] focused on the minimal transition amplitude for spot development, other studies [24–29] focused on the sustained intermittent regimes and their statistical quantification.

Experimentally the finite length of the channel sets a limitation to most statistical approaches. Numerical simulation in large domains combined with periodic boundary conditions is a well-established way to overcome such limitations. Surprisingly, despite a large number of numerical studies of transitional channel flow, investigation of spatio-temporal intermittency in large enough domains has not been possible before the availability of massive computational resources. Owing to recent numerical studies [30–32], there is currently a good consensus about a few facts concerning the transitional regime: laminar-turbulent bands with competing orientations emerge progressively as the Reynolds number is reduced below $Re_\tau \approx 100$, and their mean wavelength increases as the Reynolds number is decreased. At even lower flow rate the bands turn into isolated spots with ballistic dynamics rather than forming a seemingly robust stripe pattern [33–35]. The global centerline Reynolds number for the disappearance of the stripes is close to 660 [18,27]. However, many questions remain open. The most sensible theoretical issues revolve around the (still open) question of the universality class of the transition process (see Reference [18]), the role of the large-scale flows [23,25,36,37] in the sustainment of the stripes, or the mutual way different stripes interact together.

There is also a lack of quantitative data about the patterning regime itself. The present special issue is an opportunity to document the geometric characteristics of the stripe patterns in unconstrained settings. Moreover, there is an ongoing philosophical question about whether traces of spatio-temporal intermittency can be found in the fully turbulent regimes commonly reported at higher Reynolds numbers. In the present paper, using numerical simulation in large domains, we focus on three specific points hitherto undocumented: the angular distribution of turbulent stripes, the statistics of the laminar gaps between them, and high-order statistics of the local and instantaneous bulk velocity, wall shear stress and turbulent kinetic energy. The outline of the paper is as follows: Section 2 introduces the numerical methodology with the relevant definitions. The geometrical statistics of the stripe angles are presented in Section 3.1. The statistics of a few global quantities are presented in Sections 3.2–3.4. A discussion of the results is made in Section 4 with the conclusions and outlooks in Section 5.

2. Materials and Methods

The present section is devoted to the methodology used for the numerical simulation of pressure-driven plane channel flow. The flow is governed by the incompressible Navier Stokes equations. Channel flow is described here using the Cartesian coordinates x,y,z, respectively the streamwise, wall-normal and spanwise coordinates. The velocity field $u(x, y, z, t)$ is decomposed into the steady laminar base flow solution $\mathbf{U}(y) = (U_x, 0, 0)$ and a perturbation field $u'(x, y, z, t)$. Similarly, the pressure field is decomposed as $p(x, y, z, t) = xG + p'(x, y, z, t)$. The equation governing the steady base flow for an incompressible fluid with constant density ρ and kinematic viscosity ν is given by

$$\nu \frac{\partial^2 U_x}{\partial y^2} = \frac{1}{\rho} G \tag{1}$$

with G a constant. Together with the no-slip condition at the walls Equation (1) yields the analytic Poiseuille solution $U_x \propto 1 - (y/h)^2$. The equation governing the perturbation field involves the base flow and reads

$$\frac{\partial \mathbf{u}'}{\partial t} + \mathbf{u}' \cdot \nabla \mathbf{u}' + \mathbf{U} \cdot \nabla \mathbf{u}' + \mathbf{u}' \cdot \nabla \mathbf{U} = -\frac{1}{\rho} \nabla p' + \nu \nabla^2 \mathbf{u}' \tag{2}$$

The channel geometry is formally infinitely extended, yet in the numerical representation it is given by its extent $L_x \times 2h \times L_z$ as in Figure 1, with stationary walls at $y = \pm h$ and periodic boundary conditions in x and z.

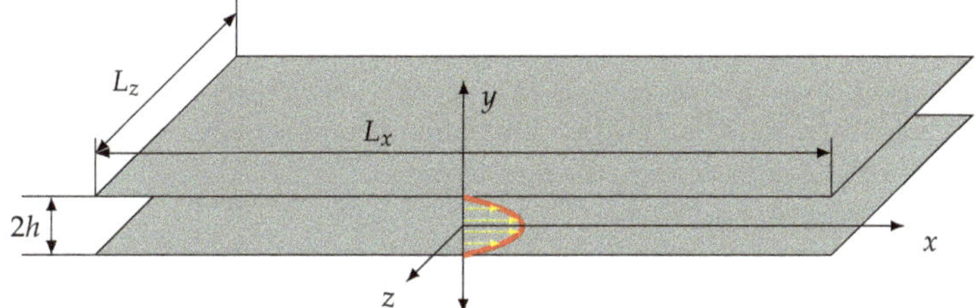

Figure 1. Schematic of the numerical domain with the laminar base flow profile (red).

The flow is driven by the imposed pressure gradient G assumed negative. The spanwise pressure gradient is explicitly constrained to be null. The centerline velocity u_{cl} of the laminar base profile with the same pressure gradient is chosen as the velocity scale (U) and the half gap h of the channel is chosen as the lengthscale used for non-dimensionalization. Time is hence expressed in units of h/U. In these units the laminar velocity profile is given by $U_x^*(y_*) = 1 - y_*^2$. From Chapter 3 onwards only dimensionless quantities will be used and the $*$ notation will be dropped from there on. Primed quantities denote perturbations to the base flow while non-primed quantities involve the full velocity field, including the laminar base flow.

In the following we shall consider, both locally and temporally fluctuating quantities, as well as their time and space averages. We denote by $\langle \bullet \rangle$ the space (x, z) average and $\overline{\bullet}$—the time average. Space-time averages are indicated by $\langle \overline{\cdot} \rangle$. More explicitly the space-average operator is defined as the discrete average over the grid points, and the time average is the discrete average sum over the total number of snapshots in the steady regime.

Different velocity scales characterize the flow. One such scale is the centerline velocity u_{cl} of the corresponding laminar flow with the same value of G. Another one is the total streamwise flow through the channel, $U_b = \langle \overline{u_b} \rangle$, where

$$u_b(x, z, t) = \frac{2}{h} \int_{-h}^{h} u_x dy \tag{3}$$

is the so-called local bulk flow. Finally, the friction velocity is defined as $U_\tau = (\langle \overline{\tau} \rangle / \rho)^{\frac{1}{2}}$, where $\tau = (\tau_t + \tau_b)/2 > 0$, with τ_t and τ_b the net shear stress on the top and the bottom wall, respectively given by:

$$\tau_{t,b}(x, z, t) = \pm \mu \frac{\partial u_x}{\partial y} \bigg|_{t,b} \tag{4}$$

where $\mu = \rho \nu$ is the dynamic viscosity of the fluid. The three Reynolds numbers arising from these velocity scales are $Re_{cl} = u_{cl}h/\nu$, $Re_b = U_bh/\nu$ and $Re_\tau = U_\tau h/\nu$. For the laminar base flow, they are inter-related as $Re_\tau^2 = 3Re_b = 2Re_{cl}$. Imposing a pressure gradient $G < 0$ translates into a fixed average shear stress $\langle \overline{\tau} \rangle$ on the walls which sets an imposed value of $Re_\tau = Re_\tau^G$ to stress that this is the control parameter.

Direct numerical simulation (DNS) of Equation (2) is carried out using the open source, parallel solver called Channelflow [38,39] written in C++. It is based on a Fourier–Chebychev discretization in space and a 3rd order semi-implicit backward difference scheme for timestepping. It makes use of the 2/3 dealiasing rule for the nonlinear terms. An influence matrix method is used to ensure the no-slip boundary condition at the walls. The numerical resolution is specified in terms of the spatial grid points (N_x, N_y, N_z) which translates into a maximum of $(N_x/2 + 1, N_z/2 + 1)$ Fourier wavenumbers and N_y Chebychev modes. Please note that the definitions of N_x and N_z take into account the aliasing modes. The domain sizes used in this study, expressed in units of h, are $L_x = 2L_z = 250$ for $55 < Re_\tau^G \leqslant 100$ and $L_x = 2L_z = 500$ for $39 \leqslant Re_\tau^G \leqslant 55$. The local numerical resolution used is $N_x/L_x = N_z/(2L_z) = 4.096$ and $N_y = 65$, comparable to that used in Reference [34]. The simulation follows an "adiabatic descent": a first simulation is carried out at sufficiently high value of Re_τ^G, known to display space-filling turbulence. After the stationary turbulent regime is reached, Re_τ^G is lowered and the simulation advanced further in time. This step-by-step reduction has been performed down to $Re_\tau^G = 39$. The initial condition for the first simulation is a random distribution of localized seeds of the kind described in Reference [40]. The time required T to reach a stationary regime gradually increases as Re_τ^G is decreased. As an order of magnitude, for $Re_\tau^G = 100$, $T \approx 1500$, while for $Re_\tau^G = 50$, $T \approx 3000$. Statistics are computed, after excluding such transients, from time series of lengths up to 2×10^4 time units.

3. Results

The entire adiabatic descent is shown using a space-time diagram of the crossflow energy shown in Figure 2a

$$E_{cf} = \frac{1}{2} \int (u_y^2 + u_z^2) dy \tag{5}$$

evaluated at an arbitrary value of z (here $z = L_z/2$). The space variable is expressed in a frame moving in the streamwise direction with the mean bulk velocity $U_b(G)$ for that particular value of Re_τ^G. Since Re_τ^G is lowered over the course of time, this allows one to capture the different flow regimes preceding full relaminarization. The intensity of turbulence, measured here by the value of E_{cf}, is seen to gradually increase as Re_τ^G is lowered. At high Re_τ^G, the so-called featureless turbulence occupies the full domain, as shown in Figure 2b at $Re_\tau^G = 100$ using isocontours of $\tau'(x,z) = \tau(x,z) - \tau_{lam}$. As Re_τ^G is lowered, turbulence self-organizes into the recognizable pattern regime [17] shown in Figure 2c for $Re_\tau^G = 80$. As Re_τ^G is further reduced the turbulent zones become sparser (see Figure 2d for $Re_\tau^G = 60$). The spatially localized turbulent regions emerge as narrow stripes throughout the process of decreasing Re_τ^G while the gaps between them constantly increase in size. The emerging patterns never feature an array of strictly parallel stripes like in former computational approaches [19,31,41], instead they feature competing orientations as in pCf [4], see Figure 2b–d. In this regime the pattern travels with a streamwise convection velocity slightly slower than $U_b(G)$. Within the quasi-laminar gaps, E_{cf} reaches very low values, at least an order of magnitude less than in the core of the turbulent stripes. The lower Re_τ^G, the lower these values. Below $Re_\tau^G = 50$ the stripe pattern eventually breaks up to form independent turbulent bands of finite length, all parallel to each other [34], as shown in Figure 2e for $Re_\tau^G = 40$. The new resulting pattern as a whole shows negligible spanwise advection, while it propagates in x with a velocity close to $\langle u_b \rangle$ [42]. The independent turbulent bands show enhanced motility in both directions x and z. This motion relative to the frame of reference causes the tilt of the stripes seen in Figure 2a for $Re_\tau^G > 50$ as well as the apparent increase of thickness.

In pipe flow it was noted recently [43] that the emergence of spatial localization does not imply the proximity to the transitional point (below which turbulence is not sustained) as long as the statistics about the size of the laminar gaps fail at displaying power-laws tails. The laminar gaps are estimated as the streamwise distance l_x between local maxima of τ (values lower than $\langle \tau \rangle + \sigma(\tau)$, with σ the standard deviation, have been discarded). The cumulative distribution (CDF) of the laminar gap size is shown in Figure 3 in lin-log coordinates. For all values of Re_τ shown, it shows an exponential tails and no algebraic part. Exponential distributions are a hallmark of spatio-temporal intermittency, unlike critical phenomena which are characterized by algebraic/power law related to the scale invariance property. The entire regime of channel flow for $39 \leqslant Re_\tau^G \leqslant 100$ can be described as being spatiotemporally intermittent, and is hence far above any critical point. Please note that the critical point of pPf is estimated to approximately $Re_{cl} = 660$ [18] i.e., $Re_\tau^G \approx 36$ and falls outside the range of parameters investigated here.

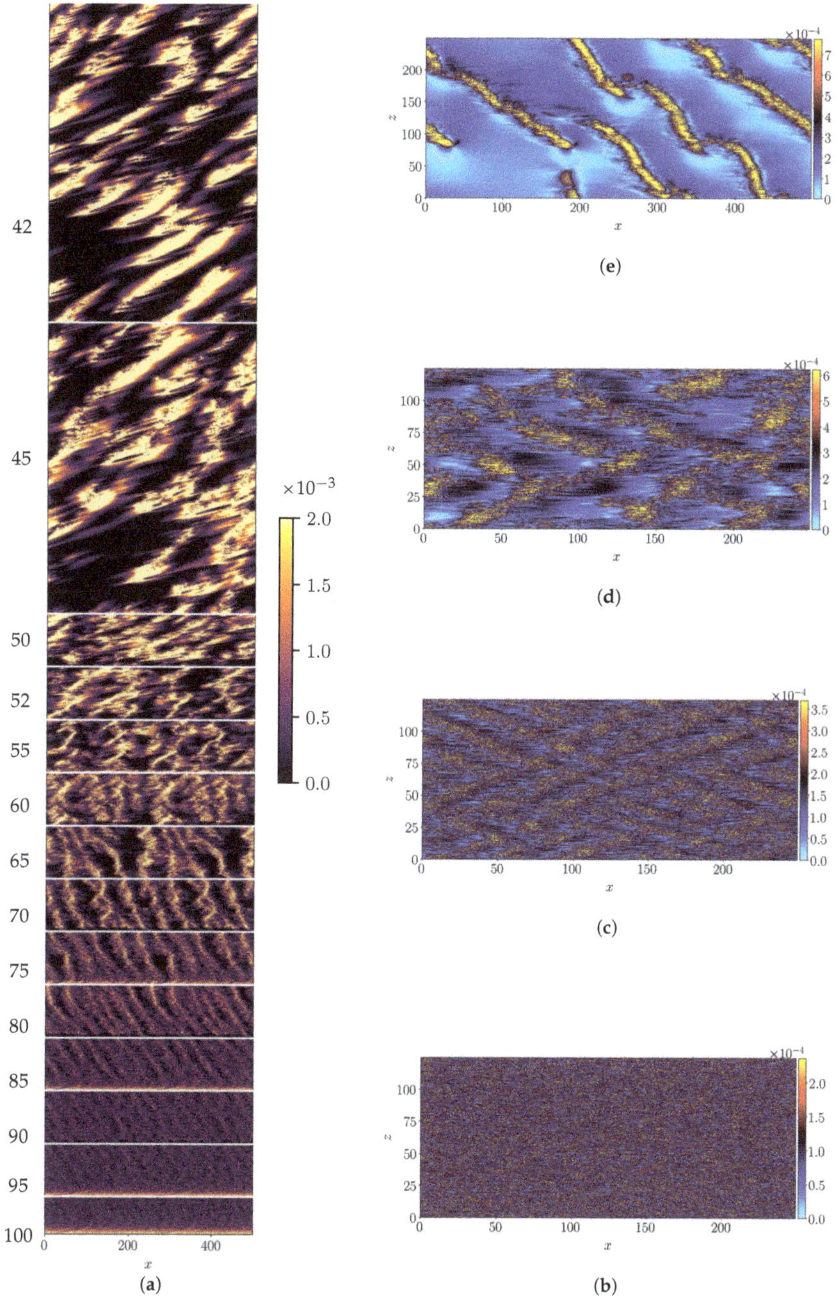

Figure 2. (**a**) Space-time diagram of $E_{cf}(x - U_b(G)\,t, t)$ for $z = L_z/2$ during the adiabatic descent protocol, in a frame travelling in the x-direction at the mean bulk velocity $U_b(G)$. Vertical axis: time with corresponding values of Re_τ^G values indicated. (**b–e**) isocontours of $\tau'(x,z)$ for $Re_\tau^G = 100, 80, 60, 40$.

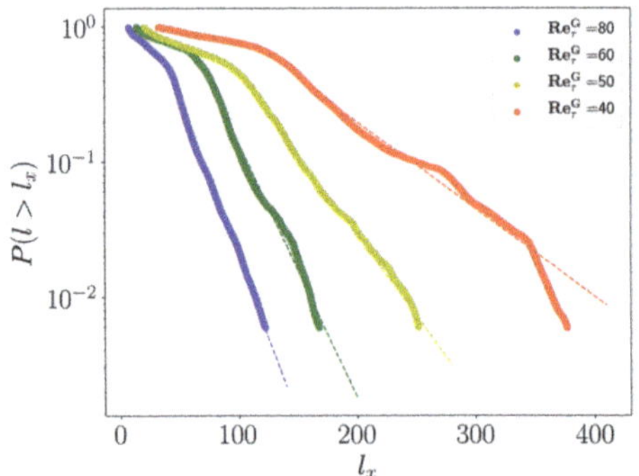

Figure 3. CDF of laminar gap size for $Re_\tau^G = 80, 60, 50, 40$.

3.1. Angular Statistics of Turbulent Bands

The self-organization of turbulence into long band–like structures, oriented with an angle with respect to the streamwise direction, is depicted in Figure 2. The (signed) angle is computed using two different methodologies. As in Duguet et al. [36] in the case of pCf, the local y–integrated velocity field is found to be parallel to the bands. The same holds for pPf, as is visible in Figure 4a,c for $Re_\tau^G = 60$ and 40, respectively. Please note that unlike Couette flow, pPf features advection with a non-zero mean bulk velocity. Hence the local velocity field is here computed by removing this mean advection velocity. A first estimation of the local and instantaneous band angle is therefore computed following Equation (6):

$$\theta_L(x, z, t) = tan^{-1}\left[\frac{\int u_z' \, dy - \langle \int u_z' \, dy \rangle}{\int u_x' \, dy - \langle \int u_x' \, dy \rangle}\right] \tag{6}$$

The second estimation is obtained from Fourier analysis and computed from Equation (7), following Reference [44] :

$$\theta_F(t) = tan^{-1}(\lambda_z/\lambda_x) \tag{7}$$

where $\lambda = 2\pi/k$, with k being the leading non-zero wavenumber identified from the power spectra (excluding the $k_x = k_z = 0$ mode). The Fourier spectrum is computed for the quantity $\tau(x, z, t)$, but similar results have been observed for other observables such as $E_{cf}(x, z, t)$ and $E_v = (1/2) \int u_y^2 \, dy$. The angles can be read directly from the Fourier spectra in polar coordinates, see Figure 4b,d for the same values of $Re_\tau^G = 60$ and 40, respectively. The mean angles $\langle \theta_L \rangle$ and $\bar{\theta}_F$ are then computed by respectively space-time-averaging and time averaging the data obtained from Equation (6) and (7).

The variation of the mean (signed) angles with Re_τ^G, computed using the two methods, is shown in Figure 5a, where the indices 1, 2 stand for the two band orientations. Both methods provide identical results. The variation of the (unsigned) angle of the band denoted by θ, computed as $\theta = |\bar{\theta}_F|$ is shown in Figure 5b. It is found that the mean angle θ of the bands remains approximately constant with $\theta = 25° \pm 2.5°$ in the range of values $60 \leqslant Re_\tau^G \leqslant 90$ and increases for lower value of $Re_\tau^G < 60$. In the patterning regime, i.e., for $Re_\tau^G \geqslant 50$, the angle of the bands is found to be distributed symmetrically with respect to zero, as a consequence of the natural symmetry $z \leftarrow -z$ of the flow. For lower Re_τ^G these quasi-regular patterns break down into individual localized structures analogous to individual puffs in cylindrical pipe flow. As the pattern dissolves, one single band orientation ends up dominating the dynamics as shown by Shimizu and Manneville [34] for a similar domain size. The angle θ further increases as the regular pattern deteriorates, with $\theta_{max} \approx 40$ at $Re_\tau^G = 39$. Previous studies [18,27] have documented that the angle of the bands approach $45°$ close to the onset of transition. The present investigation agrees well with these studies (Figure 5b) while covering a wider range in Reynolds number, highlighting the difference between the puff regime for which $\theta \approx 40 - 45°$, and the patterning regime for which θ is almost half this value (see also Figure 2).

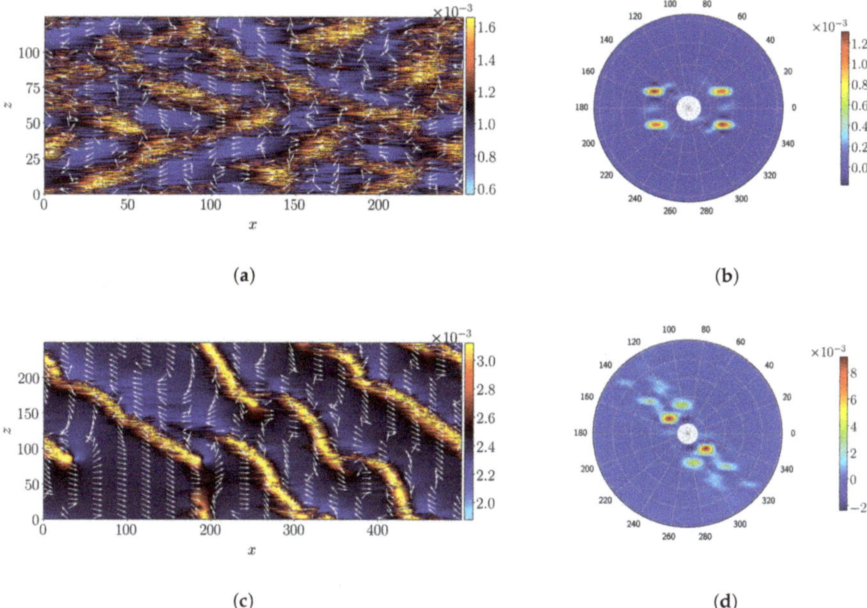

Figure 4. (**a**,**c**) Isocontours of τ' with the local velocity indicated by the normalized velocity vectors, at $Re_\tau^G = 60, 40$, respectively; (**b**,**d**) Instantaneous Fourier spectrum in polar coordinates for (**a**,**b**), respectively.

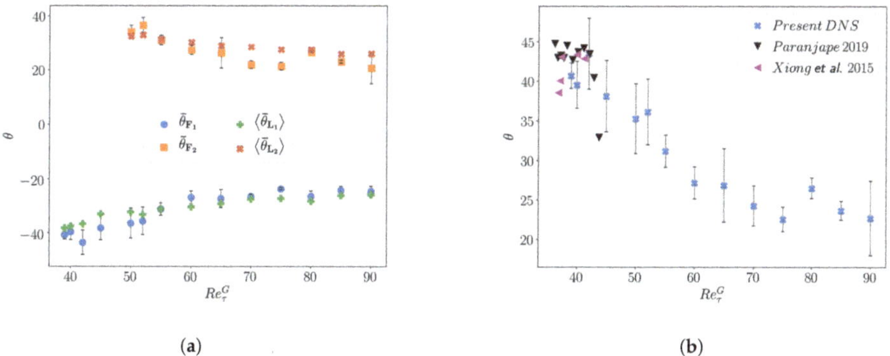

Figure 5. (**a**) Variation of the mean (signed) angle of the turbulent bands with Re_τ^G, computed from the Fourier spectra ($\overline{\theta_{F_1}}$, $\overline{\theta_{F_2}}$) and the mean (signed) angle of the local velocity ($\langle\overline{\theta_{L_1}}\rangle$, $\langle\overline{\theta_{L_2}}\rangle$) (**b**) Variation of the mean unsigned band angle θ along with the data from Reference [27,30].

Figure 4c shows that across a band, the local large-scale velocity changes orientation [41]. This property is used to sort out the local maxima of τ (higher than $\langle\tau\rangle + \sigma(\tau)$) as belonging to one band with a particular inclination. This allows one to define the respective streamwise and spanwise inter-stripe distances l_x and l_z between bands of the same orientation. Figure 6a,b displays $\langle\overline{l_x}\rangle$ and $\langle\overline{l_z}\rangle$ for orientations 1 and 2, respectively, as a function of Re_τ^G. Both increase when decreasing Re_τ^G. They vary in parallel in the patterning regime, hence the quasi-constant angle θ of the bands. When only one band orientation survives, one observes that the increase in θ amounts to the saturation of $\langle\overline{l_x}\rangle_1$, while $\langle\overline{l_z}\rangle_1$ keeps increasing.

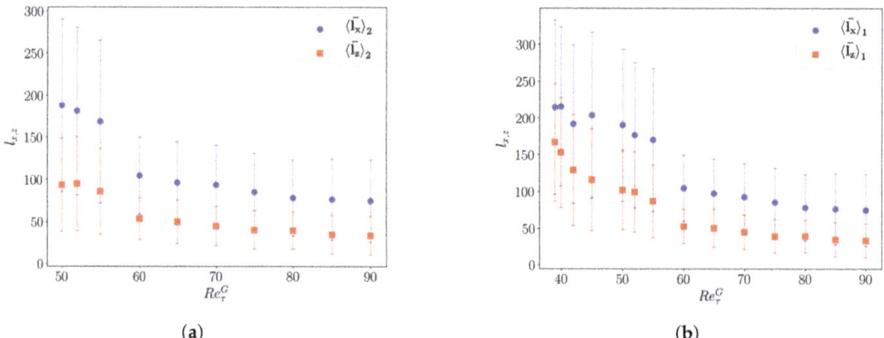

Figure 6. (**a**,**b**) Space-time-averaged inter-stripe streamwise $\overline{\langle l_x \rangle}_{1,2}$ (blue) and spanwise $\overline{\langle l_z \rangle}_{1,2}$ (red) distances for bands of orientations 1 and 2, respectively.

3.2. Global Variables: Moody Diagram

The mean velocity profile $\overline{\langle u_x^+ \rangle}$ is defined as the average of u_x over x,z and t, expressed in units of u_τ. It is shown in Figure 7 as a function of $y^+ = y u_\tau / \nu$ and compared with the classical DNS data by Kim et al. [45] obtained at higher $Re_\tau^G = 180$. The whole figure is similar to figures 3 and 10 in Reference [17,46], respectively. As expected for the present low values of Re_τ^G, the velocity field matches the linearized profile $u_x^+ = y^+$ next to the wall but does not develop a logarithmic dependence with respect to y^+.

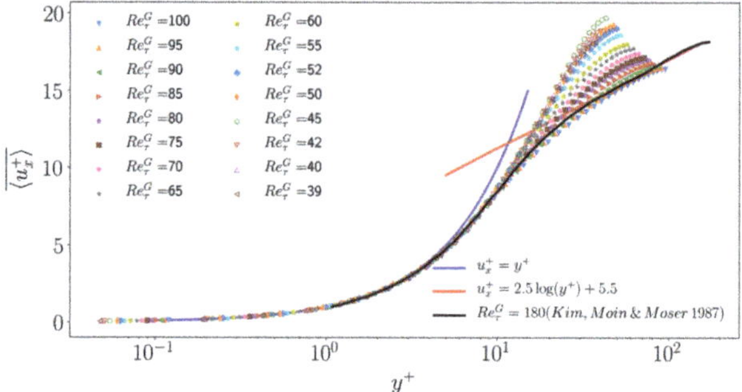

Figure 7. Mean flow profile $u_x^+(y^+)$ for Re_τ^G from 100 down to 39. Blue: law of the wall $u_x^+ = y^+$, red: logarithmic law of the wall $u_x^+ = 2.5 \log(y^+) + 5.5$, black: DNS by Kim, Moin and Moser from Reference [45].

At a global level of description, the laminar and turbulent flow are traditionally represented in the classical Moody diagram in which the Fanning friction factor C_f defined as the ratio between the pressure drop along the channel length and the kinetic energy per unit volume based on the mean bulk velocity $U_b = \overline{\langle u_b \rangle}$,

$$C_f = \frac{|\Delta p|}{1/2 \, \rho U_b^2} \frac{h}{L_x} = \frac{\overline{\langle \tau \rangle}}{1/2 \, \rho U_b^2} = \frac{2 \, Re_\tau^{G2}}{Re_b^2}, \tag{8}$$

is traditionally plotted versus Re_b as shown with plain symbols in Figure 8. Another way to express C_f is to use inner units, in which case $C_f = 2/(u_b^+)^2$, with $u_b^+ = u_b/u_\tau$. C_f is then linked only to the integral of the mean profile displayed in Figure 7.

For the laminar flow, the dependence of C_f vs. Re_b is analytically given by $Cf_{lam} = 6/Re_b$ (blue continuous line). In the featureless turbulent regime, it is known empirically as the Blasius' friction law scaling $\overline{Re_b}^{-1/4}$ (red continuous line). For intermediate values of Re_b, C_f clearly deviates from the turbulent branch, and remains far from the laminar value [47]. Here we notice, in agreement with [30] and [34] that $C_f \approx 0.01$ remains essentially constant in this transitional regime. What is remarkable is that this regime of constant C_f coincides with the patterning regime observed for $50 \leqslant Re_\tau^G \leqslant 90$, corresponding to $690 \leqslant Re_b \leqslant 1225$, *as if the respective amount of turbulent and laminar domains was precisely ensuring $C_f = cst$.* As the pattern fractures, C_f increases and approaches the laminar curve. We note that the observation of this property requires large computational domains to be observed, which explains why it had not been noticed until recently, even in experiments.

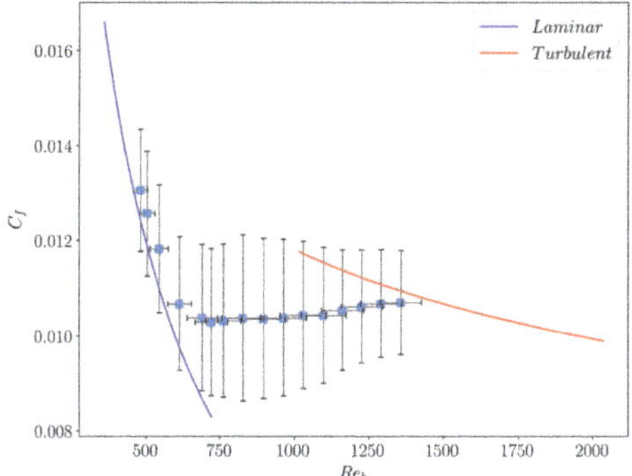

Figure 8. Friction coefficient C_f vs. Re_b, with horizontal and vertical error bars indicating the fluctuations these quantities would inherit from that of the field u_b (see text for details).

Given the complex spatio-temporal dynamics in the transitional regime, the bulk velocity u_b is expected to strongly fluctuate both in space and time. We also report in Figure 8, how these fluctuations would translate on Re_b and C_f, if the latter were computed using the locally fluctuating field u_b instead of its mean value U_b. These fluctuations are significant (up to 10–15%) and suggest to further explore them, which is the topic of the next section and the main focus of the present work.

3.3. Joint Probability Distribution of Re_τ and Re_b

Reynolds numbers such as Re_τ and Re_b are traditionally seen as global parameters characterizing the flow. They are defined based on velocity scales obtained from space-time average. It is straightforward to extend these definitions to the local fields $Re_b(x,z,t) = u_b(x,z,t)h/\nu$ and $Re_\tau = u_\tau(x,z,t)h/\nu$, with $u_\tau(x,z,t) = (\tau(x,z,t)/\rho)^{1/2}$. Please note that with this definition, $\langle Re_\tau \rangle$ is not strictly equal to the imposed Re_τ^G, because of the nonlinear relation between Re_τ and τ.

Investigation of the entire transitional regime is provided through a two-dimensional state portrait $(Re_b - Re_\tau)$ constructed from this local definition of the Reynolds number. The joint probability density distribution is constructed in this state space with the space-time data for different Re_τ^G. The state space for $Re_\tau^G = 100, 80, 60, 40$ is shown in Figure 9. The continuous blue and red lines again correspond to the scalings known analytically for the laminar flow, and empirically for featureless turbulent flows for high enough Reynolds numbers. As expected the most probable values of Re_b and Re_τ, follow the same trend as their global counterpart: they match the continuous curve in the featureless turbulent regime, and progressively depart from it to move towards the laminar branch at the lowest Re_τ^G explored here. More interesting are the distributions. First, we observe that the relative fluctuations are significantly larger for Re_τ than for Re_b, the difference being larger for the larger Re_τ^G. Secondly the distributions are not simple Gaussians. Even in the featureless turbulent regime, the marginal distribution of Re_τ is already relatively skewed (Figure 9(a_3)).

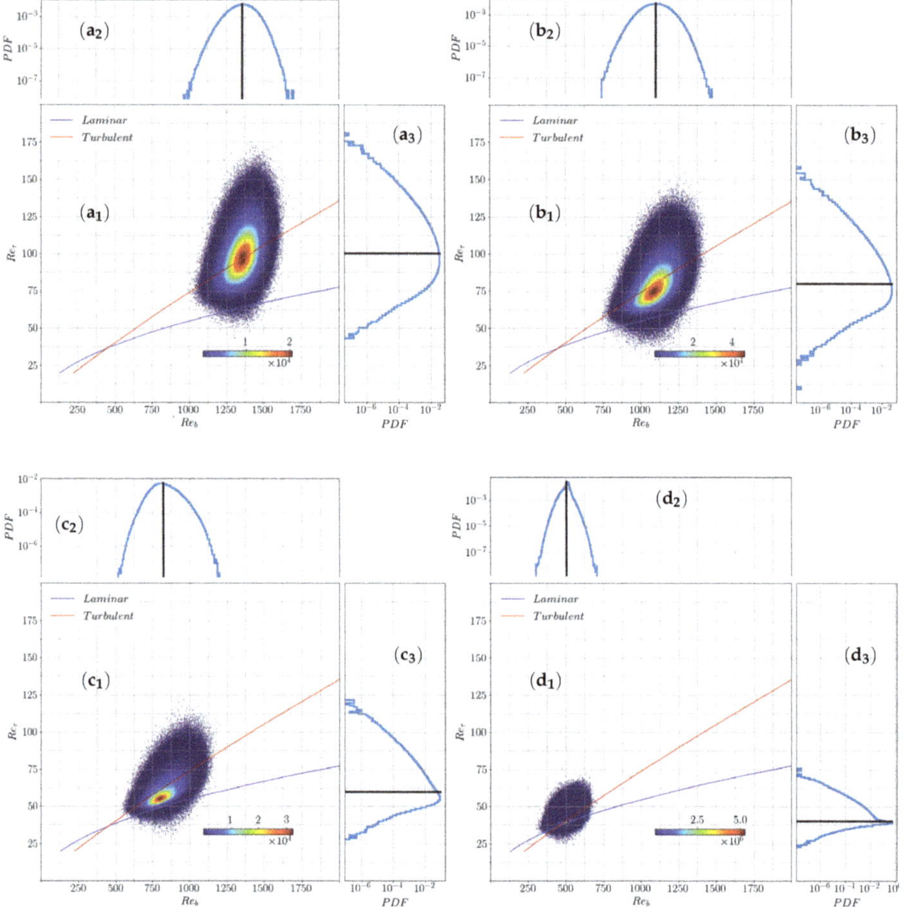

Figure 9. ($\mathbf{a_1}$) ($\mathbf{b_1}$) ($\mathbf{c_1}$) ($\mathbf{d_1}$) Joint probability distribution of the quantities Re_b and Re_τ for $Re_\tau^G = 100$, 80, 60, 40 together with their marginal distribution shown in lin-log scale for Re_b in ($\mathbf{a_2}$) ($\mathbf{b_2}$) ($\mathbf{c_2}$) ($\mathbf{d_2}$) and for Re_τ in ($\mathbf{a_3}$) ($\mathbf{b_3}$) ($\mathbf{c_3}$) ($\mathbf{d_3}$) with the mean value indicated by a vertical/horizontal black line.

As Re_τ^G is reduced, the overall width of the distribution decreases, but the shape of the marginal distributions of Re_τ differs more and more from a Gaussian. More specifically, although the distribution remains unimodal, we note that the marginal distribution of Re_τ is more and more skewed. We also note that the right wing of the distribution is not convex anymore. To further quantify these observations, a systematic analysis of the moments of this distribution is conducted in the next section.

3.4. Higher-Order Statistics

The higher-order statistics of Re_τ, Re_b and E_{cf} are presented in this section. For any field $A = A(x, z, t)$, we compute the spatio-temporal average $m = \overline{\langle A \rangle}$, the variance $\sigma^2 = \langle (A - m)^2 \rangle$ and the kth standardized higher-order moment $\langle (A - m)^k \rangle / \sigma^k$ (for $k \geqslant 3$).

Their mean values of Re_b and Re_τ (Figure 10a) simply follow the trends described above for the most probable value of the distribution, connecting the turbulent and the laminar branch, when Re_τ^G decreases. Away from the turbulent and laminar branches Re_τ is linearly related to Re_b, in agreement with the observation of a constant C_f. The standard deviation σ (Figure 10b) for Re_τ and Re_b decrease together with Re_τ^G. This decreasing trend agrees well with the experimental wall shear stress data reported in Reference [29]. The standard deviation for E_{cf} is found to increase with decreasing Re_τ^G, matching the trend reported in Reference [34].

The variation of the 3rd and 4th moments m_3 and m_4, i.e., the Skewness (S) and Kurtosis (K), versus Re_τ^G for the observable Re_τ and E_{cf} is shown in Figure 10c. These moments exhibit a strongly increasing trend with reducing Re_τ^G for both quantities. This similarity in behavior leads to $K \propto S^2$ as shown in Figure 10e. This correlation between the third and fourth statistical moments was first noted in Reference [48] for the fluctuating velocity in turbulent boundary layers at high Reynolds number. In the transitional regime, the same relationship has been found to hold in the experiments of Agrawal et al. [29] from wall shear stress data. We therefore confirm this yet-to-be-understood extension of a high Reynolds number scaling down to the spatio-temporal intermittent regime. Furthermore, we observe that the same scaling also holds for the turbulent kinetic energy E_{cf} (Figure 10e). In contrast it does not apply to Re_b (inset of Figure 10e). The reason is that while the Kurtosis follows the same trend as for the two other observables, (Figure 10d), the skewness shows a markedly different behavior: it is non-monotonous, changes sign twice and exhibit a maximum in the core of the spatio-temporal intermittent regime.

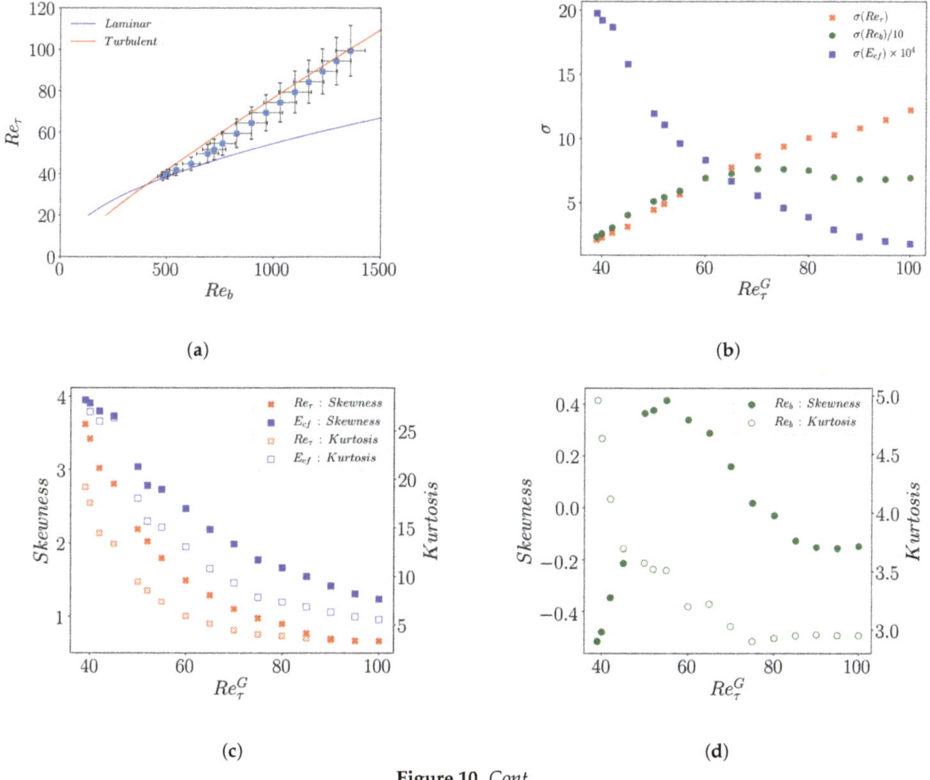

Figure 10. *Cont.*

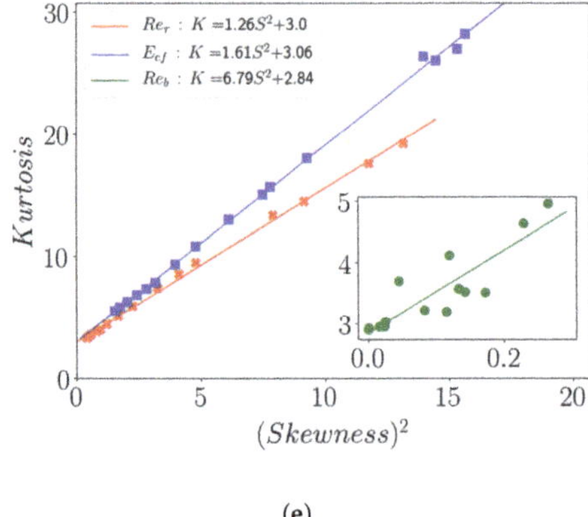

(e)

Figure 10. (**a**) Mean values (x_m) of Re_b and Re_τ. (**b**) Variation of the Standard deviation (σ) of Re_τ (red), Re_b (green), E_{cf} (blue) (indicated in the legend) vs. Re_τ^G. The $\sigma(Re_b)$ and $\sigma(Re_\tau)$ are scaled as indicated in the legend in order make them comparable. (**c**) Variation of Skewness (y-axis on left, filled symbols) and kurtosis (right y-axis, open symbols) vs. Re_τ^G for the observables Re_τ (red) and E_{cf} (blue) (**d**) Variation of Skewness (left y-axis on the left, filled symbol) and kurtosis (y-axis on right, open symbols) vs. Re_τ^G for the observable Re_b (green). (**e**) Kurtosis vs. squared skewness for Re_τ (red), Re_b (green, inset), E_{cf} (blue).

4. Discussion

The present simulations of the transitional regime of pPf confirm and extend previously documented knowledge, such as the constancy of C_f in the patterning regime and the variation of the band orientations close to the transition point.

The statistical analysis of the distribution of laminar gaps reveals that the distributions are exponentially tailed over the entire parameter range $39 \leqslant Re_\tau^G \leqslant 100$, demonstrating that even the value $Re_\tau^G = 39$ remains away from any sort of critical regime, which would be marked by algebraic distributions. This is consistent with the existing estimation of the location of the transitional critical point $Re_{cl} \approx 660$ [18,27], which translates to $Re_\tau^G \approx 36$. The entire patterning regime should thus be seen as bona fide spatio-temporal intermittency, with the critical behavior and transition point being relegated to values of $Re_\tau^G < 39$. Exploring the statistics of the flow closer to the critical point would require even larger domains and longer observation times. Such an investigation is outside the scope of the current study.

The orientation of the bands in the patterning regime for $60 \leqslant Re_\tau^G \leqslant 90$ ($1800 \leqslant Re_{cl} \leqslant 4050$) is essentially constant, with an angle $\theta = 25° \pm 2.5°$. This validates the choice of $\theta = 24°$ as a suitable value in the numerical approach of Tuckerman et al. [5,31,32], where slender computational domains are tilted at a chosen value of the angle. However, this angle of $24°$ no longer fits the mean orientation of the independent turbulent bands in the lower range $Re_\tau^G \leqslant 60$ ($Re_{cl} \leqslant 1800$), where the orientation of the bands increases by a factor close to two, with $\theta \approx 40°$ for $Re_\tau^G = 39$.

We confirm the observation of a constant C_f in the patterning regime, which also implies $\overline{\langle Re_\tau \rangle} \sim \overline{\langle Re_b \rangle}$, as reflected in Figure 10a. This constant value of C_f in the transitional regimes further enforces the long lasting analogy with first order phase transitions [49], for which the thermodynamic parameter conjugated to the order parameter remains constant while the system evolves from one homogeneous phase to the other, when a suitable control parameter is varied. At the mean-field level, a trademark of phase coexistence, is then the presence of a bimodal distribution of the order parameter in the coexistence regime. Capturing this bi-modality is however known as being a challenge, even in simulations of standard equilibrium systems: first, not all protocols allow for observing the phase coexistence; second, the order parameter must be coarse-grained on appropriate length-scales as compared to the correlation lengths such that non-mean-field effect do not dominate [50]. More than often, the bi-modality of the order parameter distribution is replaced by a mere concavity and a large kurtosis. If the two phases have very different fluctuations, as is the case here, one also expects a strong skewness of the distribution. Our observations extend the analogy, already reported at the level of the mean observable, to their fluctuations. However, a lot remain to be done to further exploit this analogy, in particular by making more precise what the relevant order and control parameters are. Let us stress that whether the analogy with a first order transition is valid or not, it does not preclude the dynamics at the spinodals from obeying a critical scenario, such as directed percolation close to the laminar phase spinodal [51] and a modulated instability of the turbulent flow close to the turbulent one [4].

Finally, the statistical moments showcased here demonstrate a correlation between the skewness and the kurtosis of both Re_τ and E_{cf}. Such a correlation, observed in both the transitional regime and higher Reynolds number turbulence but originally developed for the latter only [48], suggests a universal turbulent character, beyond the mere distinction transitional/featureless.

5. Conclusions

The transitional regime of pPf has been investigated numerically in large periodic domains. The transitional regime is composed of two sub-regimes each demarcated by a distinct behavior. The *patterning regime* is characterized, for $50 \leqslant Re_\tau^G \leqslant 90$, by a constant value of $C_f \approx 0.01$ and by a propagation downstream at approximately the mean bulk velocity $< u_b >$. For lower Re_τ^G all the way down to the critical point close to $Re_\tau^G \leqslant 36$, independent turbulent bands define a regime analogous to the puff regime of cylindrical pipe flow. The patterns are shown to exhibit a near constant angle of inclination $\theta = 25° \pm 2.5°$ for $60 \leqslant Re_\tau^G \leqslant 90$, which increases with reducing Re_τ^G. Both sub-regimes can be classified as spatiotemporally intermittent, as demonstrated by the exponential tails of the distribution of laminar gaps. The statistics of the local fields τ and u_b reinforce the feeling that a fruitful analogy with first order phase transitions could be developed, but the later remains to be made more precise and exploited.

Author Contributions: Conceptualization, Y.D. and O.D.; methodology, P.V.K., Y.D. and O.D.; data curation, P.V.K.; original draft preparation, Y.D. and P.V.K.; visualization, P.V.K.; supervision, Y.D. and O.D. All authors have read and agreed to the published version of the manuscript.

Funding: This research received no external funding.

Acknowledgments: This study was made possible using computational resources from IDRIS (Institut du Développement et des Ressources en Informatique Scientifique) and the support of its staff. We would like to acknowledge and thank the entire team of *channelflow.ch* for building the code and making it open source. The authors would also like to thank Takahiro Tsukahara, Kazuki Takeda, Jalel Chergui, Florian Reetz, Rob Poole, Rishav Agrawal, Laurette S. Tuckerman, and Sebastian Gomé for valuable discussions and technical input.

Conflicts of Interest: The authors declare no conflict of interest.

References

1. Reynolds, O., III. An experimental investigation of the circumstances which determine whether the motion of water shall be direct or sinuous, and of the law of resistance in parallel channels. *Proc. R. Soc. Lond.* **1883**, *35*, 84–99.
2. Tuckerman, L.S.; Chantry, M.; Barkley, D. Patterns in Wall-Bounded Shear Flows. *Annu. Rev. Fluid Mech.* **2020**, *52*, 343–367. [CrossRef]
3. Coles, D. Transition in circular Couette flow. *J. Fluid Mech.* **1965**, *21*, 385–425. [CrossRef]
4. Prigent, A.; Grégoire, G.; Chaté, H.; Dauchot, O.; van Saarloos, W. Large-scale finite-wavelength modulation within turbulent shear flows. *Phys. Rev. Lett.* **2002**, *89*, 014501. [CrossRef]
5. Barkley, D.; Tuckerman, L.S. Computational study of turbulent laminar patterns in Couette flow. *Phys. Rev. Lett.* **2005**, *94*, 014502. [CrossRef]
6. Duguet, Y.; Schlatter, P.; Henningson, D.S. Formation of turbulent patterns near the onset of transition in plane Couette flow. *J. Fluid Mech.* **2010**, *650*, 119–129. [CrossRef]

7. Cros, A.; Le Gal, P. Spatiotemporal intermittency in the torsional Couette flow between a rotating and a stationary disk. *Phys. Fluids* **2002**, *14*, 3755–3765. [CrossRef]

8. Ishida, T.; Duguet, Y.; Tsukahara, T. Transitional structures in annular Poiseuille flow depending on radius ratio. *J. Fluid Mech.* **2016**, *794*. [CrossRef]

9. Kunii, K.; Ishida, T.; Duguet, Y.; Tsukahara, T. Laminar-turbulent coexistence in annular Couette flow. *J. Fluid Mech.* **2019**, *879*, 579–603. [CrossRef]

10. Brethouwer, G.; Duguet, Y.; Schlatter, P. Turbulent-laminar coexistence in wall flows with Coriolis, buoyancy or Lorentz forces. *J. Fluid Mech.* **2012**, *704*, 137. [CrossRef]

11. Orszag, S.A. Accurate solution of the Orr–Sommerfeld stability equation. *J. Fluid Mech.* **1971**, *50*, 689–703. [CrossRef]

12. Carlson, D.R.; Widnall, S.E.; Peeters, M.F. A flow-visualization study of transition in plane Poiseuille flow. *J. Fluid Mech.* **1982**, *121*, 487–505. [CrossRef]

13. Alavyoon, F.; Henningson, D.S.; Alfredsson, P.H. Turbulent spots in plane Poiseuille flow–flow visualization. *Phys. Fluids* **1986**, *29*, 1328–1331. [CrossRef]

14. Henningson, D.S.; Alfredsson, P.H. The wave structure of turbulent spots in plane Poiseuille flow. *J. Fluid Mech.* **1987**, *178*, 405–421. [CrossRef]

15. Li, F.; Widnall, S.E. Wave patterns in plane Poiseuille flow created by concentrated disturbances. *J. Fluid Mech.* **1989**, *208*, 639–656. [CrossRef]

16. Henningson, D.S.; Kim, J. On turbulent spots in plane Poiseuille flow. *J. Fluid Mech.* **1991**, *228*, 183–205. [CrossRef]

17. Tsukahara, T.; Seki, Y.; Kawamura, H.; Tochio, D. DNS of turbulent channel flow at very low Reynolds Numbers. In *TSFP Digital Library Online*; Begel House Inc.: Danbury, CT, USA, 2005; pp. 935–940.

18. Tao, J.; Eckhardt, B.; Xiong, X. Extended localized structures and the onset of turbulence in channel flow. *Phys. Rev. Fluids* **2018**, *3*, 011902. [CrossRef]

19. Lemoult, G.; Shi, L.; Avila, K.; Jalikop, S.V.; Avila, M.; Hof, B. Directed percolation phase transition to sustained turbulence in Couette flow. *Nat. Phys.* **2016**, *12*, 254–258. [CrossRef]

20. Chantry, M.; Tuckerman, L.S.; Barkley, D. Universal continuous transition to turbulence in a planar shear flow. *J. Fluid Mech.* **2017**, *824*. [CrossRef]

21. Lemoult, G.; Aider, J.L.; Wesfreid, J.E. Experimental scaling law for the subcritical transition to turbulence in plane Poiseuille flow. *Phys. Rev. E* **2012**, *85*, 025303. [CrossRef]

22. Lemoult, G.; Aider, J.L.; Wesfreid, J.E. Turbulent spots in a channel: Large-scale flow and self-sustainability. *J. Fluid Mech.* **2013**, *731*. [CrossRef]

23. Lemoult, G.; Gumowski, K.; Aider, J.L.; Wesfreid, J.E. Turbulent spots in channel flow: An experimental study. *Eur. Phys. J. E* **2014**, *37*, 25. [CrossRef] [PubMed]

24. Hashimoto, S.; Hasobe, A.; Tsukahara, T.; Kawaguchi, Y.; Kawamura, H. An experimental study on turbulent-stripe structure in transitional channel flow. In Proceedings of the Sixth International Symposium on Turbulence, Heat and Mass Transfer, Rome, Italy, 14–18 September 2009; p. 10

25. Seki, D.; Matsubara, M. Experimental investigation of relaminarizing and transitional channel flows. *Phys. Fluids* **2012**, *24*, 124102. [CrossRef]

26. Sano, M.; Tamai, K. A universal transition to turbulence in channel flow. *Nat. Phys.* **2016**, *12*, 249. [CrossRef]

27. Paranjape, C. Onset of Turbulence in Plane Poiseuille Flow. Ph.D. Thesis, IST Austria, Klosterneuburg, Austria, 2019.

28. Whalley, R.; Dennis, D.; Graham, M.; Poole, R. An experimental investigation into spatiotemporal intermittencies in turbulent channel flow close to transition. *Exp. Fluids* **2019**, *60*, 102. [CrossRef]

29. Agrawal, R.; Ng, H.C.H.; Dennis, D.J.; Poole, R.J. Investigating channel flow using wall shear stress signals at transitional Reynolds numbers. *Int. J. Heat Fluid Flow* **2020**, *82*, 108525. [CrossRef]

30. Xiong, X.; Tao, J.; Chen, S.; Brandt, L. Turbulent bands in plane-Poiseuille flow at moderate Reynolds Numbers. *Phys. Fluids* **2015**, *27*, 041702. [CrossRef]

31. Tuckerman, L.S.; Kreilos, T.; Schrobsdorff, H.; Schneider, T.M.; Gibson, J.F. Turbulent-laminar patterns in plane Poiseuille flow. *Phys. Fluids* **2014**, *26*, 114103. [CrossRef]

32. Gomé, S.; Tuckerman, L.S.; Barkley, D. Statistical transition to turbulence in plane channel flow. *arXiv* **2020**, arXiv:2002.07435.

33. Kanazawa, T. Lifetime and Growing Process of Localized Turbulence in Plane Channel Flow. Ph.D. Thesis, Osaka University, Osaka, Japan, 2018. [CrossRef]
34. Shimizu, M.; Manneville, P. Bifurcations to turbulence in transitional channel flow. *Phys. Rev. Fluids* **2019**, *4*, 113903. [CrossRef]
35. Xiao, X.; Song, B. The growth mechanism of turbulent bands in channel flow at low Reynolds numbers. *J. Fluid Mech.* **2020**, *883*. [CrossRef]
36. Duguet, Y.; Schlatter, P. Oblique laminar-turbulent interfaces in plane shear flows. *Phys. Rev. Lett.* **2013**, *110*, 034502. [CrossRef] [PubMed]
37. Couliou, M.; Monchaux, R. Large-scale flows in transitional plane Couette flow: A key ingredient of the spot growth mechanism. *Phys. Fluids* **2015**, *27*, 034101. [CrossRef]
38. Gibson, J.F. Channelflow: A spectral Navier-Stokes simulator in C++. Tech. Rep. (U. New Hampshire, 2012). Available online: www.channelflow.org (accessed on 7 September 2020).
39. Gibson, J.; Reetz, F.; Azimi, S.; Ferraro, A.; Kreilos, T.; Schrobsdorff, H.; Farano, N.; Yesil, A.F.; Schütz, S.S.; Culpo, M.; et al. Channelflow2.0. Unpublished Work. 2020. Available online: www.channelflow.ch (accessed on 7 September 2020).
40. Lundbladh, A.; Johansson, A.V. Direct simulation of turbulent spots in plane Couette flow. *J. Fluid Mech.* **1991**, *229*, 499–516. [CrossRef]
41. Paranjape, C.S.; Duguet, Y.; Hof, B. Oblique stripe solutions of channel flow. *J. Fluid Mech.* **2020**, *897*. [CrossRef]
42. Fukudome, K.; Iida, O. Large-scale flow structure in turbulent Poiseuille flows at low-Reynolds Numbers. *J. Fluid Sci. Technol.* **2012**, *7*, 181–195. [CrossRef]
43. Vasudevan, M.; Hof, B. The critical point of the transition to turbulence in pipe flow. *J. Fluid Mech.* **2018**, *839*. [CrossRef]
44. Barkley, D.; Tuckerman, L.S. Mean flow of turbulent–laminar patterns in plane Couette flow. *J. Fluid Mech.* **2007**, *576*, 109–137. [CrossRef]
45. Kim, J.; Moin, P.; Moser, R. Turbulence statistics in fully developed channel flow at low Reynolds number. *J. Fluid Mech.* **1987**, *177*, 133–166. [CrossRef]
46. Tsukahara, T. Transition to/from turbulence in subcritical flows between two infinite parallel plates. In Proceedings of the Korea-Japan CFD Workshop 2010, POSCO International Center, Pohang, Korea, 19 November 2010; pp. 296–306.
47. Dean, R.B. Reynolds number dependence of skin friction and other bulk flow variables in two-dimensional rectangular duct flow. *J. Fluids Eng.* **1978**. [CrossRef]
48. Jovanović, J.; Durst, F.; Johansson, T.G. Statistical analysis of the dynamic equations for higher–order moments in turbulent wall bounded flows. *Phys. Fluids A Fluid Dyn.* **1993**, *5*, 2886–2900. [CrossRef]
49. Pomeau, Y. Front motion, metastability and subcritical bifurcations in hydrodynamics. *Phys. D Nonlinear Phenom.* **1986**, *23*, 3–11. [CrossRef]
50. Binder, K.; Landau, D.P. Finite-size scaling at first-order phase transitions. *Phys. Rev. B* **1984**, *30*, 1477–1485. [CrossRef]
51. Barkley, D. Theoretical perspective on the route to turbulence in a pipe. *J. Fluid Mech.* **2016**, *803*. [CrossRef]

Article

Low- and High-Drag Intermittencies in Turbulent Channel Flows

Rishav Agrawal [1,2], **Henry C.-H. Ng** [1], **Ethan A. Davis** [3], **Jae Sung Park** [3], **Michael D. Graham** [4], **David J.C. Dennis** [1] **and Robert J. Poole** [1,*]

[1]   School of Engineering, University of Liverpool, Liverpool L69 3GH, UK; ad5289@coventry.ac.uk (R.A.);
     hchng@liverpool.ac.uk (H.C.-H.N.); David.Dennis@liverpool.ac.uk (D.J.C.D.)
[2]   Fluid and Complex Systems Research Centre, Coventry University, Coventry CV1 5FB, UK
[3]   Department of Mechanical and Materials Engineering, University of Nebraska-Lincoln,
     Lincoln, NE 68588-0526, USA; ethan.davis@huskers.unl.edu (E.A.D.); jaesung.park@unl.edu (J.S.P.)
[4]   Department of Chemical and Biological Engineering, University of Wisconsin-Madison,
     Madison, WI 53706, USA; mdgraham@wisc.edu
*   Correspondence: robpoole@liverpool.ac.uk; Tel.: +44-151-794-4806

Received: 7 September 2020; Accepted: 29 September 2020; Published: 4 October 2020

Abstract: Recent direct numerical simulations (DNS) and experiments in turbulent channel flow have found intermittent low- and high-drag events in Newtonian fluid flows, at $Re_\tau = u_\tau h/\nu$ between 70 and 100, where u_τ, h and ν are the friction velocity, channel half-height and kinematic viscosity, respectively. These intervals of low-drag and high-drag have been termed "hibernating" and "hyperactive", respectively, and in this paper, a further investigation of these intermittent events is conducted using experimental and numerical techniques. For experiments, simultaneous measurements of wall shear stress and velocity are carried out in a channel flow facility using hot-film anemometry (HFA) and laser Doppler velocimetry (LDV), respectively, for Re_τ between 70 and 250. For numerical simulations, DNS of a channel flow is performed in an extended domain at $Re_\tau = 70$ and 85. These intermittent events are selected by carrying out conditional sampling of the wall shear stress data based on a combined threshold magnitude and time-duration criteria. The use of three different scalings (so-called outer, inner and mixed) for the time-duration criterion for the conditional events is explored. It is found that if the time-duration criterion is kept constant in inner units, the frequency of occurrence of these conditional events remain insensitive to Reynolds number. There exists an exponential distribution of frequency of occurrence of the conditional events with respect to their duration, implying a potentially memoryless process. An explanation for the presence of a spike (or dip) in the ensemble-averaged wall shear stress data before and after the low-drag (or high-drag) events is investigated. During the low-drag events, the conditionally-averaged streamwise velocities get closer to Virk's maximum drag reduction (MDR) asymptote, near the wall, for all Reynolds numbers studied. Reynolds shear stress (RSS) characteristics during these conditional events are investigated for $Re_\tau = 70$ and 85. Except very close to the wall, the conditionally-averaged RSS is higher than the time-averaged value during the low-drag events.

Keywords: hibernating turbulence; hot-film anemometry; turbulence; channel flow

---

1. Introduction

In the past few decades, the understanding of near-wall coherent structures has been greatly improved via the discovery of travelling-wave (TW) solutions [1]. These TW solutions were first obtained by Nagata [2] for plane Couette flow. They are non-trivial invariant solutions to the Navier–Stokes equation and are also sometimes called "exact coherent states (ECS)". Later, Waleffe [3,4] found ECS solutions for plane channel flow. The spatial structure of these solutions is similar to the commonly observed structure of near-wall turbulence: mean flow with counter-rotating streamwise vortices and alternating low- and high-speed streaks. Most of these ECS solutions are observed to occur in pairs at a saddle-node bifurcation point, arising at a finite value of Reynolds number. The upper branch solution has a higher fluctuation amplitude and higher drag than the lower branch solution [2–5].

One way to investigate the complex turbulent dynamics using TW solutions is to employ "minimal flow units". The minimal flow units or MFU denotes the smallest computational domain where turbulence can persist [6] at a given Reynolds number. Jiménez and Moin [6] observed a cyclic and intermittent behaviour of the fluctuations of all important quantities while employing MFU to study plane channel flow. They also observed a rapid increase in the fluctuations and wall shear stress during the "active" part of the cycle. Later, Hamilton et al. [7] and Jiménez and Pinelli [8] further studied this cycle and observed that during the time when the wall shear stress is near its lowest values the streamwise variation of the flow is also reduced. The presence of intermittency in Newtonian turbulent flow has also been investigated earlier by McComb [9]. Xi and Graham [10] carried out DNS in an MFU for low Reynolds number, $Re_\tau = u_\tau h / \nu = 85$ for both Newtonian and viscoelastic flows. Here, u_τ, h and ν are the friction velocity, channel half-height and kinematic viscosity, respectively. They observed that even in the limit of Newtonian flows, there are the moments of "low-drag" or "hibernating" turbulence, which display many similar features to MDR (a phenomenon generally associated with the polymer additives). They coined the nomenclature of a "hibernating" state when the flow was drag-reducing and resembles MDR, and "active" state for the rest of the flow. The major flow characteristics observed during hibernation were only weak streamwise vorticity and three-dimensionality, and lower than average wall shear stress. The frequency of these events increases with increasing viscoelasticity, although the events remain unchanged, i.e., they display similar flow properties as MDR. The connection between the polymeric drag reduction in turbulent flows and transition to turbulence in Newtonian flows has also been discussed earlier by Dubief et al. [11].

Xi and Graham [12] further investigated this phenomenon to provide detailed characteristics of active and hibernating turbulence in Newtonian and viscoelastic flows. They defined hibernation when the area-averaged wall shear stress was below 90% of the mean for a dimensionless time duration of $\Delta t^* = \Delta t u_\tau / h \gtrsim 3.5$, where Δt represents the dimensional time duration. Park and Graham [13] carried out DNS for MFU in a channel flow geometry, close to transition. They obtained five families of ECS solutions, which they denoted as the "P1, P2, P3, P4 and P5" solutions. Out of these five families of solutions, "P4" solution shows the most interesting behaviour. For the upper branch solutions, the velocity profile approaches the classic von Kármán log-law, while for the lower branch solutions the velocity profile approaches the Virk's MDR asymptote. They suggested that most of the time the turbulent trajectories remain at the upper-branch state (or the "active" state) with few excursions to the lower-branch state (or the hibernating state). This result provided a further verification that there are intervals of low-drag in Newtonian flows when the mean velocity profile is close to Virk's MDR profile as previously observed by Xi and Graham [10,12]. The existence of such solutions for Newtonian flows has a potential application in drag reduction, which makes it a practically significant field of research.

One major characteristic of wall-bounded turbulent flows is the so-called bursting process, which is an abrupt breaking of a low-speed streak as it moves away from the wall [14]. Itano and Toh [15] investigated the bursting process for channel flow at $Re_\tau = 130$ by computing TW solutions in a MFU using a shooting method. They observed that the bursting process is linked to the instability of the TW solution. Park et al. [16] studied the connection between the bursting process and the ECS solutions in minimal channel flow for $75 \leq Re_\tau \leq 115$. They focussed on the P4 family of ECS solutions, as identified earlier by Park and Graham [13]. To detect a hibernating event they used the criteria that the area-averaged wall shear stress should go below 90% of the mean wall shear stress and stays there for a duration of $\Delta t U_{cl,lam} / h > 65$, where $U_{cl,lam}$ is the laminar centerline velocity. This time-duration corresponds to $\Delta t^* > 3$ for $Re_\tau = 85$. They defined bursting events based on an increase in the volume-averaged energy dissipation rate by 50% of its standard deviation for a duration of $\Delta t U_{cl,lam}/h > 15$. They observed that many of the low-drag or hibernating events are followed by strong turbulent bursts. Based on this observation, they divided the turbulent bursts into two categories: weak and strong bursts, and suggested that the strong bursts are the ones which are always preceded by a hibernating event. They also investigated the possible link between the turbulent bursts and the instability of the P4-lower branch solution. Very similar trajectories were observed for the strong bursts and the lower branch of the P4 solution, which provides further evidence that the turbulent bursts are directly related to the instability of the ECS.

Initially, the investigation of these low-drag events was conducted in minimal channels, and therefore the need was to study this phenomenon for fully turbulent flow in extended domains. The relation between the minimal channels and flow in large domains was studied by Jiménez et al. [17] and Flores and Jiménez [18]. They suggested that the flow dynamics in minimal channels have many features that are representative of fully turbulent flows. It has also been seen that some of these solutions are highly localised and display the nontrivial flow only for a small region of an extended domain, whereas the rest of the flow remains laminar [19–21]. Kushwaha et al. [22] carried out an investigation into these low-drag events in an extended domain for channel flow at three Reynolds numbers, $Re_\tau = 70$, 85 and 100. The computational domain, in wall (or inner) units, was $L_x^+ \approx 3000$ and $L_z^+ \approx 800$ long in the streamwise and spanwise directions, respectively. They carried out a temporal and spatial analysis for extended domains and compared the results between the two. Regions or events of both low- and high-drag events were investigated in large domains, unlike previous MFU studies where the focus was primarily on low-drag events. To study the temporal intermittency, they employed the following criteria to detect low-drag (hibernating) or high-drag (hyperactive) events: the instantaneous wall shear stress (τ_w) should remain below 90% or above 110% of time-averaged value for a time duration of $\Delta t^* = \Delta t u_\tau/h = 3$ for low or high drag events, respectively. For studying the velocity characteristics during these low- and high-drag intervals in the flow, a conditional sampling technique was employed. They observed that, although the temporal and spatial analyses are independent of each other, the characteristics of low- and high-drag events obtained using these two methods were very similar. They found that for Re_τ between 70 and 100, the regions of low-drag in an extended domain show similar conditional mean velocity profiles as obtained from temporal interval of low-drag in minimal channels for $y^+ = y u_\tau/\nu < 30$, where y is the wall-normal distance. This showed that the spatiotemporal intermittency observed in extended channel flow is related to the temporal intermittency in a minimal channel.

Whalley et al. [23,24] carried out an experimental investigation of the low- and high-drag events in a plane channel flow at three Reynolds numbers, $Re_\tau = 70$, 85 and 100. Instantaneous velocity, wall shear stress and flow structure measurements were conducted using laser Doppler velocimetry (LDV), hot-film anemometry (HFA) and stereoscopic particle image velocimetry (SPIV), respectively. They employed the same criteria as Kushwaha et al. [22] to detect the low-drag events, but for the high-drag events, the criteria were slightly relaxed in order to obtain more events, as the high-drag events were found to occur at a lower frequency than the low-drag events. Instantaneous velocity and wall shear stress measurements were made at the same streamwise/spanwise location, enabling conditional sampling of the velocity data to be carried out. The conditionally averaged streamwise velocity and wall shear stress were found to be highly correlated until $y^+ \approx 40$ and a resemblance was observed between the conditionally sampled mean velocity profiles for $y^+ \lesssim 40$ and the lower branch of the P4 ECS solution as observed earlier in minimal channels [13]. They also observed that the fraction of time spent in hibernation (low-drag) decreases with increasing Reynolds number for $70 < Re_\tau < 100$.

Recently, Pereira et al. [25] carried out DNS in channel flow of domain size, $L_x \times L_y \times L_z = 8\,\pi h \times 2\,h \times 1.5\,\pi h$ at Re_τ between 69.26 and 180 for Newtonian flow, and at $Re_{\tau 0} = 180$ for drag-reducing flow (65% drag reduction). The flow was identified as hibernating if the spatially-averaged wall shear stress was lower than 95% of its time-averaged value and no time criteria were used (unlike previous studies where a minimum time duration was also used to detect a hibernating event, for example, in [16,22,24]). They demonstrated that the transition to turbulence in Newtonian flows shares various common features to the polymer induced drag reduction in turbulent flows.

Until now, these low- and high-drag events are investigated for $70 \leq Re_\tau \leq 100$, and therefore a natural question arises as to what are the characteristics of these events in the so-called fully-turbulent flow regime (often associated with a threshold value of $Re_\tau \geq 180$ [26]). The Reynolds shear stress characteristics during these events has been studied using the DNS in MFUs [12,13], yet there is no relevant experimental data or numerical data in extended domains available. In this paper, the low- and high-drag intermittencies are investigated using experimental and numerical techniques to answer these fundamental questions. The experiments are conducted in a channel flow facility using wall shear stress and velocity measurements. Recently, Agrawal et al. [27] observed that the flow in the present channel consists only of turbulent events beyond $Re_\tau \approx 67$ and that significant Reynolds number dependence of the skewness and flatness of wall shear stress fluctuations starts to disappear by $Re_\tau \simeq 73 - 79$. Based on these results, in this work, the intermittences associated with the turbulent flow are investigated for $Re_\tau \geq 70$. An experimental study is made for Reynolds number up to $Re_\tau = 250$, to probe the characteristics of these events for fully-turbulent channel flow. To study the Reynolds shear stress for $Re_\tau = 70$ and 85, experimental as well numerical techniques are employed.

2. Experimental Set-Up

In this study, a channel flow facility at the University of Liverpool has been utilised to carry out the experimental investigation. The same facility has been used earlier by Whalley et al. [23,24] and Agrawal et al. [27,28,29], and is shown here in Figure 1. The channel-flow facility is a rectangular duct consisting of 6 stainless steel modules and a test section. The test section is connected downstream of five stainless steel modules. Each module is of length 1.2 m and the test section has a length of 0.25 m. The width (w) and half-height (h) of the duct are 0.298 m and 0.0125 m, respectively, giving an aspect ratio ($w/2h$) of 11.92. The modules are constructed in such a manner as to ensure a hydraulically smooth transition between the modules.

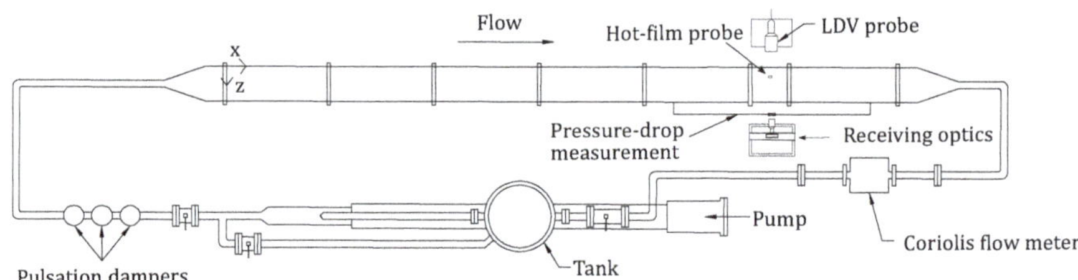

Figure 1. Schematic of channel-flow flow facility (not to scale).

The working fluid is stored in a stainless steel header tank of capacity about 500 L. A Mono type E101 progressive cavity pump is used to circulate the fluid via the tank in a closed loop. The flow loop also consists of an additional mixing loop which provides an opportunity for having lower flow rates. Three pulsation dampers are situated just after the pump, which helps in damping any pulsations in the flow before entering the channel. A Promass Coriolis flow meter is installed in the return loop to measure the mass flow rate ($\dot{m}$) of the fluid. This enables the bulk velocity (U_b) to be determined by the relation $U_b = \dot{m}/(\rho A)$, where A is the cross-sectional area of the channel and ρ is the density of the working fluid. A platinum resistance thermometer (PRT) is present in the last module of the channel which is used to measure the temperature of the working fluid. The PRT is powered by an Agilent 34,970 A switch unit, which provides temperature readings with a resolution of 0.01 °C. Throughout this study, only Newtonian fluids are used as working fluids. These are water–glycerol mixtures of different concentrations where glycerol is used to increase the viscosity to get to lower Reynolds number. For example, while studying the flow for $Re_\tau \geq 180$, water is used as the working fluid and while studying low Reynolds number flow ($Re_\tau = 70$), a 65% : 35% by weight glycerol–water mixture is used as the working fluid. The density of the working fluid is measured using an Anton Paar DMA 35 N density meter. The shear viscosity of the working fluid is measured using an Anton Paar MCR 302 rheometer. A cone and plate geometry is employed to measure shear viscosity for shear rate ($\dot{\gamma}, s^{-1}$) ranging from 10^{-2} to 10^{2}.

Pressure-drop measurements are conducted using a Druck LPX-9381 low-differential pressure transducer, which has a working range of 5 kPa with an accuracy of ± 5 Pa. A Baratron differential pressure transducer made by MKS is used to regularly calibrate the Druck pressure transducer. Instantaneous wall shear stress and velocity measurements are carried out using a hot-film anemometry (HFA) system and a laser Doppler velocimetry (LDV) system, respectively, in the test section. The side- and top-walls of the test section are made of borosilicate glass to provide optical access for the LDV measurements. A Dantec FiberFlow laser system is employed for velocity measurements which uses a 300 mW argon-ion continuous wave laser. Up to two component velocity measurements have been carried out thus requiring two pairs of laser beams of different wavelengths: blue (488 nm) and green (515.5 nm). A Bragg cell is utilised to resolve the directional ambiguity of the velocity of seeding particles by giving a frequency shift of 40 MHz to one of the laser beams. The laser beams are emitted using a transmitting optics (or laser head) which provides a beam separation of 51.5 mm and a focal length of 160 mm in air. The crossing of two beams of the same colour creates a measurement volume of 24 μm diameter and 150 μm length in air. The transmitting optics is placed on a traverse which allows movement of the measurement volume in all three directions. For the seeding particles, generally, natural particles present in the working fluid (for example, supply water) are found to be sufficient to obtain a good data rate. In cases where the natural seeding particles are found to be low, for example, when the working fluid has a high concentration of glycerol, Timiron Supersilk MP-1005, having an average size of 5 μm, are added to the working fluid. In this study, both single component and two-component velocity measurements have been carried out. In the case of two-component velocity measurements, the data are acquired in co-incident mode. This mode samples both velocity components of the same seeding particle simultaneously in the measurement volume. The LDV is operated in a forward-scatter mode and the typical data rate is around 100–500 Hz. The light scattered from the seeding particle enters the photodetector (receiving optics) which splits the laser beams based on the wavelengths. The laser beams then pass to the photomultiplier tubes (PMTs) which sends the Doppler frequencies to the flow processor, burst spectrum analyzer (BSA)-F50, made by Dantec Dynamics. The signals are converted to the corresponding velocity signals using the inbuilt signal processors in the flow processor.

Calculation of RSS requires simultaneous measurements of streamwise and wall-normal velocities, but the wall-normal velocity measurements cannot be made close to the bottom wall because of the cut-off of the laser beams [30], and therefore some modifications to the transmitting optics of the LDV set-up are made. The first modification is to rotate the laser head by 45° about the spanwise axis to get closer to the bottom wall, similarly to as previously done by Melling and Whitelaw [31], Walker and Tiederman [32] and Günther et al. [33]. Streamwise (U) and wall-normal (V) velocity components are recovered based on the coordinate transformation equation, as shown below.

$$\begin{bmatrix} U \\ V \end{bmatrix} = \begin{bmatrix} \cos 45° & \sin 45° \\ -\sin 45° & \cos 45° \end{bmatrix} \begin{bmatrix} U_1 \\ U_2 \end{bmatrix}. \tag{1}$$

Here, U_1 and U_2 are the velocity components measured by blue and green beams, respectively. This modification makes the minimum vertical height where the measurement of the wall-normal velocity component can be made reduced by a factor of $1/\sqrt{2}$. Next, an external LD1613-N-BK7 biconcave lens, made by Thorlabs, is placed in front of the laser head to increase the focal length of the laser beams. This lens has a diameter of 25.4 mm and a focal length of 100 mm. Increasing the focal length enables the measurement volume to go further into the test section from the side-wall. Therefore, if the aim is to measure at the same spanwise location in the test section, the laser head needs to be moved further back from the side-wall. This modification enables the laser beams to be closer to each other when they enter through the side-wall. The measurement volume can get closer to the bottom wall as the laser beams get closer to each other. Thus, the two-component velocity measurements can be carried out closer to the bottom-wall after the addition of a biconcave lens. The lens is connected on a lens mount which is attached to an optical post. The optical post is then attached to the traverse of the transmitting optics. Therefore, the entire lens system can be traversed with the transmitting optics. It is important that both pairs of laser beams are aligned properly to the external lens. This alignment is checked based on the high data rate of the LDV signal in co-incident mode and validating the time-averaged RSS profile against available DNS data at the same Reynolds number. By making these two modifications, the two-component velocity measurements can be conducted for $y/h \geq 0.3$ at a spanwise location of $z/h = 5$ in the channel-flow facility.

In this study, constant temperature anemometry (CTA) is employed for measuring the instantaneous wall shear stress by utilising the commercially available 55R48 glue-on hot-films probes (made by Dantec Dynamics). The hot-film sensor has a physical spanwise length (Δz) of 0.9 mm. In inner units, this corresponds to $\Delta z^+ = 18$ for $Re_\tau = 250$. In this study, the effect of measurement resolution issues due to sensor sizes are thought to be negligible as Ligrani and Bradshaw [34] considered a sensor length of about $\Delta z^+ \lesssim 20 - 25$ to be acceptable to make well-resolved turbulence measurements. In order to attach the sensor to the channel wall, removable Delrin plugs are designed and fabricated inhouse. The hot-film probes are glued on these plugs and these plugs are then inserted into the bottom wall of the test section. We ensure that the hot-films are flush with the bottom wall of the test section. A detailed description of the mounting process for the hot-film probes in the present channel has been provided in Agrawal [35]. The probe is powered by a Dantec StreamLine Pro velocimetry system. The bridge ratio and the overheat ratio of the anemometer are set at 10 and 1.1, respectively. The typical frequency response of the anemometer, against the square-wave generator is found to be around 10–30 kHz, which is generally considered sufficient for turbulence measurements [36]. The output voltage signal from the anemometer is then digitized using a 14-Bit USB6009 Multifunction A/D converter, made by National Instruments. After A/D converter, the signal is acquired using the CTA application software, StreamWare Pro, installed on the computer. In the case of simultaneous measurements of velocity and wall shear stress, the digitised voltage is sampled by the BSA flow processor which helps in the acquisition of time-synchronised velocity and wall shear stress data. The voltage output signals from the anemometer is converted to instantaneous wall shear stress signals using calibration against the mean pressure-drop obtained from the pressure transducer. The same procedure for the hot-film calibration as discussed in Agrawal et al. [27,28] has been conducted here.

In CTA, all the changes in the fluctuations in voltage output from the anemometer should be representative of fluctuations in the flow. Therefore, any change in voltage output due to thermal and non-thermal drifts need to be minimised. To minimise the thermal drift, an open-loop copper cooling coil is added to the overhead tank and the main supply water is used to control the temperature of the working fluid. Using this set-up, the temperature of the working fluid could be controlled to the precision of ± 0.01 °C for the entire experimental run of the day (typically about 6–8 h). Non-thermal drifts are also observed which are generally caused due to the contamination of the hot-films [37]. A novel nonlinear regression technique, as discussed in Agrawal et al. [28], has been employed to recover the wall shear stress signals from the drifted voltage signal.

Experiments are conducted for five Reynolds numbers: $Re_\tau = 70, 85, 120, 180$ and 250 and for each Reynolds number, wall shear stress and velocity data are acquired simultaneously in the measurement test section using HFA and LDV, respectively, at a location of $z/h = 5$ and $x/h = 496$. As discussed in Agrawal et al. [27], the spanwise location of $z/h = 5$ is observed to be devoid of side-wall effects.

Velocity acquisition is realised at various wall-normal locations, where each wall-normal location is sampled for 2 h at a typical data rate of around 300–400 Hz. Table 1 shows the Reynolds numbers, corresponding wall-normal locations studied and the parameters measured in this work. For $Re_\tau = 70$ and 85, both streamwise and wall-normal velocity components are measured simultaneously with the wall shear stress. These particular measurements have been conducted to study the RSS behaviour during the low- and high-drag events. For other Reynolds numbers, due to experimental limitations, only streamwise velocity measurements have been executed along with the wall shear stress because the near peak region of the RSS could not be measured for higher Reynolds numbers as this moves physically closer to the wall at higher Reynolds numbers where the LDV beams lose optical access.

Table 1. Reynolds numbers and various wall-normal locations studied. Parameters measured for each Reynolds numbers are also shown.

Re_τ	y^+	Parameters
70	21, 24, 28, 32, 35, 40, 46, 51, 60, 68	τ_w, U, V
85	26, 31, 36, 32, 41, 47, 54, 61, 76, 85	τ_w, U, V
120	22, 26, 30, 37, 46, 59, 71, 85, 103	τ_w, U
180	24, 30, 38, 48, 60, 75, 98, 128, 157	τ_w, U
250	35, 45, 58, 74, 94, 118, 143, 171, 202, 242	τ_w, U

The procedure described by Kline and McClintock [38] has been employed here to conduct an uncertainty analysis of the measured and calculated variables. The employed channel-flow facility is carefully machined to provide negligible relative uncertainties (~0.15%) in the channel dimensions (w and h) and the length between the pressure tappings, l. The pressure transducer has an accuracy of ± 5 Pa, and therefore the relative uncertainty in the mean wall shear stress is $\Delta \tau_w / \overline{\tau_w} = 1\text{--}3\%$. The density meter has a quoted accuracy of ± 1 kg/m^3. This gives a relative uncertainty in the density of the working fluid of $\Delta \rho / \rho = 0.09\%$. The relative uncertainty in the viscosity (μ) measurement of the working fluid using the rheometer is $\Delta \mu / \mu = 2\%$. The relative uncertainty in the friction velocity ($u_\tau = \sqrt{\tau_w / \rho}$) is $\Delta u_\tau / u_\tau = 0.5\text{--}1.5\%$. This gives an uncertainty in the friction Reynolds number ($Re_\tau = u_\tau h / \nu$) measurement of $\Delta Re_\tau / Re_\tau = 2\text{--}2.5\%$. The major sources of error in LDV data are due to velocity gradient broadening, velocity bias effect or fringe distortion [39]. These combined effects, in general, give the relative uncertainties in the mean velocity of 2–3% and the turbulent intensities of 4–6%. In inner units, the relative uncertainties in the mean velocities and turbulent intensities are $\Delta U^+ / U^+ = 2\text{--}3.5\%$ and $\Delta uv^+ / uv^+ = 4\text{--}7\%$. Here, u and v represent streamwise velocity fluctuation and wall-normal velocity fluctuation, respectively. The LDV transmitting optics traverse has a precision of 0.001 mm, providing a relative uncertainty in the wall-normal position (y) measurement, close to the wall ($y = 0.5$ mm), to be $\Delta y / y = 0.2\%$. In inner units, at this wall-normal location, y^+ has an uncertainty of $\Delta y^+ / y^+ = 2\text{--}2.5\%$.

In this study, two different ways of averaging the measured variables are carried out: time-averaging and conditional-averaging. To differentiate between these two averages the following nomenclature are used: an overbar indicates a time-averaged quantity (e.g., $\overline{U}$), and an overbar with an L or H superscripts indicates the conditionally-averaged quantity for low- and high-drag events (e.g., $\overline{U}^L, \overline{U}^H$), respectively. Similarly, friction velocities are calculated using two different wall shear stress: time-averaged wall shear stress ($\overline{u_\tau}$) and conditionally-averaged wall shear stress ($\overline{u_\tau}^L, \overline{u_\tau}^H$). Based on these definitions of the friction velocities, the wall-normal locations are also normalised in three different ways: $y^+ = y\overline{u_\tau}/\nu$, $y^{+L} = y\overline{u_\tau}^L/\nu$ and $y^{+H} = y\overline{u_\tau}^H/\nu$.

3. Numerical Procedure

We consider an incompressible Newtonian fluid in the plane Poiseuille (channel) geometry, driven by a constant volumetric flux Q. The x, y and z coordinates are aligned with the streamwise, wall-normal and spanwise directions, respectively. Periodic boundary conditions are imposed in the x and z directions with fundamental periods L_x and L_z, and a no-slip boundary condition is imposed at the walls $y = \pm h$, where $h = L_y/2$ is the half-channel height. The laminar centreline velocity for a given volumetric flux is given as $U_{cl,lam} = (3/4)Q/h$. Using the half-height h of the channel and the laminar centreline velocity $U_{cl,lam}$ as the characteristic length and velocity scales, respectively, the non-dimensionalised Navier–Stokes equations are given as

$$\nabla \cdot u = 0, \tag{2}$$

$$\frac{\partial u}{\partial t} + u \cdot \nabla u = -\nabla p + 1/(Re_c)\nabla^2 u. \tag{3}$$

Here, we define the Reynolds number for the given laminar centreline velocity as $Re_c = U_{cl,lam}h/\nu$, where ν is the kinematic viscosity of the fluid. Characteristic inner scales are the friction velocity $u_\tau = \sqrt{(\overline{\tau_w}/\rho)}$ and the near-wall length scale, or wall unit, $\delta_\nu = \nu/u_\tau$, where ρ is the fluid density and $\overline{\tau_w}$ is the time- and area-averaged wall shear stress. Quantities non-dimensionalised by the inner scales are denoted with a superscript '+'. The friction Reynolds number is then defined as $Re_\tau = u_\tau h/\nu = h/\delta_\nu$. For the current simulations, friction Reynolds numbers of $Re_\tau = 70$ and 85 are considered. Simulations are performed using the open source code ChannelFlow written and maintained by Gibson [40]. We focus on a domain of $L_x \times L_y \times L_z = 13.64\ \pi h \times 2\ h \times 3.64\ \pi h$. These dimensions correspond to $L_x^+ \times L_z^+ \approx 3000 \times 800$ for $Re_\tau = 70$, and $L_x^+ \times L_z^+ \approx 3640 \times 970$ for $Re_\tau = 85$. A numerical grid system is generated on $N_x \times N_y \times N_z$ (in x, y, and z) meshes, where a Fourier–Chebyshev–Fourier spectral spatial discretisation is applied to all variables. A resolution of $(N_x, N_y, N_z) = (196, 73, 164)$ is used for both Reynolds numbers. The numerical grid spacing in the streamwise and spanwise direction are $\Delta x_{min}^+ \approx 15.3\ (18.6)$ and $\Delta z_{min}^+ \approx 4.9(5.9)$ for $Re_\tau = 70$ and $(Re_\tau = 85)$ cases. The nonuniform Chebyshev spacing used in the wall-normal direction results in $\Delta y_{min}^+ \approx 0.07\ (0.08)$ at the wall and $\Delta y_{max}^+ \approx 3.0\ (3.7)$ at the channel centre for $Re_\tau = 70$ and $(Re_\tau = 85)$ cases. For the computation time, 50×10^3 strain times $(> 25Re_c)$ is chosen to attain meaningful statistics.

The present experiment provides temporal information for the flow, and therefore for a comparison of the DNS and experimental data, temporal information from the DNS data is extracted. To obtain reliable statistics, nine wall locations are chosen at the wall on the top and on the bottom walls of the computational domain. These locations are selected in such a way that each spatial location is not correlated with the others [22]. The streamwise/spanwise spatial locations correspond to the combinations of three x^+ locations and three z^+ locations: $x^+ \approx 505$, 1500 and 2495; $z^+ \approx 151$, 400 and 649 for $Re_\tau = 70$, and $x^+ \approx 613$, 1820 and 3027; $z^+ \approx 183$, 485 and 787 for $Re_\tau = 85$. The instantaneous wall shear stress is obtained by using the streamwise velocity gradient information at $y^+ \approx 1$, although no difference in its value was observed between $y^+ \approx 1$ and lower y^+ locations.

4. Identifying Low- and High-Drag Events

Figure 2a shows the PDF (probability density function) of wall shear stress fluctuations (τ_w') obtained at $Re_\tau = 180$ using experiments. The PDF of wall shear stress has a longer positive tail which means that the PDF is positively skewed. This shows that some of the positive fluctuations have much larger magnitude than the negative fluctuations. In the present study, the wall shear stress is representative of the skin-friction drag. Previously, Gomit et al. [41] used the PDF of wall shear stress to divide low- and high-wall shear stress events in a turbulent boundary layer. They divided the PDF into four quartiles, where each quartile contains one-fourth of the realisations. In this study, to define the low- and high-drag "events", two significant parameters are considered: the magnitude of the wall shear stress fluctuations and the duration of time the fluctuations stay below or above the time-averaged value.

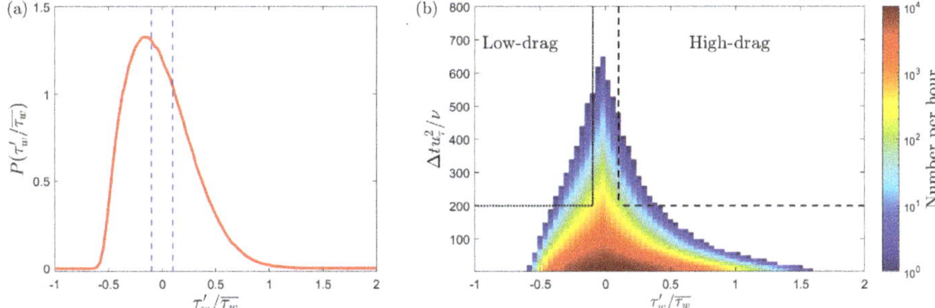

Figure 2. (**a**) PDF of wall shear stress fluctuations for $Re_\tau = 180$ obtained using experiments. Blue dashed lines represent the threshold criteria for the low- and high-drag events, i.e., $\tau_w/\overline{\tau_w} < 0.9$ and $\tau_w/\overline{\tau_w} > 1.1$, respectively. (**b**) Distribution of negative and positive wall shear stress fluctuations per hour for $Re_\tau = 180$ obtained using experiments. The black dotted lines cover the region of low-drag events based on the criteria: $\tau_w/\overline{\tau_w} < 0.9$ and $\Delta t_{cr}^+ = 200$ and the black dashed lines cover the region of high-drag events based on the criteria: $\tau_w/\overline{\tau_w} > 1.1$ and $\Delta t_{cr}^+ = 200$.

The PDF of wall shear stress fluctuations, as shown in Figure 2a, provides statistical information about the magnitude of the fluctuations but information regarding the time-duration of the fluctuations cannot be inferred. Therefore, it is necessary to find a way to visualise all the positive and negative fluctuations as a function of the magnitude and time-duration. This is carried out by calculating the distribution of all the fluctuations (τ_w') about the time-averaged value ($\overline{\tau_w}$) with their corresponding time durations (Δt). Figure 2b shows this distribution for $Re_\tau = 180$. Here, inner scaling (u_τ^2/ν) is used to scale the time-duration of the negative and positive wall shear stress fluctuations. The strength of the wall shear stress fluctuations is given by $\tau_w'/\overline{\tau_w}$. The number of these fluctuations is higher for the lower strengths and lower time-durations.

In this study, to detect a low-drag or a high-drag event, a magnitude threshold criterion and a time duration criterion are employed on the wall shear stress signals. For the threshold criteria, values less than $0.9\tau_w$ for the low-drag events and greater than $1.1\tau_w$ for the high-drag events have been typically employed previously by Kushwaha et al. [22]. Whalley et al. [24] used the same threshold criteria for the low-drag events, but for the high-drag events they employed a less stringent criteria of greater than $1.05\tau_w$, in order to obtain more data points to carry out the statistical analysis. In the present study, the same values for the threshold criteria as used by Kushwaha et al. [22] are employed to detect the conditional events; however, the effect of varying the threshold criteria will also be discussed. For the time-duration criteria, Kushwaha et al. [22] and Whalley et al. [23,24] employed a mixed scaling ($\Delta t^* = \Delta t u_\tau/h$) to detect conditional events in channel flows. They typically used $\Delta t^* = 3$ as the time-duration criterion while discussing the sensitivity of the value of the time-duration criterion on the conditional quantities. Unlike these previous studies, in the present investigation, an inner scaling is used for the time-duration criterion for the conditional events: $\Delta t^+ = 200$ is used as the minimum time-duration to detect conditional events. The reasons for, and implications of, choosing this scaling will be discussed in detail in the next section. The effect of varying the length of the time-duration criterion on the conditional quantities will be discussed in Section 6. To further understand the definition of these conditional events, examples of instantaneous wall shear stress signals meeting the above-mentioned criteria for the low-drag and the high-drag events are shown in Figure 3. This figure shows the instantaneous normalised wall shear stress during the low-drag (Figure 3a) and the high-drag (Figure 3b) events. In Figure 3, the acquisition time of the wall shear stress is shifted such that $t^+ = 0$ indicates the beginning of a low- or a high-drag event. Each event is shown to act longer than the minimum time duration (for "low-drag" ~230 units and for "high-drag" ~320 units).

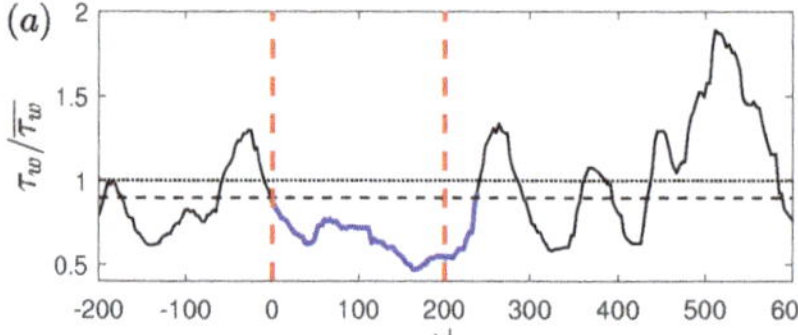 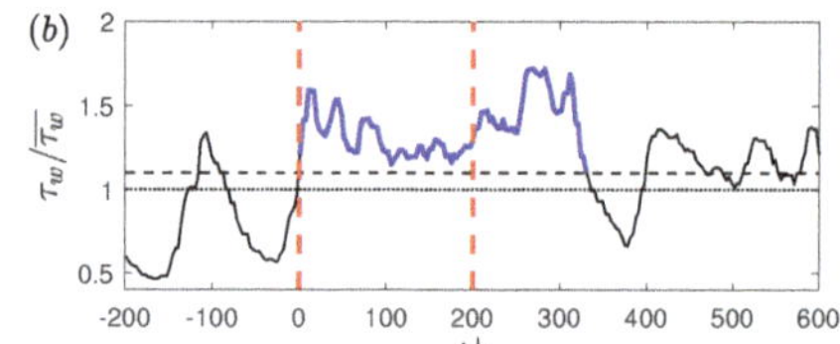

Figure 3. Time history of normalised wall shear stress at $Re_\tau = 180$ during (**a**) a low-drag and (**b**) a high-drag event obtained using experiments. Blue solid lines highlight the low-drag and the high-drag events in panels (**a**,**b**), respectively. Black dotted lines show mean value of normalised wall shear stress $\tau_w/\overline{\tau_w} = 1$. Black dashed lines show $\tau_w/\overline{\tau_w} = 0.9$ and $\tau_w/\overline{\tau_w} = 1.1$ in panels (**a**,**b**), respectively. Red dashed line indicates the time-duration criteria of $\Delta t_{cr}^+ = 200$. In panels (**a**,**b**), t^+ is shifted such that $t^+ = 0$ indicates the beginning of a conditional event.

5. Time Spent in Low- and High-Drag Events

Here we study the effect of three different scalings, i.e., inner scaling, mixed scaling and outer scaling for the time-duration criteria to detect a conditional event. Outer scaling is simply $\Delta t U_b/h$. Inner scaling ($\Delta t^+ = \Delta t \frac{u_\tau^2}{\nu}$) and the mixed scaling ($\Delta t^* = \Delta t \frac{u_\tau}{h}$) are related by the following relation.

$$\Delta t^+ = Re_\tau \Delta t^*. \tag{4}$$

From Equation (4), it can be observed that with increasing Reynolds numbers, the Δt^+ value increases for the same Δt^* value. Whalley et al. [24] studied the fraction of time spent in low- and high-drag events with changing Reynolds numbers where the time-duration criterion was kept constant in mixed scaling. They observed that with increasing Reynolds number between $70 \leq Re_\tau \leq 100$, the fraction of time spent in low-drag events decreases by approximately 500% while increasing the Re_τ from 70 to 100. The effect of other scalings has not been considered previously.

The fraction of time spent in the conditional events is investigated for $Re_\tau = 70, 85, 120, 180$ and 250 using all three scalings. For $Re_\tau = 70$, $\Delta t u_\tau/h = 3$ corresponds to about $t u_\tau^2/\nu = 200$ and $t U_b/h = 42$. Based on this information, three values are chosen for each scaling to study the effect of Reynolds number on the fraction of time spent in the conditional events. For the mixed scaling, $\Delta t u_\tau/h = 1, 2$ and 3, for outer scaling, $t U_b/h = 15, 30$ and 45, and for the inner scaling, $t u_\tau^2/\nu = 100, 200$ and 300 are used. For the low-drag events the threshold criterion is kept constant as $\tau_w/\overline{\tau_w} < 0.9$ and for the high-drag events the threshold criterion is kept constant as $\tau_w/\overline{\tau_w} > 1.1$.

Figure 4 shows the fraction of time spent in low- and high-drag for different Reynolds numbers and the time-duration criteria. Results are shown for both the experimental as well as DNS data. It can be observed that the fraction of time spent in low-drag or high-drag decreases with increasing Reynolds numbers when mixed or outer scaling is used for the time duration criteria. This is similar to the result obtained using the mixed scaling for the time-duration criteria by Whalley et al. [24]. However, the fraction of time spent in the conditional events remains almost independent of the Reynolds number for $70 \leq Re_\tau \leq 250$ for the experimental data, when the time-duration criteria is kept constant in inner units. DNS data shows a qualitatively consistent behaviour (i.e., show a similar trend for all three scalings) in the fraction of the conditional events compared to the experimental data although for a smaller range of Reynolds numbers. One possibility for the differences observed between DNS and experiments here is that these very rare low- or high-drag events involve flow structures that are much longer in the streamwise direction than usual, and that a domain size that is adequate for the vast majority of the turbulent dynamics might not be long enough to quantitatively capture the frequency of these rare events. Alternatively, subtle differences caused by the finite aspect ratio of the experimental set-up in comparison to the periodic boundary conditions used in the simulations, or the inherent uncertainties associated with the calibration of the hot-film signals maybe the cause of these differences. Based on this observation, inner scaling is chosen for the time-duration criteria in the remainder of this paper. Figure 4e,f also shows that increasing the value of the time-duration criteria ($100 \leq \Delta tu_\tau^2/\nu \leq 300$) decreases the fraction of time spent in these conditional events. The fraction of time spent in the intervals of low-drag is found to be greater than the intervals of high-drag for the same values of the time-duration criteria for $100 \leq \Delta tu_\tau^2/\nu \leq 300$, and where the threshold criteria is kept the same in terms of the magnitude ($\tau_w/\overline{\tau_w} < 0.9$ for the low-drag events and $\tau_w/\overline{\tau_w} > 1.1$ for the high-drag events).

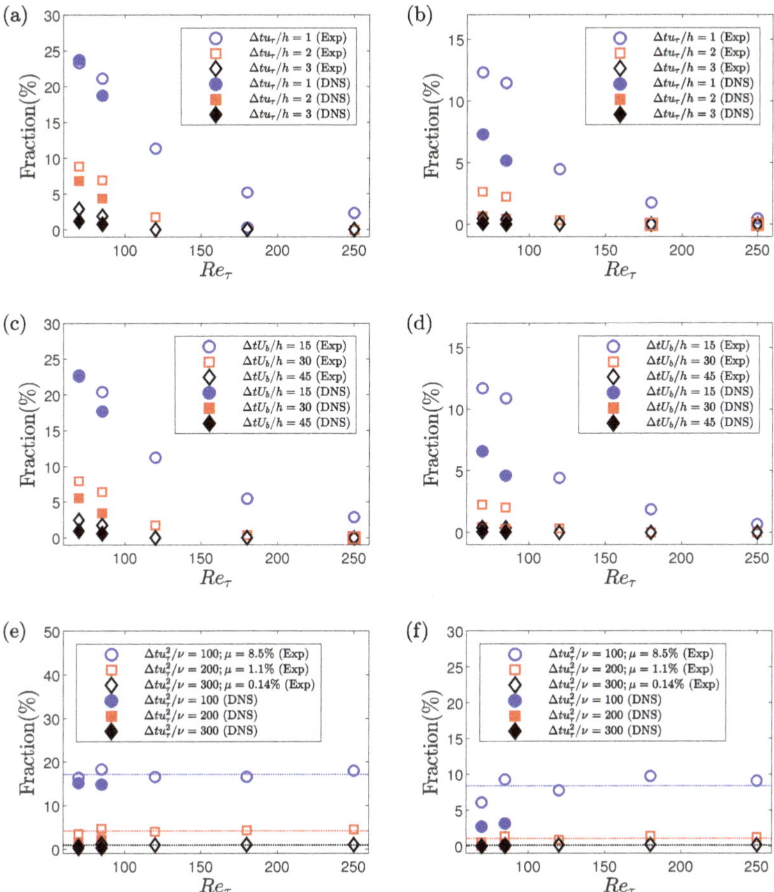

Figure 4. Reynolds number variation of fraction of time spent in low-drag events with using (**a**) mixed scaling, (**c**) outer scaling and (**e**) inner scaling for the time-duration criteria. Reynolds number variation of fraction of time spent in high-drag events with using (**b**) mixed scaling, (**d**) outer scaling and (**f**) inner scaling for the time-duration criteria. Open symbols represent the experimental data and filled symbols represent the DNS data. The threshold criteria to detect a low- and high-drag event are $\tau_w/\overline{\tau_w} < 0.9$ and $\tau_w/\overline{\tau_w} > 1.1$, respectively. Note that the y-axis is not the same between low- and high-drag data. Error bars obtained by dividing the sample size into two halves and calculating the respective fraction are found to be within the size of the symbols and are therefore removed to avoid cluttering of data. Dotted lines in panels (**e**,**f**) highlight the average value of fraction (%) for $70 \leq Re_\tau \leq 250$ at different values of $\Delta tu_\tau^2/\nu$ obtained using experiments.

A similar observation was also made previously by Whalley et al. [24] while using mixed scaling for the time-duration criterion.

Figure 4 shows that the fraction of time spent in the conditional events decreases with increasing the value of the time-duration criterion. A further investigation of this phenomenon is made by studying the dependence of the occurrence of conditional events as a function of their durations. Figure 5 shows the distribution of the occurrence of low- and high-drag events as a function of Δt^+ for $Re_\tau = 180$. The threshold criteria to detect a low- and high-drag events are $\tau_w/\overline{\tau_w} < 0.9$ and $\tau_w/\overline{\tau_w} > 1.1$, respectively. The probability of occurrence of both low- and high-drag events decreases almost exponentially (as the y-axis is in log scale) with increasing Δt^+. For $\Delta t^+ \gtrsim 400$, $P(\Delta t^+)$ does not seem to be well resolved because of the lower occurrence of low- and high-drag events for higher Δt^+, thus leading to lower number of events to carry out the statistical analysis. The distribution of high-drag events is observed to be different to the distribution of low-drag events. There is a higher probability of occurrence of high-drag events for lower Δt^+ as compared to the low-drag events and vice versa. The crossover Δt^+, where the behaviour of the low- and high-drag events becomes opposite, is about 60. The decay of the probability of the low- and high-drag events is then fitted with an exponential relationship for $100 \leq \Delta t^+ \leq 300$, given by $P(\Delta t^+) = Ae^{-\lambda \Delta t^+}$. Here, λ indicates the rate of decay. The decay rate is calculated for all the Reynolds numbers. Exponential distributions like this arise in so-called Poisson processes, also called memoryless processes. The exponential decay implies that the probability of the interval ending between time Δt^+ and time $\Delta t^+ + d(\Delta t^+)$ is independent of Δt^+, i.e., the probability of the low- or high-drag intervals ending are independent of how long they have lasted. Avila et al. [42] observed a similar memoryless process with regards to puff splitting during transition in a pipe flow. After an initial formation time, the distribution of puff splitting were exponential and therefore memoryless, thus showing that the probability of a puff splitting does not depend on its age. Table 2 shows the rate of decay obtained for low- and high-drag events at various Reynolds numbers. The rate of decay is found to be almost independent of the Reynolds numbers for both low- and high-drag events, and the λ values are lower for the low-drag than the high-drag for the $100 \leq \Delta t^+ \leq 300$. A slight discrepancy is observed for $Re_\tau = 70$, which can be attributed to the presence of transitional effects at this Reynolds number, as discussed in Agrawal et al. [27]. These results are also consistent with the results shown in Figure 4e,f that the fraction of the conditional events are almost independent of the Reynolds number and the fraction of time spent in low-drag events is higher than for the high-drag events. This is the first evidence that the "low-drag" hibernating turbulent events exist significantly above the Reynolds numbers close to transition [24] and well into the regime where the flow is usually considered to be "fully-turbulent", i.e., $Re_\tau \geq 180$ [26].

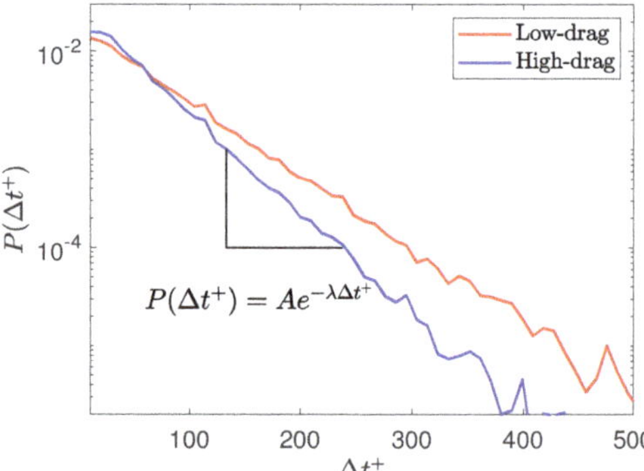

Figure 5. PDF of occurrence of low- and high-drag events as a function of Δt^+ for $Re_\tau = 180$ where the threshold criteria for low- and high-drag events are $\tau_w/\overline{\tau_w} < 0.9$ and $\tau_w/\overline{\tau_w} > 1.1$, respectively. Here, x-axis (Δt^+) represents the lifetime or duration of a conditional event.

Table 2. Rate of decay (λ) of the PDF of occurrence of conditional events for $100 \leq \Delta t^+ \leq 300$ at various Reynolds numbers. Numbers in brackets correspond to the R^2 value. The threshold criteria for low- and high-drag events are $\tau/\overline{\tau_w} < 0.9$ and $\tau/\overline{\tau_w} > 1.1$, respectively.

Re_τ	Low-Drag	High-Drag
70	0.0192 (0.89)	0.0258 (0.90)
85	0.0181 (0.95)	0.0252 (0.97)
120	0.0185 (0.98)	0.0245 (0.97)
180	0.0185 (0.96)	0.0251 (0.97)
250	0.0183 (0.99)	0.0245 (0.99)

6. Wall Shear Stress Statistics during Conditional Events

To study the statistics of the conditional wall shear stress, the instantaneous wall shear stress during the low-drag or high-drag events are ensemble-averaged. Figure 6 shows the instantaneous and ensemble averaged wall shear stress fluctuations during low- and high-drag events for $Re_\tau = 180$. The ensemble averaging is executed in two ways: by shifting all the instantaneous low- and high-drag events such that $t^+ = 0$ indicates the beginning of a conditional event (shown in Figure 6a,c), and by shifting all the instantaneous low- and high-drag events such that $t^+ = 0$ indicates the end of a conditional event (shown in Figure 6b,d). This has been done to study the time evolution of the ensemble-averaged wall shear stress with respect to the start and the end of a conditional event. It can be seen that during the low-drag events, the ensemble averaged wall shear stress drops approximately 35% below the time-averaged value. During the high-drag events, the ensemble averaged wall shear stress rises approximately 45% above the time-averaged value. This figure also highlights that although the time-duration criteria for the conditional events is $\Delta t_{cr}^+ = 200$, these events can last up to $\Delta t^+ \geq 400$.

The effect of the time-duration and magnitude threshold criteria on the conditional wall shear stress is investigated for $Re_\tau = 180$. For the time-duration criterion, Δt_{cr}^+ is varied between 150 and 250 while keeping the threshold criteria constant as $\tau_w/\overline{\tau_w} < 0.9$ and $\tau_w/\overline{\tau_w} > 1.1$ for the low- and high-drag events, respectively. Figure 7a–d shows the ensemble-averaged wall shear stress for the low- and high-drag events at $Re_\tau = 180$ for various time-duration criteria. The figure shows the ensemble-averaged wall shear stress for the conditional events for both methods of ensemble averaging, i.e., $t^+ = 0$ indicates either the start or end of a conditional event. The plateau of the ensemble-averaged wall shear stress during the low- and high-drag events is observed to be insensitive to the time-duration criteria when varying Δt^+ from 150 to 250, but the duration of these conditional events itself becomes smaller when making the criteria less stringent. A spike in the ensemble-averaged wall shear stress can be observed near the start and end of the low-drag events and similarly, a dip can be seen near the start and end of the high-drag events. Analogous results corresponding to the ensemble-averaged wall shear stress during the low-drag events were also obtained by Kushwaha et al. [22] in channel flow using DNS for $Re_\tau = 100$. They employed mixed scaling ($\Delta t^* = 2$ and 3) as the time-duration criteria to detect low-drag events. Similar results were obtained for the other Reynolds numbers studied here and are not shown for brevity.

It can be said that the time-duration criteria, either based on mixed or inner scaling (for the range studied), does not affect the strength of the low- or high-drag events. For the rest of this paper, the time-duration criteria for the both low- and high-drag events is fixed at $\Delta t_{cr}^+ = 200$ unless stated otherwise. Next, the effect of changing the threshold criteria on the conditional wall shear stress is investigated while keeping the time-duration criterion constant at $\Delta t_{cr}^+ = 200$. The threshold criteria used for low-drag events are $\tau_w/\overline{\tau_w} < 0.8$, $\tau_w/\overline{\tau_w} < 0.9$ and $\tau_w/\overline{\tau_w} < 1$, and for the high-drag events are $\tau_w/\overline{\tau_w} > 1$, $\tau_w/\overline{\tau_w} > 1.1$ and $\tau_w/\overline{\tau_w} > 1.2$. The most stringent limits for the strength in the threshold criteria are chosen based on the availability of a sufficient number of conditional events to obtain well-resolved ensemble-averaged wall shear stress results. As the threshold criterion is made more stringent, for the low-drag events (shown in Figure 7e,f), the lower plateau of the ensemble-averaged wall shear stress decreases. Similarly, for the high-drag events (shown in Figure 7g,h), the upper plateau of the ensemble-averaged wall shear stress increases. Similar results were observed for low-drag events only by Kushwaha et al. [22] at $Re_\tau = 100$. The results are shown only for $Re_\tau = 180$ as very similar results were obtained for the other Reynolds numbers studied.

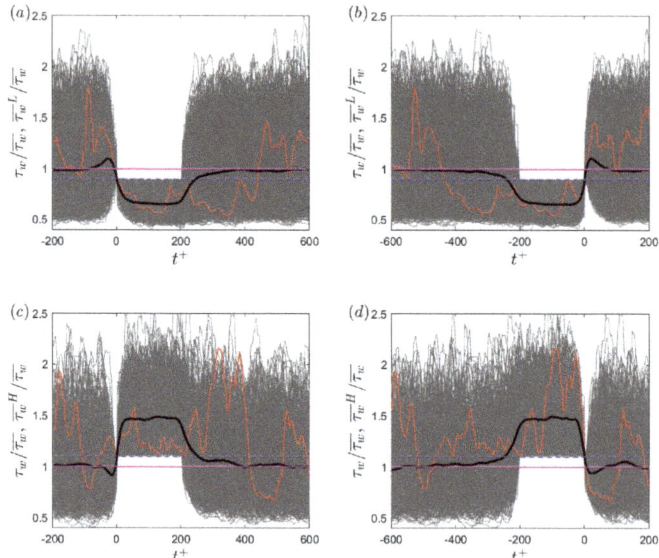

Figure 6. (**a**,**b**) Instantaneous normalised wall shear stress (thin grey lines) and ensemble-averaged wall shear stress (thick black line) during the low-drag events for $Re_\tau = 180$ where $t^+ = 0$ indicates (**a**) start of a low-drag event and (**b**) end of a low-drag event. Red line highlights an instantaneous low-drag event with a duration of $\Delta t^+ \approx 410$. Purple line and dashed blue line represent the time-averaged value and the threshold value of $\tau_w / \overline{\tau_w} < 0.9$, respectively. (**c**,**d**) Instantaneous normalised wall shear stress (thin grey lines) and ensemble-averaged wall shear stress (thick black line) during the high-drag events for $Re_\tau = 180$ where $t^+ = 0$ indicates (**a**) start of a high-drag event and (**b**) end of a high-drag event. Red line highlights an instantaneous low-drag event with a duration of $\Delta t^+ \approx 400$. Purple line and dashed blue line represent the time-averaged value and the threshold value of $\tau_w / \overline{\tau_w} > 1.1$, respectively.

Interestingly, as can be seen from Figure 7e–h, the spike in the ensemble-averaged wall shear stress for the low-drag events and dip in the ensemble-averaged wall shear stress for the high-drag events seems to be less significant with increasingly strict threshold criteria. Kushwaha et al. [22] mentions that they have no physical explanation for the existence of the spike or dip in the ensemble-averaged wall shear stress data. To investigate the reason for the spike or dip in the ensemble-averaged data during the conditional events, two artificially generated time series have been produced where one signal is Gaussian and the other signal has the same first four moments as the wall shear stress moments for $Re_\tau = 180$ obtained in the present experiment. The Gaussian signal has a rms value the same as the wall shear stress for $Re_\tau = 180$. This has been conducted to understand if the reason for the spike or the dip is unique to the wall shear stress signals or is merely a statistical artefact of the conditioning. An equal number of samples ($N = 2 \times 10^8$) are generated for both of the artificially generated signals using the inbuilt MATLAB function: "pearsrnd".

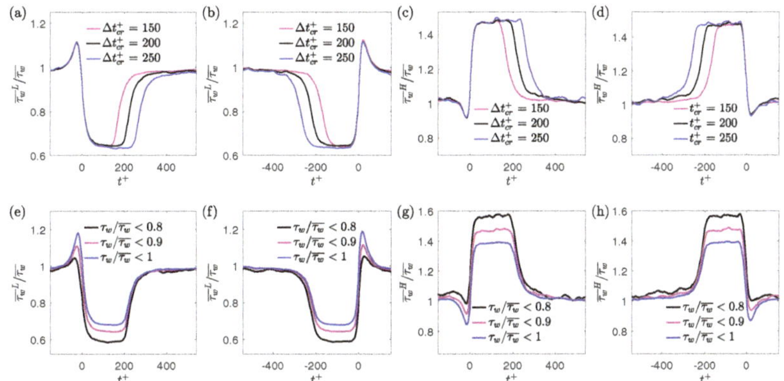

Figure 7. Ensemble-averaged wall shear stress for various time-duration criteria at $Re_\tau = 180$ for (**a**) start and (**b**) end of low-drag events, and (**c**) start and (**d**) end of high-drag events. The threshold criteria to detect a low- and high-drag event are $\tau_w\,/\overline{\tau_w} < 0.9$ and $\tau_w\,/\overline{\tau_w} > 1.1$, respectively. Ensemble-averaged wall shear stress for various threshold criteria at $Re_\tau = 180$ for (**e**) start and (**f**) end of low-drag events, and (**g**) start and (**h**) end of high-drag events. The time-duration criteria to detect a low-drag or a high-drag event is kept constant at $\Delta t_{cr}^+ = 200$.

A comparison of the ensemble averaged data during the conditional events is made between the two artificially generated signals. The time duration is kept the same as $\Delta t_{cr}^+ = 200$ to detect the low- and high-drag events. The threshold criteria are varied to study their effect on the ensemble averaged values. For the low-drag events, the threshold criteria are $\tau_w/\overline{\tau_w} < 0.925$, $\tau_w/\overline{\tau_w} < 0.95$, $\tau_w/\overline{\tau_w} < 0.975$ and $\tau_w/\overline{\tau_w} < 1$, and for the high-drag events, the threshold criteria are $\tau_w/\overline{\tau_w} > 1$, $\tau_w/\overline{\tau_w} > 1.025$, $\tau_w/\overline{\tau_w} > 1.05$ and $\tau_w/\overline{\tau_w} > 1.075$. Figure 8 shows the ensemble averaged wall shear stress during low- and high-drag events obtained from the two artificially generated signals. There is a spike (and dip) in the ensemble-averaged wall shear stress near the start of the low-drag (and high-drag) events for both artificially generated signals. The existence of spikes or dips in the ensemble-averaged data from the artificially-generated signals, even in the limit of a Gaussian signal, suggest that these are artefacts of the conditional sampling and ensemble averaging and are not unique to the wall shear stress signals. It is also seen that the spikes (and dips) in the ensemble-averaged data from the low-drag events (and high-drag events) becomes less significant when making the threshold criteria more stringent. This further reinforces the idea that these spikes and dips in the ensemble averaged data are the consequence of the conditional sampling of any time-series signal. Thus, these spikes or dips cannot be used to identify the onset/footprint of low- or high-drag events. Park et al. [16], using MFU simulations, observed that many of the low-drag events are followed by strong turbulent bursts which were detected based on an increase in the volume-averaged energy dissipation rate. There may exist a relation between these turbulent bursts and spikes in the ensemble-averaged wall shear stress data after low-drag events which needs further investigation.

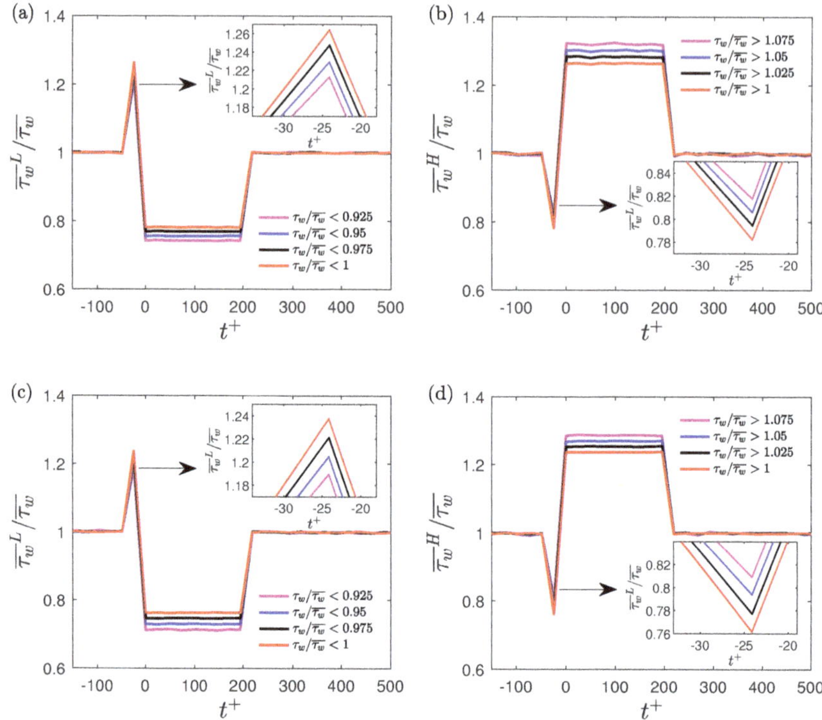

Figure 8. Ensemble-averaged wall shear stress during (**a**) low-drag events and (**b**) high-drag events for the artificially generated wall shear stress signal with same first four moments as one measured for $Re_\tau = 180$. Ensemble-averaged wall shear stress during (**c**) low-drag events and (**d**) high-drag events for a Gaussian signal. The time-duration criteria to detect a low-drag or a high-drag event is kept constant at $\Delta t_{cr}^+ = 200$. Inset plots show the same data as the main plot but only near the spike or dip in the ensemble averaged data.

7. Velocity Characteristics during Conditional Events

As mentioned in Section 2, simultaneous measurements of velocity using LDV above the hot-film are made for various wall-normal locations (shown in Table 1) at every Reynolds numbers studied. However, the wall-normal velocities were measured only for $Re_\tau = 70$ and 85, due to the limited access of the laser beams for LDV measurements closer to the bottom wall of the channel. Velocity information is also obtained using DNS in large computation domains (discussed in Section 3) for $Re_\tau = 70$ and 85. In this section, the criteria for conditional events are kept constant at $\Delta t_{cr}^+ = 200$ and $\tau_w/\overline{\tau_w} < 0.9$ for the low-drag events, and $\Delta t_{cr}^+ = 200$ and $\tau_w/\overline{\tau_w} > 1.1$ for the high-drag events, unless stated otherwise. To carry out the conditional sampling of the velocity data, we ensured that there are a sufficient number of conditional events ($\sim$100) to obtain well-converged results. For the DNS, the number of high-drag events obtained were quite few in number, between 10 and 20 for both $Re_\tau = 70$ and 85. Therefore, the characteristics of only low-drag events are studied for the DNS data, whereas characteristics of both low-drag and high-drag events are studied using the experimental data.

7.1. Streamwise Velocity

The conditional sampling of the velocity data and their ensemble-averaging is conducted in a similar manner as has been conducted earlier by Whalley et al. [24] and Kushwaha et al. [22]. For the low-drag events, the drop in the ensemble averaged velocities is observed to be more significant near the wall, with the effect disappearing near the centreline. For the high-drag events, an analogous behaviour to low-drag events is observed. Figure 9 shows an example of the ensemble averaged streamwise velocities for various wall-normal locations at $Re_\tau = 180$ during the low- and high-drag events. Here, the ensemble-averaged streamwise velocities ($\overline{U}^L, \overline{U}^H$) are normalised by $\overline{u_\tau}$. Very similar results were observed for other Reynolds numbers and therefore are not shown. This behavior of the ensemble averaged streamwise velocities is similar to those previously obtained by Whalley et al. [24] and Kushwaha et al. [22] for $70 \leq Re_\tau \leq 100$. Therefore, it can be said that the ensemble-averaged streamwise velocity during the low- and high-drag events, which were previously observed for $70 \leq Re_\tau \leq 100$, shows similar characteristics even for the flow in the fully-turbulent regime.

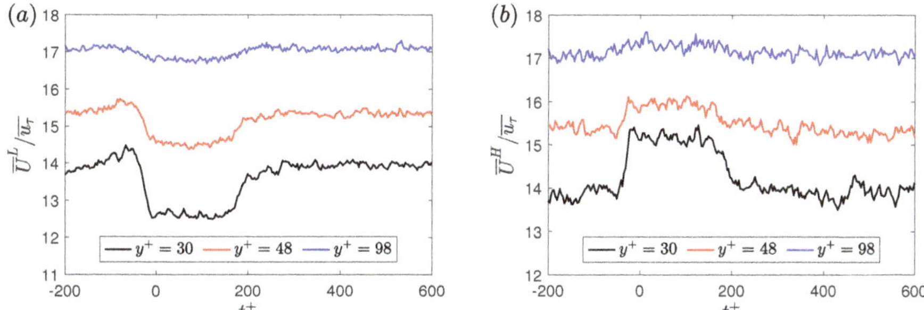

Figure 9. Ensemble-averaged streamwise velocity for $Re_\tau = 180$ during (**a**) low-drag events and (**b**) high-drag events. Here, $t^+ = 0$ indicates the beginning of a low-drag or a high-drag event. The criteria to detect a low-drag event are $\Delta t_{cr}^+ = 200$ and $\tau_w / \overline{\tau_w} < 0.9$, and a high-drag event are $\Delta t_{cr}^+ = 200$ and $\tau_w / \overline{\tau_w} > 1.1$.

Figure 10 shows the unconditional and conditionally-averaged streamwise velocity profiles for $Re_\tau = 70, 85, 120, 180$ and 250 obtained using experiments, and $Re_\tau = 70$ and 85 obtained using DNS. Here, the normalisation of the unconditional velocity and the corresponding wall-normal locations are carried out using the time-averaged friction velocity ($\overline{u_\tau}$). The conditionally-averaged streamwise velocities and the corresponding wall-normal locations are normalised by the conditionally-averaged friction velocities ($\overline{u_\tau}^L$ for low-drag and $\overline{u_\tau}^H$ for high-drag). Before studying the profiles during the conditional events, we first focus on the unconditional (time-averaged) profiles. Experimental and DNS results are in good agreement for $Re_\tau = 70$ and 85. The unconditional profile obtained for $Re_\tau = 180$ is also in good agreement with the DNS profile obtained by [26] for $Re_\tau = 180$, and the velocity profiles for Re_τ of 180 and 250 approximately collapses on the log-law profile ($U^+ = 2.5$ $\ln y^+ + 5.5$) for $y^+ \geq 30$.

The velocity statistics during the conditional events is investigated in such a way that only the upper (for high-drag) or lower plateau (for low-drag) of the instantaneous wall shear stress and velocity are considered for the conditional sampling. This is done to avoid any transient behaviours (start and end of conditional events) affecting the result. Therefore, only wall shear stress and velocity data between $30 < t^+ < t_{end}^+ - 30$ are used for conditional sampling, where t_{end}^+ indicates the end of a low-drag or a high-drag event. For $y^+ \lesssim 10$, the unconditional and conditional profiles for $Re_\tau = 70$ and 85 obtained using DNS almost collapse on each other. For $y^+ \gtrsim 10$, the conditionally averaged velocity profiles are closer to Virk's MDR asymptote than their time-averaged values (for all the Reynolds numbers studied). Previously, Kushwaha et al. [22] and Whalley et al. [24] showed that at $70 \leq Re_\tau \leq 100$, the low-drag velocity profiles get closer to the Virk's MDR and the lower-branch of the nonlinear TW solutions (as obtained by Park and Graham [13]) for similar wall-normal locations, $y^+ \lesssim 35$. Therefore, the present result confirms the validity of this phenomenon for Reynolds numbers in the fully-turbulent regime. There is a very good agreement between the experimental and DNS results for the velocity profiles during the low-drag events at $Re_\tau = 70$ and 85. For higher wall-normal locations the conditional velocity profiles start to deviate from Virk's MDR profile, and for $y^+ \gtrsim 100$, the conditional velocity profiles have a slightly higher slope as compared to the Prandtl-von Kármán log-law, as seen for $Re_\tau = 180$ and 250. For the high-drag events, the conditional velocity profiles are lower than the unconditional profiles for all the Reynolds numbers.

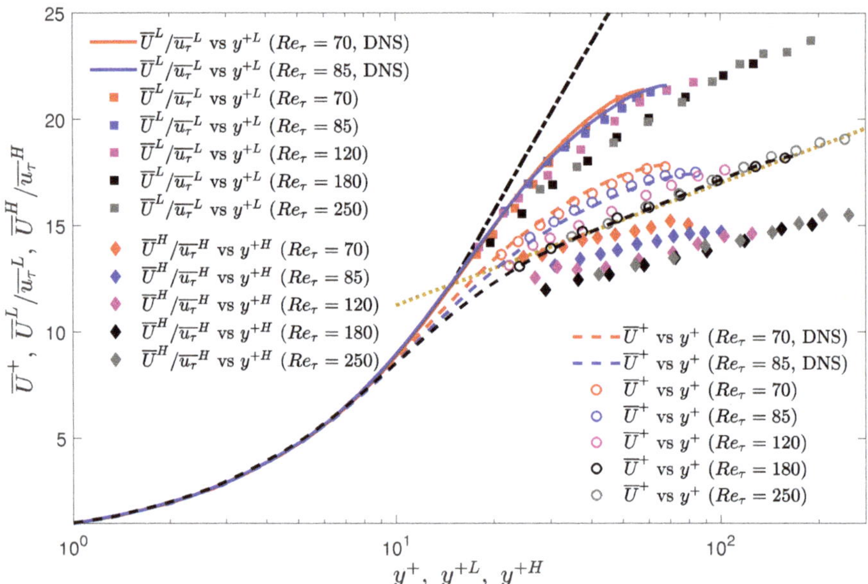

Figure 10. Unconditional and conditionally averaged streamwise velocity profiles for $Re_\tau = 70, 85, 120,$ 180 and 250 during low-drag and high-drag events. All the symbols represent the experimental data. Here, the conditionally averaged streamwise velocity data is normalised using conditionally averaged friction velocity. Yellow dotted line represents the Prandtl-von Kármán log-law: $U^+ = 2.5 \ln y^+ + 5.5$ and the black dash-dotted line represents the lower end of the 95% confidence interval of the Virk's MDR asymptote: $U^+ = 11.4 \ln y^+ - 18.5$ [43]. Black dashed line represents the time-averaged velocity profile obtained using DNS at $Re_\tau = 180$ by Kim et al. [26].

To further investigate the slope of the conditional velocity profiles, the so-called indicator function is calculated, which is generally used to study the logarithmic dependence of the mean velocity profile [44]. For the unconditional velocity data, the indicator function is given by: $\bar{\zeta} = y^+ d\overline{U}^+/dy^+$. For the conditional velocity data, the indicator functions are given by $\bar{\zeta}^L = y^{+L} d\overline{U}^{+L}/dy^{+L}$ and $\bar{\zeta}^H = y^{+H} d\overline{U}^{+H}/dy^{+H}$ for the low- and high-drag events, respectively. The profiles of indicator function are shown in Figure 11. It can be seen that that for $Re_\tau = 70$ and 85, the $\bar{\zeta}$ profiles do not exhibit a logarithmic dependence. For $Re_\tau = 120, 180$ and 250, the $\bar{\zeta}$ profiles approximately collapse on the value of $1/\kappa = 2.5$ for $y^+ \geq 30$, thus suggesting a logarithmic dependence. Here, κ is the von Kármán constant. It is observed from Figure 11a,b that the $\bar{\zeta}^L$ profiles at all Reynolds numbers are closer to the Virk's MDR ($1/\kappa = 11.7$) for $y^+ \leq 30$. For $Re_\tau = 120, 180$ and 250, the $\bar{\zeta}^L$ profiles remain above the unconditional profiles for $y^+ \geq 30$, thus showing that the slope of the low-drag velocity profiles is slightly higher than the unconditional profiles in the log-law region. Figure 11c,d shows that the $\bar{\zeta}^H$ profiles at $Re_\tau = 70$ and 85, are lower than the $\bar{\zeta}$ profiles (except close to the centreline), with the effect being more significant for $y^+ \leq 30$. For $Re_\tau = 120, 180$ and 250, the slope of the $\bar{\zeta}^H$ profiles is slightly lower than the $\bar{\zeta}$ profiles for all wall-normal locations.

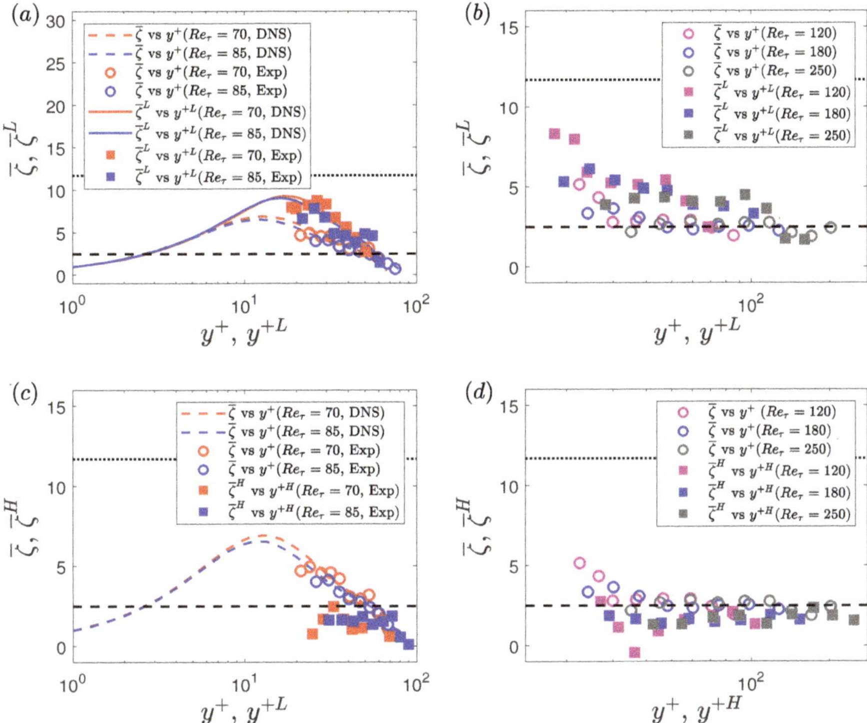

Figure 11. Unconditional (open circles) and conditionally averaged (closed squares) indicator functions for (**a**) $Re_\tau = 70$ and 85, and for (**b**) $Re_\tau = 120$, 180 and 250 during low-drag events. Unconditional (open circles) and conditionally averaged (closed squares) indicator functions for (**c**) $Re_\tau = 70$ and 85, and for (**d**) $Re_\tau = 120$, 180 and 250 during high-drag events. The criteria to detect a low-drag event is $\Delta t_{cr}^+ = 200$ and $\tau_w/\overline{\tau_w} < 0.9$, and a high-drag event is $\Delta t_{cr}^+ = 200$ and $\tau_w/\overline{\tau_w} > 1.1$. Dashed lines represent 2.5 and dotted lines represent 11.7.

7.2. Similarity between Turbulent Drag Reduction and Low-Drag Events in Newtonian Turbulence

To quantify the "drag reduction" during the low-drag events a percentage decrease in the wall shear stress, during these low-drag events, is calculated. The comparison with the drag-reduction literature is carried out only for $Re_\tau = 180$ and 250. It is found that the percentage drag reduction is about 36% for $Re_\tau = 180$ and 250 when calculated using Equation (5).

$$\%DR = \frac{\overline{\tau_w} - \overline{\tau_w}^L}{\overline{\tau_w}} \approx 36\%\,(Re_\tau = 180 \text{ and } 250). \tag{5}$$

This level of drag reduction is similar to some of the other techniques employed previously to reduce drag in channel flows. For example, when using polymer additives at low concentration, the low-drag reduction (LDR) regime is observed [45,46]. A comparison is made with the experimental data obtained by Warholic et al. [45] at $Re_h \approx 20{,}000$ for the case where a drag reduction of about 33% was observed. Drag reduction due to superhydrophobic surfaces were investigated by Min and Kim [47]. They conducted DNS in a channel flow for $Re_\tau = 180$ (for DR = 0) and by using streamwise slip, they obtained a maximum drag reduction of 29%. Choi et al. [48] implemented DNS in a channel flow at $Re_\tau = 180$ (for DR = 0) to numerically study the effect of blowing and suction on the skin-friction drag. They employed out-of-phase boundary conditions for the spanwise and wall-normal velocities to simulate the blowing and suction effects on the channel, and obtained a drag reduction of about 26% by applying spanwise control.

In Figure 12, a comparison is shown between the streamwise velocity profiles obtained using these three techniques for turbulent drag reduction and the conditional streamwise velocity profile obtained in the present experiment at $Re_\tau = 180$ and 250.

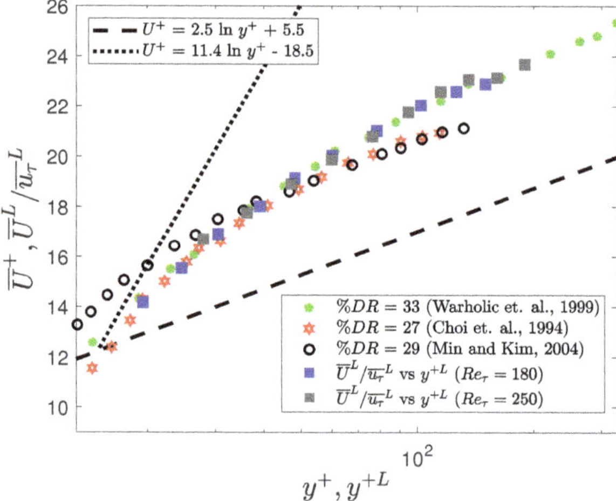

Figure 12. Conditional streamwise velocity profiles for $Re_\tau = 180$ and 250 during the low-drag events. Streamwise velocity profiles, where different drag reduction mechanisms are employed previously: Warholic et al. [45] used polymeric additive, Min and Kim [47] used hydrophobic surface in the form of slip-boundary condition for the streamwise direction and Choi et al. [48] applied out-of-phase boundary condition to the spanwise velocity at the surface. Dashed line represents the Prandtl-von Kármán log-law: $U^+ = 2.5 \ln y^+ + 5.5$ and dotted line represents the lower end of the 95% confidence interval of the Virk's MDR asymptote: $U^+ = 11.4 \ln y^+ - 18.5$ [43].

A good agreement can be seen between the conditionally averaged profile for $Re_\tau = 180$ and 250 and the profile obtained by Warholic et al. [45] for $DR = 33\%$ using polymer additives. The profiles obtained by Min and Kim [47] and Choi et al. [48], and the present experiment are also in relatively good agreement with the obvious difference arising due to the lower levels of drag reduction reported in these cases. One major difference in the result obtained by Min and Kim [47] is that the velocity profile shifts upwards even closer to the wall which is the consequence of the slip boundary condition. Therefore, it suggests that for the fully-turbulent flows ($Re_\tau = 180$ and 250), the conditional streamwise velocity for $y^+ \gtrsim 20$ during the low-drag events mimics the flow as observed during the LDR phenomenon due to polymer addition or the drag reduction due to spanwise oscillation. For the case of superhydrophobicity, this similarity between the velocity profiles can be observed approximately in the log-law region. Thus, if a method could be found to encourage the turbulent state to enter the low-drag "hibernating" state more often, a significant time-averaged drag reduction would be achievable.

7.3. Reynolds Shear Stress

DNS studies by Park and Graham [13] and Xi and Graham [12], using MFU at $Re_\tau = 85$, showed that the Reynolds shear stress drops to a very low value during the low-drag events. There is still no information in the prior literature regarding the RSS characteristics, during the conditional events, from either physical experiments or from DNS in extended domains. For the experiments (discussed in Section 2), two-component (streamwise and wall-normal) velocity measurements have been made for $Re_\tau = 70$ and 85 to study the behaviour of the Reynolds shear stress during the conditional events. To carry out the conditional sampling, each wall-normal location is sampled for 2 h while simultaneously measuring the wall shear stress using HFA. DNS study is conducted for $Re_\tau = 70$ and 85 which provides the streamwise and wall-normal velocity information for various wall-normal locations (discussed in Section 3).

To calculate the conditional RSS, the streamwise velocity fluctuations and the wall-normal velocity fluctuations during the conditional events are calculated by subtracting their time-averaged values from the instantaneous conditional values. Figure 13 shows the ensemble averaged wall-normal velocities ($\overline{V}^L$) and ensemble averaged Reynolds shear stress ($-\overline{uv}^L$). All the quantities are normalised by the time-averaged friction velocity ($\overline{u_\tau}$). The threshold and time-duration criteria to detect a low-drag events are $\tau_w/\overline{\tau_w} < 0.9$ and $\Delta t^+_{cr} = 200$, respectively. For $y^+ < 21$, experimental data are not available and therefore only DNS results are shown. A fairly good agreement between the experimentally and numerically obtained ensemble-averaged wall-normal velocity and RSS is observed. From continuity, the time-averaged wall-normal velocity must be zero, as can be observed from the DNS data. There is a slight discrepancy in the time-averaged values for the experimental data which is attributed to the error associated with the LDV measurements (discussed in Section 2). The conditionally averaged wall-normal velocity is higher than the time-averaged value during the low-drag events.

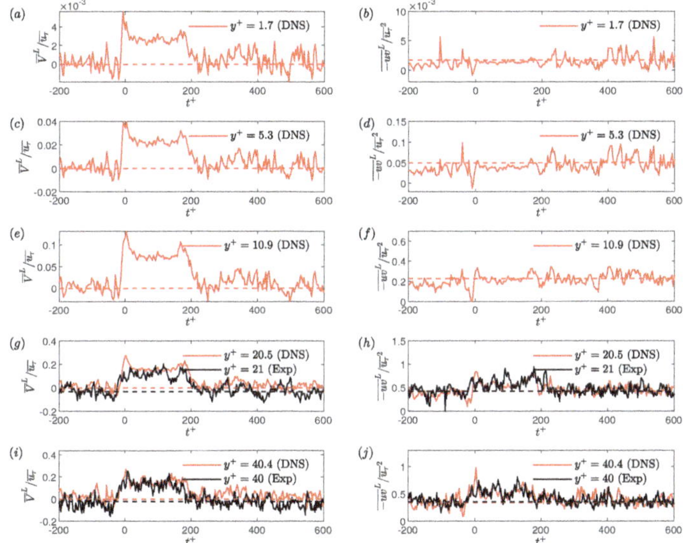

Figure 13. Ensemble-averaged wall-normal velocities (**a,c,e,g,i**) and Reynolds shear stresses (**b,d,f,h,j**) obtained using DNS (red solid lines) and experiment (black solid lines) during low-drag events for $Re_\tau = 70$. Here, $t^+ = 0$ indicates start of low-drag events. The time-averaged values for the corresponding wall-normal locations are shown using red dashed lines (obtained using DNS) and black dashed lines (obtained using experiment). The criteria to detect a low-drag event is $\Delta t^+_{cr} = 200$ and $\tau_w/\overline{\tau_w} < 0.9$.

The ensemble averaged streamwise velocities have already been shown previously in Section 7.1. Based on the conditionally-averaged streamwise and wall-normal velocities, it can be said that the low-drag events form a subset of so-called Q2 events, i.e., $u < 0$ and $v > 0$. Figure 14 shows the ensemble-averaged wall-normal velocity and RSS during the high-drag events for $y^+ = 21$ and 40. The ensemble averaged wall-normal velocity is lower than the time-averaged wall-normal velocity whereas the ensemble averaged RSS is unchanged. Again, based on the conditionally-averaged streamwise and wall-normal velocities, it can be said that the high-drag events form a subset of Q4 events, i.e., $u > 0$ and $v < 0$. This behaviour will be further investigated in the following discussions.

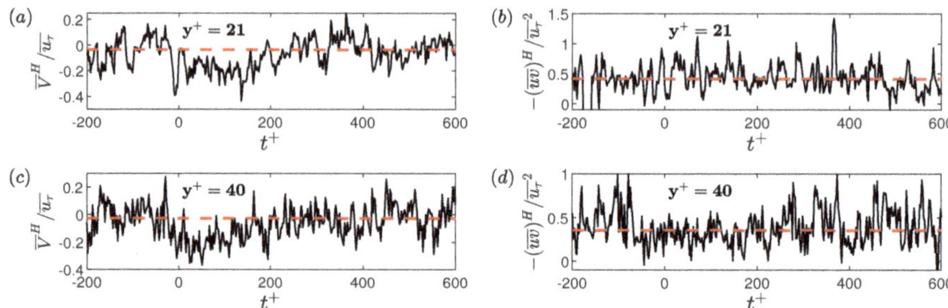

Figure 14. Ensemble-averaged wall-normal velocities (**a,c**) and Reynolds shear stresses (**b,d**) obtained using experiment during high-drag events for $Re_\tau = 70$. Here, $t^+ = 0$ indicates start of high-drag events. The time-averaged values for the corresponding wall-normal locations are shown using red dashed lines (obtained using experiment). The criteria to detect a high-drag event is $\Delta t_{cr}^+ = 200$ and $\tau_w / \overline{\tau_w} > 1.1$.

The unconditional and conditionally-averaged RSS profiles, obtained for these two Reynolds numbers, shown in Figure 15. A good agreement can be observed between the experimental and DNS unconditional profiles. The conditionally-averaged data are normalised using $\overline{u_\tau}^2$, are shown in Figure 15a,b, for low- and high-drag events, respectively. For the low-drag case, both experimental and DNS results are shown, and for high-drag case only experimental results are shown. A good agreement is observed between the conditionally averaged profiles obtained using experiments and DNS, with a slight discrepancy observed for the $Re_\tau = 85$ results. As seen in Figure 15a, the conditionally averaged profiles have slightly lower values than the unconditional profiles for $y^+ \lesssim 10$. For $y^+ \gtrsim 10$ the conditionally averaged profiles are higher than the unconditional profiles with the effect being more significant for y^+ between 20 and 40. For the high-drag case, as seen in Figure 15b, the conditionally-averaged RSS profiles almost collapse onto the unconditional profiles for all the wall-normal locations measured. This result suggests that the Reynolds shear stress is more affected by the low-drag events compared to the high-drag events. A sensitivity check has been executed to study the effect of changing the criteria for conditional events on the conditional RSS profiles for $Re_\tau = 70$. No significant dependence of the RSS profiles is observed for the different values of criteria studied here.

A quadrant analysis is conducted to calculate the contribution to the Reynolds shear stress from various turbulent events [49]. In quadrant analysis, the Reynolds shear stress is divided into four quadrants based on the signs of the streamwise and wall-normal velocity fluctuations: Q1 $(+u, +v)$, Q2 $(-u, +v)$, Q3 $(-u, -v)$ and Q4 $(+u, -v)$. The Q2 and Q4 events are generally related to the ejection and sweep events, respectively [49]. Here, the normalisation of both unconditional and conditional velocity fluctuations is based on the time-averaged friction velocity $(\overline{u_\tau})$. For unconditional velocity fluctuations, the time-averaged velocities are subtracted from the instantaneous velocities, and for the conditional velocity fluctuations, the time-averaged velocities are subtracted from the instantaneous conditional velocities during the low- or high-drag events. Figure 16a,d shows the jpdfs (joint probability density functions) of the unconditional streamwise and wall-normal velocity fluctuations for $Re_\tau = 70$ obtained using the experiment (at $y^+ = 24$) and DNS (at $y^+ = 25$). The shape of the unconditional jpdfs are roughly elliptical with their major axes tilted in the direction of Q2 and Q4 motions. During the low-drag events the jpdf shifts towards the Q2 quadrant, whereas during the high-drag events the jpdf shifts towards the Q4 quadrant.

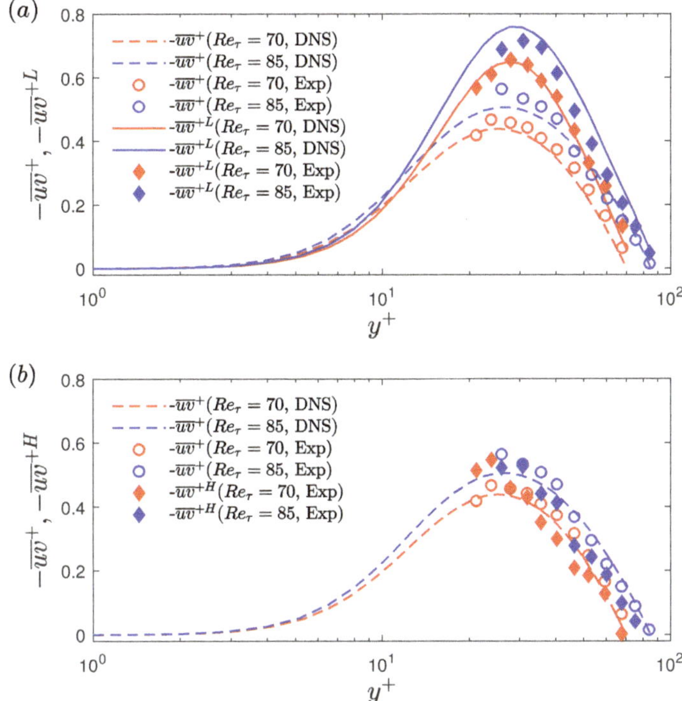

Figure 15. (**a**) Unconditional and conditionally averaged RSS profiles for $Re_\tau = 70$ and 85 during low-drag events. (**b**) Unconditional and conditionally averaged RSS profiles for $Re_\tau = 70$ and 85 during high-drag events.

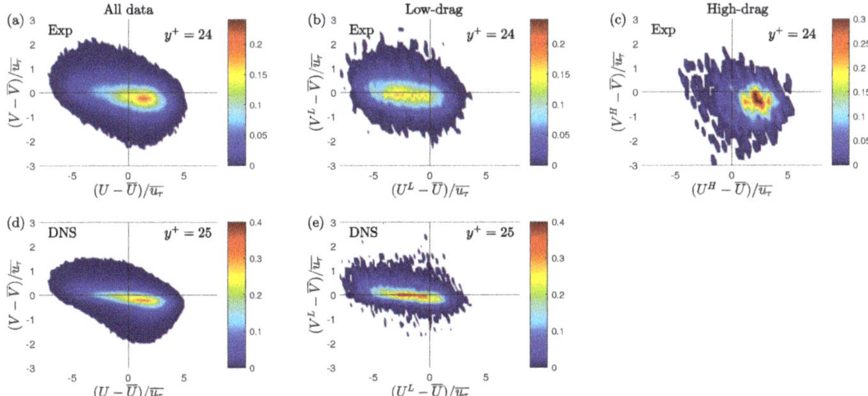

Figure 16. Unconditional (**a**), low-drag (**b**) and high-drag (**c**) jpdfs of streamwise and wall-normal velocity fluctuations for $y^+ = 24$ at $Re_\tau = 70$ using experiments. Unconditional (**d**) and low-drag (**e**) jpdfs of streamwise and wall-normal velocity fluctuations for $y^+ = 25$ at $Re_\tau = 70$ using DNS. Unconditional and conditional velocity fluctuations are normalised using the time-averaged $\overline{u_\tau}$.

This observation is consistent with the previous results where it is shown that during the low-drag events the ensemble-averaged streamwise decreases and wall-normal velocities increases for $y^+ \approx 20$–40, whereas the opposite is true for high-drag events.

Figure 17a,b shows the unconditional and conditional (low-drag) profiles of contribution from the various quadrants in the Reynolds shear stress for $Re_\tau = 70$ and $Re_\tau = 85$, respectively. Figure 16b,d shows the joint distribution of streamwise and wall-normal velocity fluctuations during the low-drag events for $Re_\tau = 70$, obtained using experiments (at $y^+ = 24$) and DNS (at $y^+ = 25$), respectively. Figure 16c shows the joint distribution during the high-drag events for $Re_\tau = 70$ at $y^+ = 24$, obtained using experiments. A good qualitative agreement is observed between the experimental and DNS results for the unconditional data. It can be seen that the major contributors to the Reynolds shear stress are the Q2 and Q4 motions, which explains the reason for the tilted shape of the jpdf shown in Figure 16a,d. These two quadrants are considered to be responsible for the turbulence production [50,51]. It is also observed that the Q4 motions or the "sweep" type motions are the most dominant motions for $y^+ \lesssim 20$ and for the higher wall-normal locations Q2 motions or the "ejection" type motions are the most dominant. For the low-drag case, the Q2 events contribute more than the other quadrants for all the wall-normal locations at both $Re_\tau = 70$ and 85. Another interesting observation is that the Q4 events contribution decreases to a very low value during these low-drag events. This further reinforces the hypothesis that the low-drag events are composed of low-streamwise speed and upwash motions. There is a good qualitative and also fairly good quantitative (for $y^+ \gtrsim 30$–40) agreement between the experimental and DNS results. The discrepancies between the experimental and DNS data in the conditional data are aligned with their unconditional values, which suggests that these slight variations are the result of noise in the measurement rather than different physical observations.

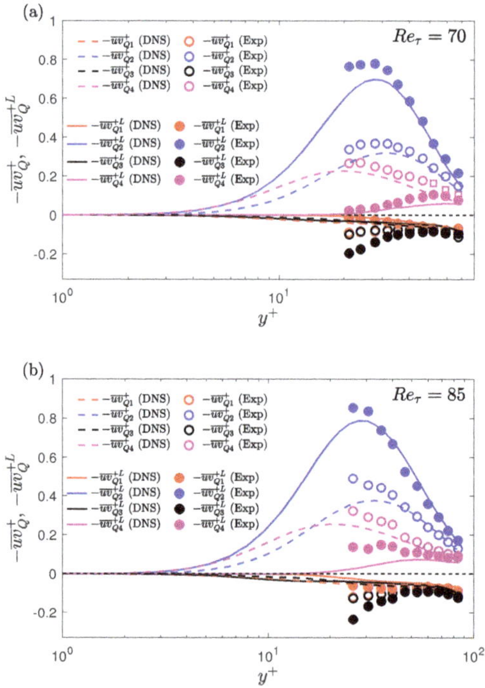

Figure 17. (a) Contribution to $-\overline{uv}$ from different quadrants for the unconditional case and during the low-drag events for (a) $Re_\tau = 70$ and (b) $Re_\tau = 85$. The criteria to detect a low-drag event is $\Delta t_{cr}^+ = 200$ and $\tau_w/\overline{\tau_w} < 0.9$. Thin black dashed line represents a constant value of zero.

The observation from the quadrant contributions is consistent with the previous numerical findings by Kushwaha et al. [22] where it is shown that the low-wall shear stress events are associated with counter-rotating streamwise vortex pairs transferring momentum away from the wall. Park et al. [16] showed in MFU simulations that the low-drag event is the precursor to a strong bursting event which is again consistent with the present result. The low-speed fluid moves away from the wall (ejection process) during these low-drag events which ultimately undergo a bursting process. The ejection and bursting processes are well studied in the past in regards to the low-speed streaks moving away from the wall and bursting in the buffer layer region (for more details, see in [52,53]). Adrian et al. [54] provided a hairpin vortex model in an effort to unify the various previous findings related to the coherent structures observed in the turbulent boundary layer. It was stated that the hairpin vortex originates from the wall inducing a region of low speed between two legs of the vortex which then lifts up by ejection process. The present work suggests that the low wall shear stress events are representative of low-speed regions which are generally observed between the legs of the hairpin vortices in wall-bounded turbulent flows [54,55]. Although it should be noted that the present work employs a different criterion to detect these low-drag events ($\tau_w/\overline{\tau_w} < 0.9$ and $\Delta t^+ > 200$) and therefore these conditional events form only a subset of the low-speed streaks/events observed in the past [55].

Results for the high-drag events are shown in Figure 18a,b for $Re_\tau = 70$ and 85, respectively. It can be observed that during the high-drag events, the Q4 events are the significant contributor to the Reynolds shear stress. This is again expected based on the ensemble-averaged data, i.e., high-drag events are composed of high-speed and downwash motions for $y^+ \geq 20$.

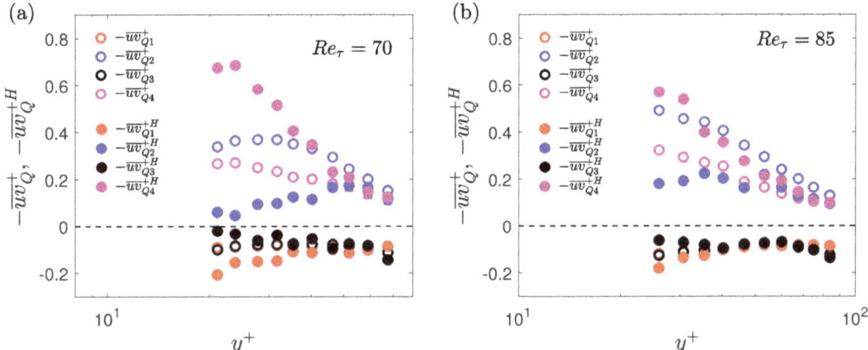

Figure 18. (a) Contribution to $-\overline{uv}$ from different quadrants for the unconditional case and during the high-drag events for (a) $Re_\tau = 70$ and (b) $Re_\tau = 85$. The criteria to detect a high-drag event is $\Delta t_{cr}^+ = 200$ and $\tau_w/\overline{\tau_w} < 0.9$. Thin black dashed line represents a constant value of zero.

8. Summary

An investigation into the intermittencies associated with the low- and high-drag events in turbulent channel flow has been conducted using experiments and DNS. For experiments, simultaneous measurements of streamwise velocity and wall shear stress are carried out to detect and characterise these intermittencies for Re_τ between 70 and 250. DNS is carried out in a large computational box for $Re_\tau = 70$ and 85. The fraction of time spent in the intervals of low- and high-drag is found to be roughly independent of the Reynolds number for $70 \leq Re_\tau \leq 250$ when the criteria for minimum time-duration is kept constant in inner units. The low- and high-drag events exhibit an exponential distribution of the frequency of their occurrence when studied as a function of the duration of their intervals. It is found that even for artificially constructed signals (up to the limit of Gaussian signal), there is a presence of spikes and dips in the ensemble-averaged data, if the same criteria is applied as used to detect a low- or high-drag event in the wall shear stress signals. This suggests that these spikes (or dips) might be the consequence of the conditional averaging of a time series data.

Streamwise velocity profiles, conditionally sampled during the low-drag events, get closer to Virk's MDR profile and the lower-branch of the nonlinear TW solutions for $y^+ \approx 20\text{--}35$ at all studied Reynolds numbers. For $120 \leq Re_\tau \leq 250$, in the log-law region, the conditional velocity profile is higher than the unconditional velocity profile with the slope of the profile higher during the low-drag events. Similarly, the conditional velocity profile is lower than the unconditional velocity profile with the slope of the profile being slightly lower during the high-drag events. A comparison of the conditional streamwise velocity profiles at $Re_\tau = 180$ and 250 with other drag reduction techniques is made. A good agreement between the profiles in the log-law region is observed. For $Re_\tau = 70$ and 85, in addition to the streamwise velocity, wall-normal velocity is also measured to investigate the behaviour of RSS. There is found to be an increase in the conditionally averaged RSS for $y^+ \gtrsim 10$ during the low-drag events. This is observed to be due to a significant increase in the turbulence-generating Q2 motions during these low-drag events. The high-drag events are found to be associated with the Q4 events, although the RSS during these events remain fairly similar to the unconditional profile for $y^+ \gtrsim 20$.

Author Contributions: Conceptualisation, R.J.P., D.J.C.D. and M.D.G.; methodology, R.J.P., D.J.C.D. and H.C.-H.N.; software, R.A. and E.A.D.; validation, R.A.; formal analysis, R.A.; investigation, R.A.; resources, R.J.P., J.S.P.; data curation, R.A.; writing—original draft preparation, R.A. and E.A.D.; writing—review and editing, R.J.P., D.J.C.D., M.D.G., H.C.-H.N. and J.S.P.; visualisation, R.A.; supervision, R.J.P., D.J.C.D. and H.C.-H.N.; project administration, R.J.P., D.J.C.D. and H.C.-H.N.; funding acquisition, R.J.P. and J.S.P. All authors have read and agreed to the published version of the manuscript.

Funding: This research was funded by the Air Force Office of Scientific Research through grant FA9550-16-1-0076.

Acknowledgments: This research was funded by the Air Force Office of Scientific Research (AFOSR) through grant FA9550-16-1-0076. M.D.G. acknowledges the financial support obtained from AFOSR through grant FA9550-18-1-0174. E.A.D. and J.S.P. gratefully acknowledge the financial support in part from the National Science Foundation through a grant OIA-1832976. We would like to thank Kevin Zeng (University of Wisconsin-Madison) for helping in transferring the flow fields computed in large computational boxes.

Conflicts of Interest: The authors declare no conflict of interest.

References

1. Graham, M.D.; Floryan, D. Exact coherent states and the nonlinear dynamics of wall-bounded turbulent flows. *Annu. Rev. Fluid Mech.* **2021**, *53*, 227–253. [CrossRef]
2. Nagata, M. Three-dimensional finite-amplitude solutions in plane Couette flow: Bifurcation from infinity. *J. Fluid Mech.* **1990**, *217*, 519–527. [CrossRef]
3. Waleffe, F. Exact coherent structures in channel flow. *J. Fluid Mech.* **2001**, *435*, 93–102. [CrossRef]
4. Waleffe, F. Homotopy of exact coherent structures in plane shear flows. *Phys. Fluids* **2003**, *15*, 1517–1534. [CrossRef]
5. Nagata, M.; Deguchi, K. Mirror-symmetric exact coherent states in plane Poiseuille flow. *J. Fluid Mech.* **2013**, *735*, R4. [CrossRef]
6. Jiménez, J.; Moin, P. The minimal flow unit in near-wall turbulence. *J. Fluid Mech.* **1991**, *225*, 213–240. [CrossRef]
7. Hamilton, J.M.; Kim, J.; Waleffe, F. Regeneration mechanisms of near-wall turbulence structures. *J. Fluid Mech.* **1995**, *287*, 317–348. [CrossRef]
8. Jiménez, J.; Pinelli, A. The autonomous cycle of near-wall turbulence. *J. Fluid Mech.* **1999**, *389*, 335–359. [CrossRef]
9. McComb, W. *The Physics of Fluid Turbulence*; Oxford University Press: Oxford, UK, 1990.
10. Xi, L.; Graham, M.D. Active and hibernating turbulence in minimal channel flow of Newtonian and polymeric fluids. *Phys. Rev. Lett.* **2010**, *104*, 218301. [CrossRef]
11. Dubief, Y.; White, C.M.; Shaqfeh, E.S.; Terrapon, V.E. Polymer maximum drag reduction: A unique transitional state. *arXiv* **2011**, arXiv:1106.4482.
12. Xi, L.; Graham, M.D. Intermittent dynamics of turbulence hibernation in Newtonian and viscoelastic minimal channel flows. *J. Fluid Mech.* **2012**, *693*, 433–472. [CrossRef]

13. Park, J.S.; Graham, M.D. Exact coherent states and connections to turbulent dynamics in minimal channel flow. *J. Fluid Mech.* **2015**, *782*, 430–454. [CrossRef]
14. Kline, S.J.; Reynolds, W.C.; Schraub, F.; Runstadler, P. The structure of turbulent boundary layers. *J. Fluid Mech.* **1967**, *30*, 741–773. [CrossRef]
15. Itano, T.; Toh, S. The dynamics of bursting process in wall turbulence. *J. Phys. Soc. Jpn.* **2001**, *70*, 703–716. [CrossRef]
16. Park, J.S.; Shekar, A.; Graham, M.D. Bursting and critical layer frequencies in minimal turbulent dynamics and connections to exact coherent states. *Phys. Rev. Fluids* **2018**, *3*, 014611. [CrossRef]
17. Jiménez, J.; Kawahara, G.; Simens, M.P.; Nagata, M.; Shiba, M. Characterization of near-wall turbulence in terms of equilibrium and "bursting" solutions. *Phys. Fluids* **2005**, *17*, 015105. [CrossRef]
18. Flores, O.; Jiménez, J. Hierarchy of minimal flow units in the logarithmic layer. *Phys. Fluids* **2010**, *22*, 071704. [CrossRef]
19. Brand, E.; Gibson, J.F. A doubly localized equilibrium solution of plane Couette flow. *J. Fluid Mech.* **2014**, *750*, R3. [CrossRef]
20. Chantry, M.; Willis, A.P.; Kerswell, R.R. Genesis of streamwise-localized solutions from globally periodic traveling waves in pipe flow. *Phys. Rev. Lett.* **2014**, *112*, 164501. [CrossRef]
21. Zammert, S.; Eckhardt, B. Streamwise and doubly-localised periodic orbits in plane Poiseuille flow. *J. Fluid Mech.* **2014**, *761*, 348–359. [CrossRef]
22. Kushwaha, A.; Park, J.S.; Graham, M.D. Temporal and spatial intermittencies within channel flow turbulence near transition. *Phys. Rev. Fluids* **2017**, *2*, 024603. [CrossRef]
23. Whalley, R.D.; Park, J.S.; Kushwaha, A.; Dennis, D.J.C.; Graham, M.D.; Poole, R.J. Low-drag events in transitional wall-bounded turbulence. *Phys. Rev. Fluids* **2017**, *2*, 034602. [CrossRef]
24. Whalley, R.; Dennis, D.; Graham, M.; Poole, R. An experimental investigation into spatiotemporal intermittencies in turbulent channel flow close to transition. *Exp. Fluids* **2019**, *60*, 102. [CrossRef]
25. Pereira, A.S.; Thompson, R.L.; Mompean, G. Common features between the Newtonian laminar–turbulent transition and the viscoelastic drag-reducing turbulence. *J. Fluid Mech.* **2019**, *877*, 405–428. [CrossRef]
26. Kim, J.; Moin, P.; Moser, R. Turbulence statistics in fully developed channel flow at low Reynolds number. *J. Fluid Mech.* **1987**, *177*, 133–166. [CrossRef]
27. Agrawal, R.; Ng, H.C.H.; Dennis, D.J.; Poole, R.J. Investigating channel flow using wall shear stress signals at transitional Reynolds numbers. *Int. J. Heat Fluid Flow* **2020**, *82*, 108525. [CrossRef]
28. Agrawal, R.; Whalley, R.D.; Ng, H.C.H.; Dennis, D.J.C.; Poole, R.J. Minimizing recalibration using a non-linear regression technique for thermal anemometry. *Exp. Fluids* **2019**, *60*, 116. [CrossRef]
29. Agrawal, R.; Ng, H.C.H.; Whalley, R.D.; Dennis, D.J.; Poole, R.J. In search of low drag events in Newtonian turbulent channel flow at low Reynolds number. In Proceedings of the 11th International Symposium on Turbulence and Shear Flow Phenomena (TSFP11), Southampton, UK, 30 July–2 August 2019.
30. Poggi, D.; Porporato, A.; Ridolfi, L. An experimental contribution to near-wall measurements by means of a special laser Doppler anemometry technique. *Exp. Fluids* **2002**, *32*, 366–375. [CrossRef]
31. Melling, A.; Whitelaw, J. Turbulent flow in a rectangular duct. *J. Fluid Mech.* **1976**, *78*, 289–315. [CrossRef]
32. Walker, D.; Tiederman, W. Turbulent structure in a channel flow with polymer injection at the wall. *J. Fluid Mech.* **1990**, *218*, 377–403. [CrossRef]
33. Günther, A.; Papavassiliou, D.; Warholic, M.; Hanratty, T. Turbulent flow in a channel at a low Reynolds number. *Exp. Fluids* **1998**, *25*, 503–511. [CrossRef]
34. Ligrani, P.; Bradshaw, P. Spatial resolution and measurement of turbulence in the viscous sublayer using subminiature hot-wire probes. *Exp. Fluids* **1987**, *5*, 407–417. [CrossRef]
35. Agrawal, R. Intermittencies in Transitional and Turbulent Channel Flow. Ph.D. Thesis, University of Liverpool, Liverpool, UK, 2020.
36. Bruun, H.H. *Hot-Wire Anemometry: Principles and Signal Analysis*, 1st ed.; Oxford University Press: New York, NY, USA, 1995.
37. Bruun, H. Hot-film anemometry in liquid flows. *Meas. Sci. Technol.* **1996**, *7*, 1301. [CrossRef]
38. Kline, S.; McClintock, F. Describing uncertainties in single-sample experiments. *Mech. Eng.* **1953**, *75*, 3–8.
39. Zhang, Z. *LDA Application Methods: Laser Doppler Anemometry for Fluid Dynamics*; Springer Science & Business Media: Berlin/Heidelberg, Germany, 2010.

40. Gibson, J.F. Channelflow: A Spectral Navier-Stokes Simulator in C++. Technical Report, U. New Hampshire, 2014. Available online: Channelflow.org (accessed on 31 January 2018).

41. Gomit, G.; De Kat, R.; Ganapathisubramani, B. Structure of high and low shear-stress events in a turbulent boundary layer. *Phys. Rev. Fluids* **2018**, *3*, 14609. [CrossRef]

42. Avila, K.; Moxey, D.; de Lozar, A.; Avila, M.; Barkley, D.; Hof, B. The onset of turbulence in pipe flow. *Science* **2011**, *333*, 192–196. [CrossRef]

43. Graham, M.D. Drag reduction and the dynamics of turbulence in simple and complex fluids. *Phys. Fluids* **2014**, *26*, 625–656. [CrossRef]

44. White, C.; Dubief, Y.; Klewicki, J. Re-examining the logarithmic dependence of the mean velocity distribution in polymer drag reduced wall-bounded flow. *Phys. Fluids* **2012**, *24*, 021701. [CrossRef]

45. Warholic, M.; Massah, H.; Hanratty, T. Influence of drag-reducing polymers on turbulence: Effects of Reynolds number, concentration and mixing. *Exp. Fluids* **1999**, *27*, 461–472. [CrossRef]

46. White, C.M.; Mungal, M.G. Mechanics and prediction of turbulent drag reduction with polymer additives. *Annu. Rev. Fluid Mech.* **2008**, *40*, 235–256. [CrossRef]

47. Min, T.; Kim, J. Effects of hydrophobic surface on skin-friction drag. *Phys. Fluids* **2004**, *16*, L55–L58. [CrossRef]

48. Choi, H.; Moin, P.; Kim, J. Active turbulence control for drag reduction in wall-bounded flows. *J. Fluid Mech.* **1994**, *262*, 75–110. [CrossRef]

49. Wallace, J.M. Quadrant analysis in turbulence research: History and evolution. *Ann. Rev. Fluid Mech.* **2016**, *48*, 131–158. [CrossRef]

50. Kim, H.; Kline, S.; Reynolds, W. The production of turbulence near a smooth wall in a turbulent boundary layer. *J. Fluid Mech.* **1971**, *50*, 133–160. [CrossRef]

51. Wallace, J.M.; Eckelmann, H.; Brodkey, R.S. The wall region in turbulent shear flow. *J. Fluid Mech.* **1972**, *54*, 39–48. [CrossRef]

52. Dennis, D.J.C. Coherent structures in wall-bounded turbulence. *An. Acad. Bras. Ciênc.* **2015**, *87*, 1161–1193. [CrossRef]

53. Jiménez, J. Coherent structures in wall-bounded turbulence. *J. Fluid Mech.* **2018**, *842*, 1. [CrossRef]

54. Adrian, R.J.; Meinhart, C.D.; Tomkins, C.D. Vortex organization in the outer region of the turbulent boundary layer. *J. Fluid Mech.* **2000**, *422*, 1–54. [CrossRef]

55. Adrian, R.J. Hairpin vortex organization in wall turbulence. *Phys. Fluids* **2007**, *19*, 041301. [CrossRef]

Article

Intermittency, Moments, and Friction Coefficient during the Subcritical Transition of Channel Flow

Jinsheng Liu [1], Yue Xiao [2], Mogeng Li [3], Jianjun Tao [2,*] and Shengjin Xu [1,*]

[1]   Applied Mechanics Laboratory, School of Aerospace Engineering, Tsinghua University, Beijing 100084, China; ljs16@mails.tsinghua.edu.cn
[2]   Center for Applied Physics and Technology, Department of Mechanics and Engineering Science, College of Engineering, Peking University, Beijing 100871, China; xiaoyue904@pku.edu.cn
[3]   Department of Mechanical Engineering, University of Melbourne, Victoria 3010, Australia; mogengl@student.unimelb.edu.au
*   Correspondence: jjtao@pku.edu.cn (J.T.); xu_shengjin@tsinghua.edu.cn (S.X.)

Received: 26 October 2020; Accepted: 8 December 2020; Published: 11 December 2020

Abstract: The intermittent distribution of localized turbulent structures is a key feature of the subcritical transitions in channel flows, which are studied in this paper with a wind channel and theoretical modeling. Entrance disturbances are introduced by small beads, and localized turbulent patches can be triggered at low Reynolds numbers (Re). High turbulence intensity represents strong ability of perturbation spread, and a maximum turbulence intensity is found for every test case as $Re \geq 950$, where the turbulence fraction increases abruptly with Re. Skewness can reflect the velocity defects of localized turbulent patches and is revealed to become negative when Re is as low as about 660. It is shown that the third-order moments of the midplane streamwise velocities have minima, while the corresponding forth-order moments have maxima during the transition. These kinematic extremes and different variation scenarios of the friction coefficient during the transition are explained with an intermittent structure model, where the robust localized turbulent structure is simplified as a turbulence unit, a structure whose statistical properties are only weak functions of the Reynolds number.

Keywords: subcritical transition; channel flow; turbulence fraction; moment

---

1. Introduction

Plane Poiseuille flow (PPF), the flow driven by a pressure gradient between two parallel plates, displays a parabolic velocity profile at its laminar state and becomes linearly unstable when the Reynolds number is larger than the critical value, $Re_c = 5772$ [1]. The Reynolds number (Re) is defined as $1.5U_b^* h^*/v^*$, where U_b^* is the bulk velocity, h^* is the half-channel height, and v^* is the kinematic viscosity of the fluid. In practice, PPF may become turbulent at much lower Reynolds numbers than Re_c due to the subcritical transition, where the finite-amplitude disturbances are necessary and the nonlinear effect cannot be ignored [2–4]. Davies and White [5] measured the friction coefficient of PPF with different aspect ratios of the cross-sections in a wide range of Reynolds numbers. It was shown that the critical Reynolds number of the subcritical transition increases with the ratio between the entrance length and the channel height, and it remains at 667.5 when the entrance length is larger than 108h. Patel and Head [6] found experimentally that PPF remained laminar as $Re < 1035$, and intermittent bursts occurred as $1035 < Re < 1350$. Later experiments by Nishioka and Asai [7] confirmed that the turbulent state could hardly be sustained as $Re < 1000$. Based on flow visualizations, Carlson et al. [8] found that the orifice jet on the wall can trigger turbulent spots when Re is about 1000, and when $Re < 840$, the turbulent spots cannot be formed completely and decay eventually. Later experimental, theoretical and numerical works were mainly focused on the turbulent spots as $Re > 1000$ [9–14]. According to the experiments of Alavyoon, et al. [15], the complete spot cannot be triggered by orifice jet if $Re < 1100$. Recently, turbulent stripes or bands were revealed by numerical simulations for $Re \geq 1070$ [16,17] and were observed by flow visualizations [18]. It was found experimentally that the turbulent bands would break as $Re < 1275$, and the flow remained stable and laminar at $Re = 975$ [19].

Based on numerical simulations within a tilted long and narrow domain, Tuckerman found turbulent band structures as $Re > 850$ [20]. By applying entrance disturbances and flow visualization techniques, Sano and Tamai [21] obtained the turbulence fraction at a range of Reynolds numbers and defined a threshold of 830 for the transition by fitting the data with the Directed Percolation (DP) model. According to their experimental data, however, the turbulence fractions are not zero as $Re < 830$. Recent numerical simulations revealed that the DP power law is retrieved only when Re is above 924, and relaminarization will occur in the long-time limit as $Re < 700$ [22]. Numerical simulations in large domains showed that localized turbulent bands can be obtained when Re is reduced to 720 [23]. Further numerical investigations illustrated that the isolated turbulent band, a single banded coherent structure surrounded by a large laminar region, can obliquely extend at moderate Reynolds numbers but will decay eventually as $Re < 665$ [24]. This threshold Reynolds number, in fact, agrees with the experimental observation by Davies and White [5]. It is tested that the periodic turbulent band can sustain as $Re < 750$, though band breaking and band reconnection may occur [25]. Recently, the turbulent bands were observed at $Re = 750$ by flow visualization [26], and the mean growth rate of turbulence fraction was found to become positive at $Re \approx 650$ [27,28]. Therefore, in the literature, there have been discrepancies on the threshold Reynolds number for sustained turbulence in channel flows.

Besides the turbulence fraction, other statistical parameters are studied as well for the transitional channel flows. Turbulence intensities at the channel center are measured and are found to increase rapidly around $Re = 1050$, reach a peak at $Re = 1140$, and then gradually decrease with increasing Re [29]. The intermittent low- and high-drag events are investigated numerically and experimentally [30–32], and it is found that the conditionally averaged Reynolds shear stress is higher than the mean value during the low-drag events [33]. Based on simulations of channel flows with constant pressure gradients, a linear correlation for the wall shear stress is observed between its kurtosis and its skewness squared [34]. It is known that high-order moments of velocity derivatives are important to understand the non-Gaussian behavior of turbulence [35], and the intermittency is a key concept to develop turbulence model for the transitions of incompressible, supersonic, and hypersonic boundary layer flows [36]. However, the study on the relation between the turbulence fraction and the high order moments of velocities in the transitional channel flows is still rudimental.

In this paper, a wind channel with a large width-to-height ratio is used to study the subcritical transition of PPF, and its configuration is introduced in Section 2. In Section 3, it is revealed that the turbulence intensity and the kurtosis of midplane streamwise velocity reach their maxima while the skewness has a negative minimum during the transition. Furthermore, an intermittent structure model is constructed to describe the velocity features of localized turbulent structures and derive theoretically the high-order moments of midplane velocity and the friction coefficient, which are shown to be consistent with the experimental data. In Section 4, conclusions are presented.

2. Experimental Apparatus and Methods

2.1. Wind Channel

The open-circuit wind channel used in the experiment is shown in Figure 1. The length, width, and height $2h^*$ of the working section are 4.5, 1.0, and 0.01 m, respectively. The flow is driven by three centrifugal fans with 1.5 kW induction motors, and the midplane velocity in the working section is controlled by a frequency converter to vary between 0.4 and 28 m·s^{-1}. In order to isolate the vibration noise generated by the centrifugal fans, a soft connection is attached just in front of the expansion section. A perforated screen and 5 stainless-steel screens are mounted near the honeycomb layer to stabilize the flow and decrease the turbulence intensity. Two contractions with the contraction ratios of 4:1 and 9:1 are used to further reduce the turbulence intensity to a level less than 0.2%.

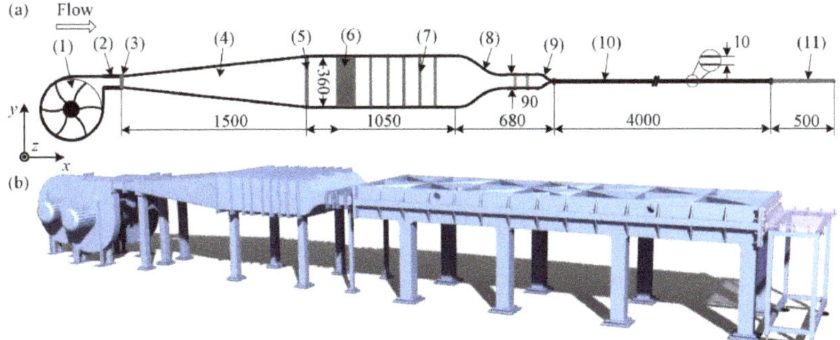

Figure 1. Sketch of the wind channel. (**a**) The components of the wind channel: (1) centrifugal fan, (2) soft connection, (3) fine damping screen, (4) expansion section, (5) perforated screen, (6) honeycomb, (7) screen, (8) first contraction, (9) second contraction, (10) first working section (steel), and (11) second working section (tempered glass). Unit of length, mm; (**b**) 3D drawing of the wind channel.

The channel walls of the first working section are polished to achieve a surface roughness less than 15 μm and are supported by steel frames, to avoid deflection. According to the finite element analyses, the maximum deflection of the whole test section is less than 3.7 μm. The second part is a transparent test section with a length of 0.5 m, granting optical access to the Particle Image Velocimetry (PIV) setup. Two 10-mm-thick side walls are sandwiched between the top and the bottom walls, and the error of channel height in the working section is less than 0.01 mm. In all experiments, the ambient temperature variation is less than 2 degrees centigrade. For non-dimensionalization, the half channel height h^* and the time averaged velocity at the midplane U_c^* are chosen as the characteristic length and velocity, respectively, and the dimensionless parameters have no superscript. For laminar flows, $U_c = 1.5U_b$. The origin of the coordinates lies at the entrance center of the working section, and the dimensionless x, y, and z represent the streamwise, the wall-normal, and the spanwise directions, respectively.

2.2. Experimental Methods and Validations

Eighteen static pressure holes with 0.5 mm diameter are drilled on the lower wall along the line $z = 0$ with an interval of $l = 200$ mm, and the first hole is located at 300 mm from the entrance of the working section. Consequently, the pressure gradient along the streamwise direction can be monitored by using micro differential pressure transducers (Alpha M168, range: 0~25 Pa, accuracy: ±0.25% FS). A low-noise hot-wire anemometer (HWA, Dantec StreamLine Pro.) with 3 channels is used to measure the velocity with a relative error less than 1.5%. The stainless-steel probe stem is mounted on a two-dimensional traversing mechanism with a positioning resolution of 5 μm. In order to minimize the interference, the probes are inserted through the outlet of the working section.

We checked that, except the region very close to the entrance, the streamwise pressure gradients remained constant at low Reynolds numbers and agreed with the theoretical values for laminar PPF as reflected by the friction coefficients, which are discussed in Section 3.1. As shown in Figure 2a, the uniform distribution of, U_c^*, in the spanwise direction indicates that the velocity field in the central part of the cross-section is hardly affected by the sidewalls. When the flow is laminar at $Re = 1096$, it is shown in Figure 2b that the velocity profiles at five different spanwise positions agree well with the theoretical parabolic distribution. When Re is increased to 7543, the time averaged velocity profiles are all close to the 1/8 power law curve, confirming that the sidewall effect is still negligible in the central region. Without the entrance artificial disturbances, it is checked that the flow can remain laminar for Re up to 3500, and hence the present setup is appropriate to study the subcritical transition of PPF.

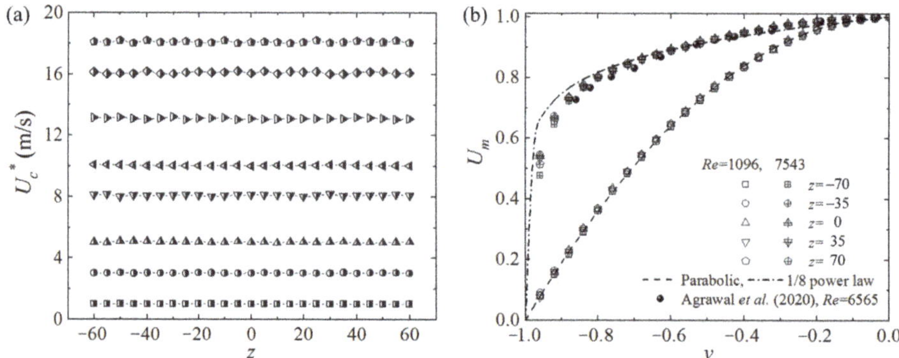

Figure 2. Streamwise velocities measured at $x = 780$. (**a**) Spanwise distributions of the time averaged velocity in the midplane U_c^*, and (**b**) the time averaged velocity profiles at different spanwise positions. The measurements of Reference [33] are added in (**b**) as references.

Nine plastic beads evenly spaced with an interval of 100 mm along a thin iron wire are placed at the centerline of channel inlet to introduce entrance disturbances. Different bead diameters, D^*, and wire diameters, d^*, are used in four cases and are listed in Table 1.

Table 1. Dimensions of the entrance disturbances.

	Baseline	Case_1	Case_2	Case_3
D^* (mm)	/	8	6	8
D^*/h^*	/	1.6	1.2	1.6
d^* (mm)	/	0.2	0.2	1.5
d^*/h^*	/	0.04	0.04	0.3

3. Results and Discussions

3.1. Friction Coefficient

The friction coefficient $C_f = 8\left(h^* \frac{dP^*}{dx^*}\right)/\left(9\rho^* U_b^{*2}\right)$ is measured at different Reynolds numbers, with different entrance disturbances, where dP^*/dx^* is the mean pressure gradient calculated based on the pressure difference between $x = 660$ and 740, and the bulk velocity, U_b^*, is obtained from the mean velocity profile. C_f is calculated for every 10-s sample, and the averaged C_f for 20 samples (totally $10^4 \sim 10^5$ time units at the transition stage) are shown in Figure 3, where the error bars represent the standard deviation. It is shown that when $Re < 600$ or there are no entrance artificial disturbances (Baseline), the present experimental data agree well with the laminar value $C_f = 4/Re$. The previous results [5,6,22,24] are shown as well for references. When Re is greater than 1750, C_f data for different entrance disturbance cases tend to agree with the "optimum log-law" labeled by the dashed line for developed turbulence, where $Re = \sqrt{\frac{2}{C_f}} \exp\left[0.41\left(\sqrt{\frac{8}{9C_f}} - 2.4\right)\right]$ [22,37]. During $950 < Re < 1010$, C_f in three disturbed cases increases abruptly, reflecting a strong development of turbulence. As shown in the inset of Figure 3b, such an abrupt increase of C_f occurs as well in the previous direct numerical simulations, where the turbulent band split occurs, i.e., parallel split to form a new band parallel to the original one and transverse split to sprout new branch (as shown by Figure 6 of Reference [24]). Recent systematical simulations [22] revealed that the transition from "one-sided" (all localized turbulent bands point to the same direction) to "two-sided" (the bands may grow in different directions) propagations takes place at $Re \approx 924$. By simulations in tilted slender domains, a critical Reynolds number is defined as 950, where the statistically estimated mean lifetimes for band decay and splitting coincide with each other [38]. All of these numerical results explain, to some degree, why C_f increases abruptly as $Re > 950$.

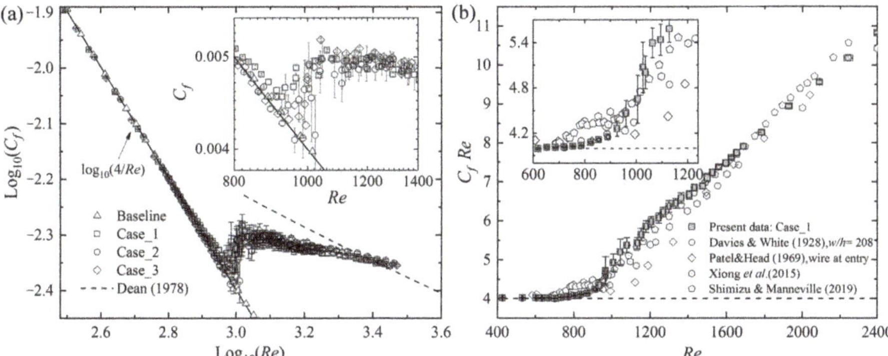

Figure 3. (**a**) The friction coefficient, C_f, as a function of Re. The previous experimental and numerical data are illustrated in (**b**) for references.

3.2. Turbulence Intensity and Pressure Turbulence Intensity

The time series of the streamwise velocity, U, obtained at the midplane by HWA are just straight lines superimposed by background noise at low Reynolds numbers, e.g., $Re = 652$ in Figure 4a. When a turbulent band or spot passes through the measuring point, the time series show a velocity defect, i.e., the midplane streamwise velocity decreases first along with the time, then oscillates strongly with high frequencies before increasing abruptly to recover its laminar level. The velocity fields of the spots and turbulent bands are measured by PIV, and their consistencies with the direct numerical simulations are confirmed and shown in [39]. The present study mainly focuses on the statistical kinematic and dynamic properties of the transitional flow. It is shown in Figure 4d that the widths and amplitudes of the velocity defects are comparable for different entrance disturbances and different Reynolds numbers, indicating that the statistical properties of localized structures are weak functions of Re and external disturbances during the transition. Such a streamwise velocity defect appears more and more frequently with the increase of Re, as shown in Figure 4.

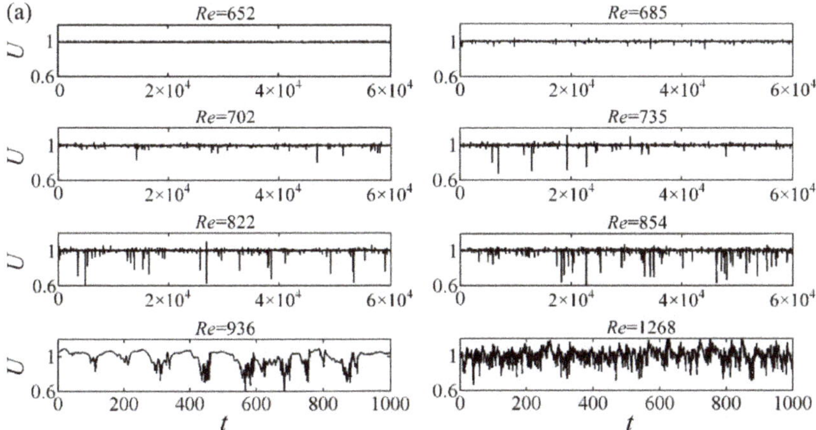

Figure 4. *Cont.*

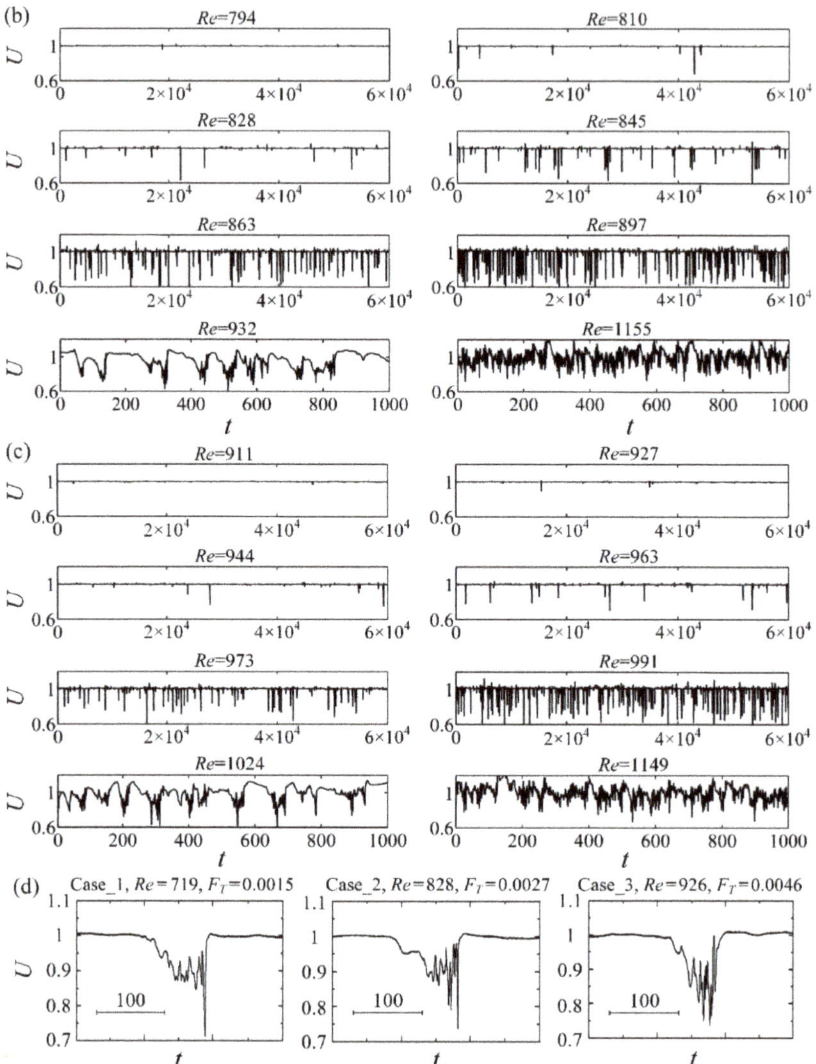

Figure 4. The time series of velocity, U, measured at $(x, z) = (780, 0)$ for (**a**) Case_1, (**b**) Case_2, and (**c**) Case_3. Typical signals of localized turbulent structures for different cases at different Re and turbulence fraction, F_T, are shown in (**d**).

The development of turbulence may be described by the turbulence intensity of streamwise velocity $I_u = \langle u^2 \rangle^{1/2} = \langle (U - U_c)^2 \rangle^{1/2}$ at the midplane ($y = 0$) and the pressure turbulence intensity $I_P = P_{rms} / (dP/dx) - [P_{rms} / (dP/dx)]_r$, where $\langle \, \rangle$ means the time averaged quantity, and the subscripts r and rms represent a reference value and the root mean square. In this paper, $[P_{rms} / (dP/dx)]_r$ is the value at $Re = 600$, corresponding to a laminar flow with background noise. When Re is smaller than 850, I_P remains a small value and is almost independent of the entrance disturbances, the downstream position, and the Reynolds number as shown in Figure 5a. When Re is larger than 850, I_P of Case_1 increases obviously and reaches a peak at about $Re = 950$ before decreasing. The corresponding Re of I_P peaks for Case_2 and Case_3 is around 980 and 1020, respectively. In the right column of Figure 5, it is shown that the turbulence intensity, I_u, has peak values at the same *Re* as I_P for all three cases. The existence of these peaks is explained in Section 3.5, with an intermittent structure model.

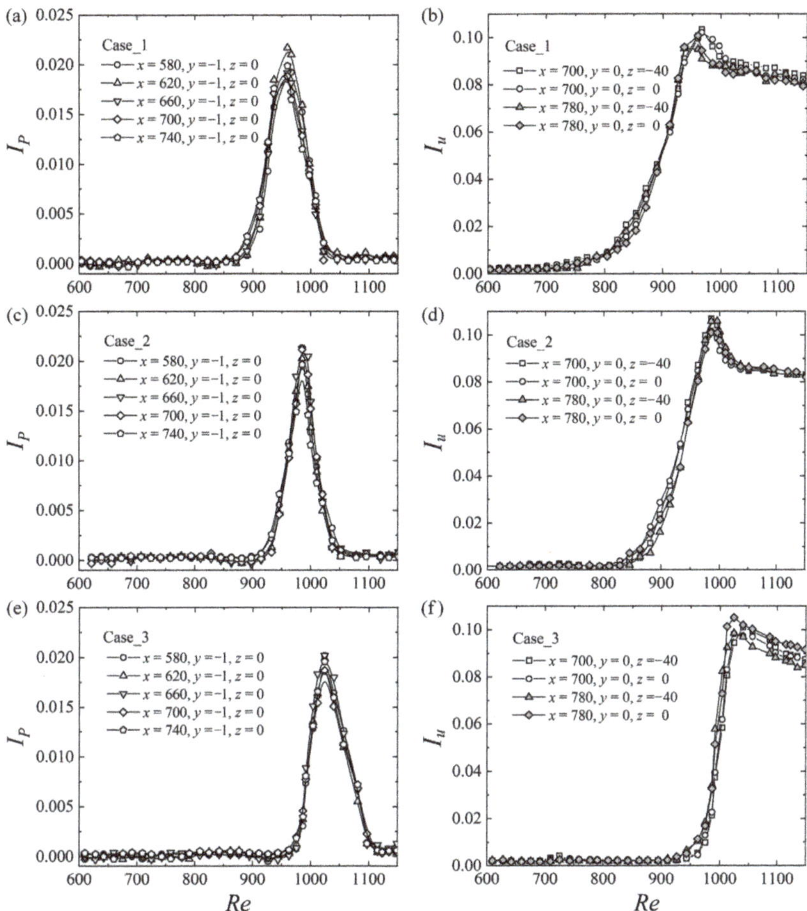

Figure 5. Pressure turbulence intensity, I_P (left column), and turbulence intensity, I_u (right column) measured at different locations. (**a,b**), (**c,d**), and (**e,f**) are for Case_1, Case_2, and Case_3, respectively.

3.3. Skewness and Kurtosis

Though I_P and I_u reflect the mean levels of fluctuation amplitudes or strengths, they cannot describe the intermittency and asymmetry of the signals. In this subsection, the skewness $S(u) = \langle u^3 \rangle / \langle u^2 \rangle^{3/2}$ is calculated based on the streamwise fluctuation velocity, u, measured at the midplane, representing the asymmetric distribution of the velocity. The kurtosis or flatness $F(u) = \langle u^4 \rangle / \langle u^2 \rangle^2$ is computed as well, reflecting the intermittency and the deviation from the random distribution. At low Reynolds numbers, the laminar velocity signal mixed with the background white noise conforms to the normal distribution, and hence $S(u) = 0$ and $F(u) = 3$. When the localized turbulent spots or bands emerge intermittently in the flow, the velocity defects appear, leading to a negative skewness and a positive flatness, e.g., *Re* < 700 for Case_1 shown in Figure 6, while the corresponding turbulence intensity (Figure 5) and the friction coefficient (Figure 3) remain nearly unchanged. Specially, it is shown in

Figure 6 that the skewness and the kurtosis reach a minimum and a maximum during the transition, respectively, and the corresponding underlying mechanisms are discussed in Section 3.5.

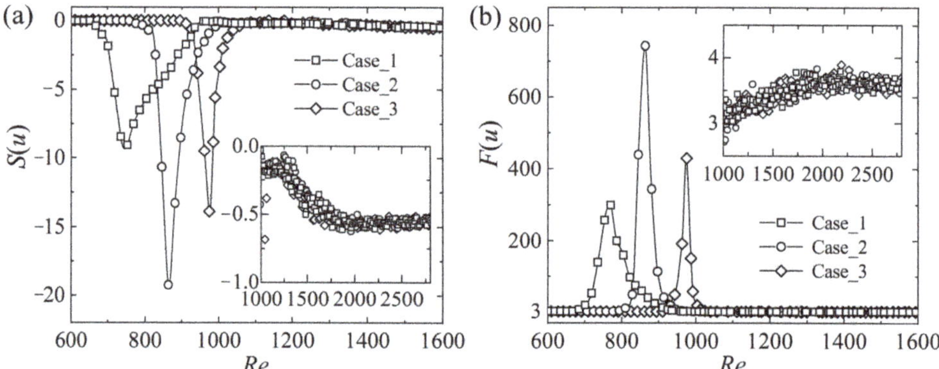

Figure 6. (a) Skewness and (b) kurtosis of the streamwise velocity measured at $(x, y, z) = (780, 0, 0)$ for different disturbance cases and Reynolds numbers.

The transition process is triggered by the entrance disturbances, the abundant vortex structures shed from the beads placed at the inlet. It has been shown that, at $Re_D = 3700$ (based on the free-stream velocity and the sphere diameter D), the turbulence intensity, I_u, along the wake centerline of a sphere quickly reduces to 0.05 at $x/D = 12$ [40]. Based on the centerline velocities measured for $Re = 600\sim1200$, the corresponding Re_D for the present inlet beads can be estimated to be 720~1920. Considering that the working section is $500D\sim666D$ long, the strong turbulence intensity, I_u, around 0.1, as shown in Figure 5, should be caused by the localized turbulent patches triggered by the remnants of the bead wakes rather than the remnants themselves. According to Figure 6, the Reynolds number intervals where the skewness and the kurtosis deviate from the normal distribution are [660,960], [780,1000], and [910,1060] for Case_1, Case_2, and Case_3, respectively. It is interesting to note that the upper limits of these Re intervals are close to the corresponding peak Res for I_P and I_u shown in Figure 5. The lower limits indicate the onset of turbulence, and the minimum lower limit of tested cases is about 660, which is consistent with the threshold determined numerically for the oblique turbulent bands [24,25] and the value obtained by flow visualization [27]. In numerical simulations, the computation may last long enough, e.g., $\sim10^4$ time units, to observe the transient growth and eventual decay of the patterns near the critical state, while, in experiments, the channel length is limited and the traveling turbulent patches may grow transiently but have no time to experience the final decay. This factor may cause a mild underestimate of the threshold value in experiments. It is shown in the insets of Figure 6 that, when $Re > 1100$ and F_T is close to 1, the skewness and the kurtosis of streamwise velocity continue to evolve, deviating from 0 and 3 (the values for white Gaussian noise) and remain at about -0.5 and 3.5 after $Re > 1750$, respectively, the values for fully developed turbulence [41]. Consequently, the threshold for fully developed turbulence may be defined as $Re \approx 1750$.

3.4. Turbulence Fraction

An important parameter to describe the pattern evolution and intermittency during the subcritical transition is the turbulence fraction, F_T, whose determination relies on the identification of the boundaries between the laminar and the turbulent regions. Different from the previous experiments, where F_T was mostly calculated based on flow visualization images, in this paper, the time series of velocity are used to define F_T as $F_T = \sum t_T/t_{Total}$, where t_T and t_{Total} are the turbulent period and the total sampling time, respectively. As shown in Figure 7a, the time series of the midplane streamwise velocity includes many velocity defects, which correspond to the traveling localized turbulent patches and include high-frequency components, as illustrated by the wavelet power spectrum shown in Figure 7b. Consequently, high-pass filtering is used to extract these components, as shown in Figure 7c, whose time intervals are defined as the turbulent period, t_T. Different cutoff frequencies, f_c, are tested, and the corresponding F_T values vary in the same trend, as shown in Figure 8a, though a higher f_c leads to a lower F_T. By comparing Figure 7a,c, the cutoff frequency of 45 Hz is found to capture the turbulent periods reasonably well, and hence is used in the following analyses.

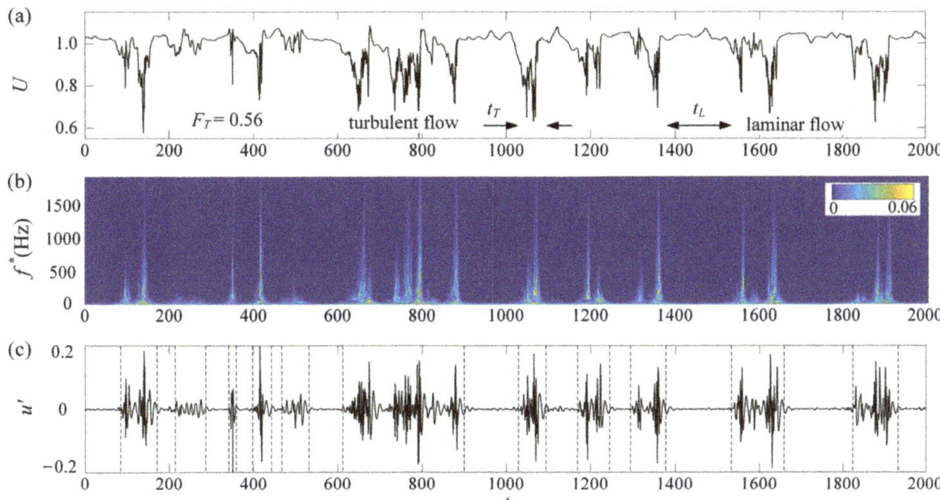

Figure 7. (**a**) The time series of streamwise velocity U measured at $(Re, x, y, z) = (935, 780, 0, 0)$ for Case_1 and (**b**) its wavelet power spectrum. (**c**) The high-frequency component, u', after high-pass filtering of the signal shown in (**a**). Localized turbulent patches are marked with shadowed areas in (**a**,**c**).

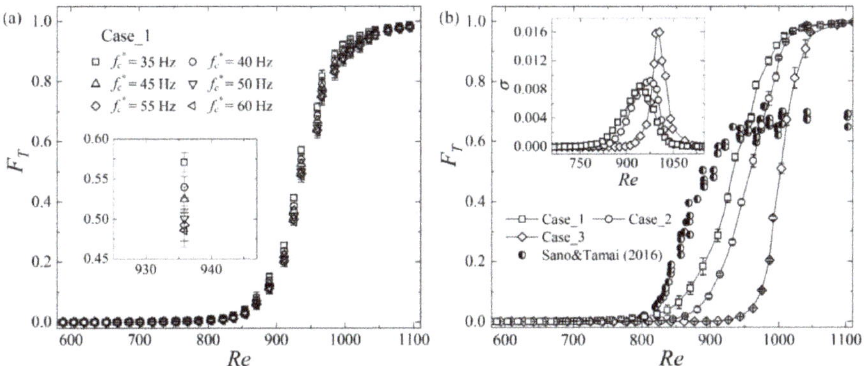

Figure 8. (**a**) F_T calculated with different cutoff frequencies, f_c^*, for Case_1, and (**b**) data calculated with $f_c^* = 45$ Hz for different entrance disturbances. Inset of (**b**): the growth steepness σ versus Re.

F_T shown in Figure 8 is computed from the midplane streamwise velocity signals sampled at six locations, i.e., $(x, z) = (700, -40)$, $(700, -20)$, $(700, 0)$, $(780, -40)$, $(780, -20)$, and $(780, 0)$. Each time series lasts 2000 s ($10^5 \sim 10^6$ time units at the transition stage), and the error bar represents the standard deviation. As $Re < 850$, the localized patches are far from each other, as shown in Figure 4, and F_T increases slowly with Re and is less than 0.1 for all three cases. When Re is larger than 1050, the localized turbulent structures almost occupy the whole flow field and are arranged nearly side by side, as shown by the case of $Re = 1155$ in Figure 4b, and hence F_T is close to 1, as shown in Figure 8. The growth steepness $\sigma = dF_T/dRe$ is calculated and is found to reach its maxima (as shown in the inset of Figure 8b) at $Re = 950, 975$, and 1005 for Case_1, Case_2, and Case_3, respectively, where F_T is around 0.6. It is interesting to note that the Reynolds numbers of the σ peaks are almost the same as those of the I_P and I_u peaks shown in Figure 5, confirming the intrinsic relation between the turbulence intensity and the growth steepness of the turbulence fraction.

According to Table 1, the beads' diameters are different for Case_1 and Case_2, representing different localized disturbance intensities, and the wire diameter of Case_3 is about one order larger than that of Case 1, denoting different entrance disturbance forms, i.e., the entrance disturbances of Case_3 are more uniform in the spanwise direction due to the vortex shedding of the thicker wire. As shown in Figure 8b, F_T data for different entrance disturbances vary in the same manner but do not collapse with each other as $850 < Re < 1050$, reflecting the sensitivity of transition to the external forcing, and the reason lies in several aspects. Firstly, F_T data collapse will occur when F_T is a single valued function of Re, e.g., at laminar state or the equilibrium state, which is found to be retrieved only as $Re > 924$ in long-term simulations [22]. In other words, when the upstream or initial disturbances are different, F_T may be different from case to case as $Re < 924$ even for simulations with the same computational configurations, e.g., domain size and mode numbers. Secondly, in reality, the lengths of experimental channels are finite, and at moderate Reynolds numbers, the turbulent structures may have no enough time to spread completely before leaving the outlet. Consequently, F_T will depend on the entrance disturbances. Thirdly, the effectiveness to trigger the transition are different for different types of perturbations. The turbulence fractions obtained based on flow visualization by Sano and Tamai [21] are shown in Figure 8b, as well, and are different from the present data: F_T does not increase with Re as $Re > 1000$ but maintain at about 0.7. In Sano and Tamai's experiments, turbulent flow was excited in a buffer box by a grid and injected from the inlet, and hence the entrance perturbations occupied the span of the channel and are different from the localized disturbances used in this paper. In addition, different approaches applied to identify the laminar–turbulent boundaries and different data (e.g., the two-dimensional images of flow visualization and the one-dimensional velocity series measured by HWA) may lead to different F_T values, as well.

3.5. Intermittent Structure Model

In order to understand the peaks and valleys of turbulence intensity and high-order moments during the transition, an intermittent structure model is constructed as follows. For convenience, the characteristic velocity is chosen as $1.5U_b^*$ instead of U_c^* in this subsection. The velocity during the turbulent period is decomposed into two parts: the turbulent mean velocity, U_T, representing the behavior of low-frequency and large-scale structures, and the turbulent perturbation velocity, u_T (relative to U_T), denoting the high-frequency and small-scale components. $U = U_T + u_T$, and it is assumed that u_T satisfies Gaussian distribution, i.e., the time averaged values $\langle u_T \rangle = 0$, $\langle u_T^3 \rangle = 0$, and $\langle u_T^4 \rangle = 3\langle u_T^2 \rangle^2$, but its temporal and spatial distribution is strongly asymmetric and aperiodic just like the measured velocity (gray curve) shown in Figure 9a. Assuming that U_T and $\langle u_T^2 \rangle$ are the same for all localized turbulent patches in a given case and F_T is known, it can be derived that the mean velocity $U_c = U_0 - F_T(U_0 - U_T)$ and the fluctuation velocity relative to U_c is as follows:

$$u = U - U_c = \begin{cases} F_T(U_0 - U_T), & \text{laminar periods,} \\ (U_0 - U_T)(F_T - 1) + u_T, & \text{turbulent periods.} \end{cases} \tag{1}$$

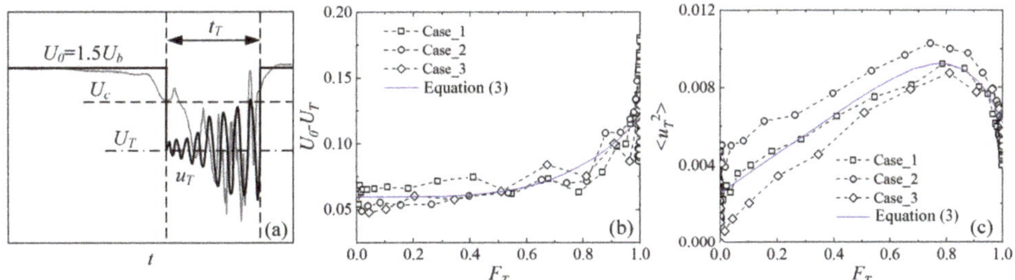

Figure 9. (**a**) The simplified velocity signal (thick solid line) of the intermittent structure model at midplane, and the time averaged (**b**) $U_0 - U_T$ and (**c**) $\langle u_T^2 \rangle$ sampled at the midplane during the turbulent periods. A measured midplane velocity signal is shown in (**a**) by the gray curve for a reference.

Consequently, the turbulence intensity and the high-order moments can be derived as follows:

$$
\begin{cases}
I_u = \dfrac{\sqrt{\langle u^2 \rangle}}{U_c} = \dfrac{\sqrt{F_T(1-F_T)(U_0-U_T)^2 + \langle u_T^2 \rangle F_T}}{U_0 - F_T(U_0 - U_T)} \\
\langle u^3 \rangle = 3F_T(U_0 - U_T)\langle u_T^2 \rangle(F_T - 1) - F_T(U_0 - U_T)^3\left(2F_T^2 - 3F_T + 1\right) \\
\langle u^4 \rangle = F_T(1 - F_T)\left(1 - 3F_T + 3F_T^2\right)(U_0 - U_T)^4 + 3F_T\langle u_T^2 \rangle^2 + 6F_T\langle u_T^2 \rangle(F_T - 1)^2(U_0 - U_T)^2
\end{cases}
\tag{2}
$$

U_T is estimated by the mean value of low-pass filtered midplane velocity during the turbulent periods at each *Re*, and the cutoff frequency, f_c, used for the filtering is the same as those used for calculating F_T. It is shown in Figure 9 that $U_0 - U_T$ increases with F_T, while the variance $\langle u_T^2 \rangle$ increases first then decreases with the growth of F_T, reflecting the fact that the localized turbulent structures are influenced to some degree by the entrance disturbances, F_T, and then *Re*. $U_0 - U_T$ and $\langle u_T^2 \rangle$ may be fitted as follows:

$$
U_0 - U_T = 0.06\left(1 + F_T^4\right), \quad \langle u_T^2 \rangle = 0.0026 + 0.01\left(F_T - 0.64F_T^7\right),
\tag{3}
$$

which are shown in Figure 9b,c as solid curves.

According to the previous studies [42], the characteristics of localized turbulent bands, e.g., the band's tilt angle, width, and convection velocity, do not change much during the transition. Similar properties are shown in Figure 4d, as well: The midplane velocity defects of localized turbulent structures are similar and not very sensitive to the Reynolds number, the entrance disturbances, and the turbulence fractions. Therefore, these localized turbulent structures may be simplified to a unified structure, whose statistical dimensionless properties are independent of time, F_T, and the initial or upstream disturbances. This unified structure is referred as turbulence unit hereafter. Consequently, $U_0 - U_T$ and u_T^2 are chosen for mature structures and are set as the values when F_T reaches 1, and then Equation (3) is simplified as follows:

$$
U_0 - U_T = 0.12, \quad \langle u_T^2 \rangle = 0.006.
\tag{4}
$$

For all three test cases, it is shown in Figure 10a–i by the solid lines that the main features of the second-, third-, and forth-order moments predicted by the model are consistent acceptably with the experimental results when the relations between F_T and *Re* shown in Figure 8b are applied. The variance of the midplane streamwise velocity $\langle u^2 \rangle$ is $F_T(1 - F_T)(U_0 - U_T)^2 + \langle u_T^2 \rangle F_T$, where the contribution of fluctuations (the second term) increases with F_T, while the first term increases first and then decreases with F_T due to the fact that the mean velocity, U_c, leaves U_0 for U_T, leading to a peak value of $\langle u^2 \rangle$. Consequently, there exist peak values of I_u and $\langle u^4 \rangle$ during the transition. Furthermore, when F_T is close to 1 and the flow field is nearly fully occupied by the localized turbulent structures, U_c is almost as low as U_T, and $\langle u^2 \rangle$ and $\langle u^3 \rangle$ are close to $\langle u_T^2 \rangle$ and $\langle u_T^3 \rangle$, respectively. Therefore, at the late transition stage, $\langle u^3 \rangle$ should be close to zero again, and then there must exist a minimum $\langle u^3 \rangle$ during the transition. Similarly, the asymptotic values for I_u and $\langle u^4 \rangle$ should be finite ($\sqrt{\langle u_T^2 \rangle}/U_T$ and $3\langle u_T^2 \rangle^2$ in the model), just as shown by the experimental data in Figure 10. The consistencies of the model curves with the experimental data indicate that, not only the turbulence fraction, but also the characteristics of localized structures is required in order to describe properly the statistical properties of transitional flows.

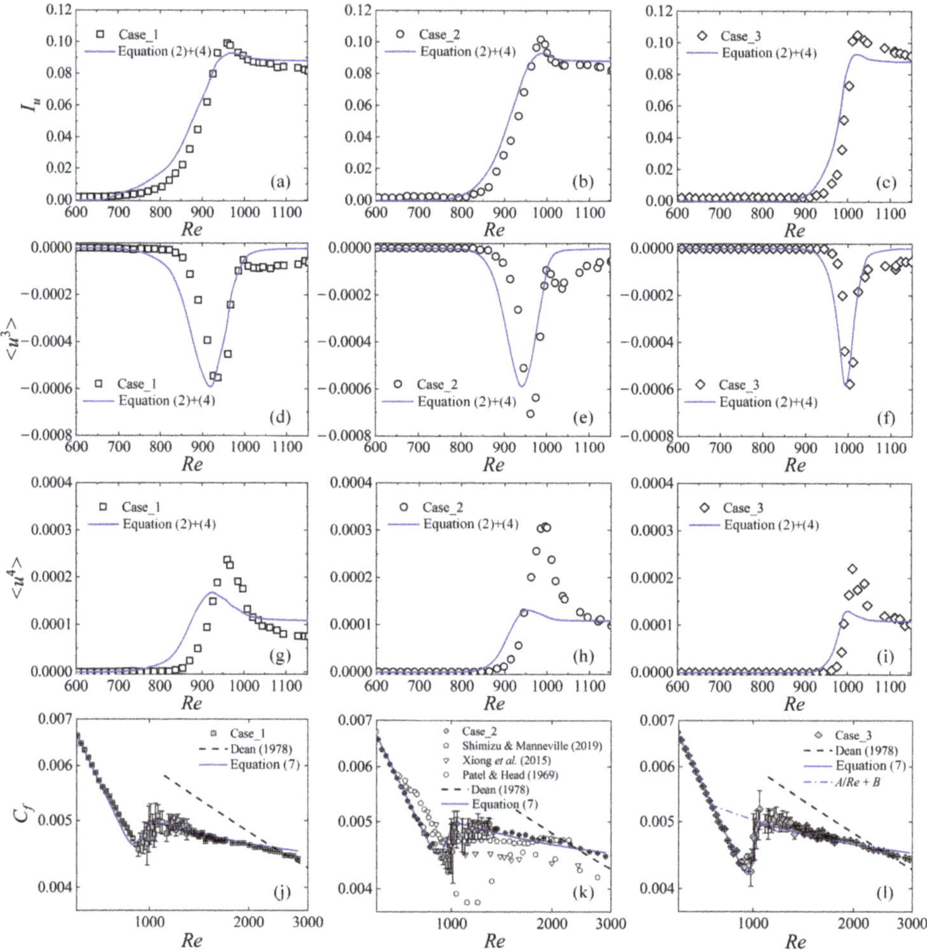

Figure 10. Turbulence intensity (**a–c**), the third (**d–f**) and the fourth (**g–i**) order moments of the midplane velocity, and the friction coefficient (**j–l**) for different disturbance cases. The symbols of different cases shown in (**a–i**) are experimental data measured at $(x, y, z) = (780, 0, 0)$, and C_f symbols shown in (**j–l**) are the same as those shown in Figure 3a. The solid curves are the results of the intermittent structure model.

Recently, it is found that, for a channel flow with constant pressure gradient, the kurtosis of the bulk velocity, which fluctuates during the transition and is represented by Re_b in the simulations [34], increases abruptly as the Reynolds number decreases to the threshold value. However, the kurtosis obtained in experiments is close to zero near the onset of turbulence, as shown in

Figure 6. This discrepancy may be explained to some degree with the present model. Considering that, in simulations, the velocities in the laminar periods are as clean as the present model and have no background random noise, an inevitable factor in experiments, then when F_T is close to 0, $\langle u^4 \rangle \sim F_T$ while $\langle u^2 \rangle^2 \sim F_T{}^2$ according to Equation (2), and hence the kurtosis will increase sharply.

Next, we use this model to study the dynamic property. Considering a turbulence unit with volume, V, mean velocity, $U_T(y)$, and mean pressure, P_T, the perturbation velocities are u_T, v_T, and w_T, and then the volume averaged friction coefficient is obtained from the mean x-momentum equation:

$$C_{fT} = -\frac{2}{V} \int \frac{\partial P_T}{\partial x} \mathrm{d}V = -\frac{2}{ReV} \int \frac{d^2 U_T}{dy^2} \mathrm{d}V + \frac{2}{V} \int \left[\frac{\partial \langle u_T^2 \rangle}{\partial x} + \frac{\partial \langle u_T w_T \rangle}{\partial z} \right] \mathrm{d}V. \tag{5}$$

Note that $\int_{-1}^{1} \frac{\partial \langle u_T v_T \rangle}{\partial y} dy = 0$. Since the velocity fluctuations are strongly asymmetric and there is nearly a velocity discontinuity at the later edge of time series (upstream edge) of the structure and the present model (Figure 9a), the Reynolds stresses, e.g., $\langle u_T^2 \rangle$, are different at the upstream and the downstream edges of the turbulence unit. In fact, the Reynolds stresses of a localized turbulent band are aperiodic in both the streamwise and the spanwise directions, as shown by the disturbance velocity structures in Figure 2b of Reference [23], due to its oblique manner. Since the transition occurs at relatively high Reynolds numbers and the properties of turbulence unit are assumed to be weak functions of Re, $-\frac{2}{ReV} \int \frac{d^2 U_T}{dy^2} \mathrm{d}V$ may be expanded with $1/Re$ as $\frac{4}{Re} - \frac{2}{Re}\left(A_0 + A_1 \frac{1}{Re} + A_2 \frac{1}{Re^2} + \ldots \right)$, where $\frac{4}{Re}$ corresponds to the laminar state, and the constants A_i represent the contribution of mean flow modification. Similarly, the Reynolds stress term (the second term on the right hand side of Equation (5)) is expanded as $B_0 + B_1 \frac{1}{Re} + B_2 \frac{1}{Re^2} + \ldots$, where the constants B_i reflect the aperiodicity of the Reynolds stress. Consequently, Equation (5) can be expressed as follows:

$$C_{fT} = B_0 + \frac{1}{Re}(4 - 2A_0 + B_1) + \frac{1}{Re^2}(B_2 - 2A_1) + \ldots = B + \frac{A}{Re} + O\left(\frac{1}{Re^2}\right), \tag{6}$$

where A and B are constants for the turbulence unit. For a transitional flow with a turbulence fraction, F_T, the total friction coefficient can be obtained as follows, after ignoring the higher orders terms in Equation (6):

$$C_f = (1 - F_T)\frac{4}{Re} + C_{fT}F_T = \left(1 - F_T + \frac{A}{4}F_T\right)\frac{4}{Re} + F_T B. \tag{7}$$

It is shown in Figure 10j–l and that Equation (7) describes well the variations of C_f data for different entrance disturbance cases when the measured relation between F_T and Re are applied. A and B are determined by fitting the data between Re = 1300 and 2000 as 0.78 and 0.00426, respectively.

At the initial and middle stages of transition, C_f may have different variation scenarios. If the external disturbances are not effective to trigger the turbulent patches and the transition starts at high Reynolds numbers, $\left(1 - F_T + \frac{A}{4}F_T\right)\frac{4}{Re}$ may become smaller than $F_T B$ after a short Re range, and then there will be a stage where C_f increases with F_T and Re, as shown in Figure 10. Note that $A < 4$ and $\left(1 - F_T + \frac{A}{4}F_T\right)\frac{4}{Re}$ decreases with the increase of F_T and Re. Consequently, there will be a maximum of C_f during the transition as illustrated by the present data shown in Figure 10l and the data of Patel and Head [6] shown in Figure 10k. If the transition begins at low Reynolds numbers, the variation of $\left(1 - F_T + \frac{A}{4}F_T\right)\frac{4}{Re}$ may be comparable with that of $F_T B$. Depending on the variation feature of F_T, the stage of C_f growth may be short or even disappear, and a C_f plateau may appear, where C_f remains nearly constant in a finite range of Re. The C_f plateaus were observed in the previous numerical simulations [22,24,34] and are shown in Figure 10k for references. According to Equation (7), provided that the decrease of $\left(1 - F_T + \frac{A}{4}F_T\right)\frac{4}{Re}$ is balanced by the rise of $F_T B$, C_f will keep constant, though this constant value may be different for different entrance or initial disturbances, domain sizes, and computational periods. At the late stage of transition, F_T tends to 1, and C_f is close to $A/Re + B$ according to Equation (7) and then decreases with Re. The dashed lines in Figure 10j–l,

$Re = \sqrt{\frac{2}{C_f}} \exp\left[0.41\left(\sqrt{\frac{8}{9C_f}} - 2.4\right)\right]$, represent the fully developed turbulence [22,37], where the Reynolds stresses are assumed to be uniform in the streamwise direction. According to the experiments, F_T is close to 1 as $Re > 1100$, but C_f still deviates from the dashed line as $Re < 1750$, indicating a moderately developed turbulent state. By extrapolating $A/Re + B$ to the laminar value $4/Re$, as shown by the dot-dash line in Figure 10l, we get $Re = 756$, corresponding to an asymptotic threshold for the moderately developed turbulence.

4. Conclusions

In this paper, the subcritical transition of channel flow is studied experimentally and theoretically. A pressure turbulence intensity is defined to describe the pressure fluctuations, and it is found that both the pressure and the velocity turbulence intensities reach maxima at the same Reynolds number during the transition, where the turbulence fraction is about 0.6 and both the friction coefficient and the turbulence fraction increase abruptly with Re. The velocity defect of localized turbulent structure leads to a negative skewness, and for all tested cases, the smallest Re where the skewness of the midplane velocity starts to be negative is about 660. Since the onset of turbulence depends on not only the intensities but also the forms of initial or upstream disturbances, the high-order moments of fluctuations are better markers for the start of transition than the turbulence intensity or fluctuation kinetic energy, and hence should be considered in the future transition control strategies.

According to the experimental data, there exist maxima of the turbulence intensity and the forth-order moment of the midplane streamwise velocity and a negative minimum for the third-order moment. At the late stage of transition, the third-order moment decreases to a low level, and the turbulence intensity and the forth-order moment remain finite values. These phenomena are explained with an intermittent structure model, where the robust localized turbulent structure is simplified as a turbulence unit. In addition, different variation behaviors of the friction coefficient are explained by this model, as well, mainly in terms of the turbulence fraction and the aperiodic distribution of Reynolds stress in the localized turbulent structures, and the latter factor should be considered in the future transition modelling.

Author Contributions: Conceptualization, J.T. and S.X.; methodology, S.X. and J.T.; validation, J.L., Y.X., and M.L.; formal analysis, J.L., J.T., and Y.X.; data curation, J.L. and Y.X.; original draft preparation, J.T. and J.L.; supervision, J.T. and S.X. All authors have read and agreed to the published version of the manuscript.

Funding: This research was funded by the National Natural Science Foundation of China (Grants Nos. 91752203, 11772173, and 11490553).

Acknowledgments: The authors would like to thank many cited authors for insightful discussions.

Conflicts of Interest: The authors declare no conflict of interest.

References

1. Orszag, S.A. Accurate solution of the Orr-Sommerfeld stability equation. *J. Fluid Mech.* **1971**, *50*, 689–703. [CrossRef]
2. Manneville, P. Understanding the sub-critical transition to turbulence in wall flows. *Pramana J. Phys.* **2008**, *70*, 1009–1021. [CrossRef]
3. Eckhardt, B. A critical point for turbulence. *Science* **2011**, *333*, 165–166. [CrossRef]
4. Tuckerman, L.S.; Kreilos, T.; Schrobsdorff, H.; Schneider, T.M.; Gibson, J.F. Turbulent-laminar patterns in plane Poiseuille flow. *Phys. Fluids* **2014**, *26*, 114103. [CrossRef]
5. Davies, S.; White, C. An experimental study of the flow of water in pipes of rectangular section. *Proc. R. Soc. Lond. Ser. A* **1928**, *119*, 92–107.
6. Patel, V.; Head, M. Some observations on skin friction and velocity profiles in fully developed pipe and channel flows. *J. Fluid Mech.* **1969**, *38*, 181–201. [CrossRef]
7. Nishioka, M.; Asai, M. Some observations of the subcritical transition in plane Poiseuille flow. *J. Fluid Mech.* **1985**, *150*, 441–450. [CrossRef]

8. Carlson, D.R.; Widnall, S.E.; Peeters, M.F. A flow-visualization study of transition in plane Poiseuille flow. *J. Fluid Mech.* **1982**, *121*, 487–505. [CrossRef]

9. Henningson, D.S.; Alfredsson, P.H. The wave structure of turbulent spots in plane Poiseuille flow. *J. Fluid Mech.* **1987**, *178*, 405–421. [CrossRef]

10. Henningson, D.S.; Spalart, P.; Kim, J. Numerical simulations of turbulent spots in plane Poiseuille and boundary-layer flow. *Phys. Fluids* **1987**, *30*, 2914–2917. [CrossRef]

11. Li, F.; Widnall, S.E. Wave patterns in plane Poiseuille flow created by concentrated disturbances. *J. Fluid Mech.* **1989**, *208*, 639–656. [CrossRef]

12. Henningson, D.S.; Kim, J. On turbulent spots in plane Poiseuille flow. *J. Fluid Mech.* **1991**, *228*, 183–205. [CrossRef]

13. Henningson, D.S.; Johansson, A.V.; Alfredsson, P.H. Turbulent spots in channel flows. *J Eng. Math.* **1994**, *28*, 21–42. [CrossRef]

14. Lemoult, G.; Gumowski, K.; Aider, J.-L.; Wesfreid, J.E. Turbulent spots in channel flow: An experimental study. *Eur. Phys. J.* **2014**, *37*, 25. [CrossRef]

15. Alavyoon, F.; Henningson, D.S.; Alfredsson, P.H. Turbulent spots in plane Poiseuille flow-flow visualization. *Phys. Fluids* **1986**, *29*, 1328–1331. [CrossRef]

16. Tsukahara, T.; Seki, Y.; Kawamura, H.; Tochio, D. DNS of Turbulent Channel Flow at Very Low Reynolds Numbers. In Proceedings of the Fourth International Symposium on Turbulence and Shear Flow Phenomena, Williamsburg, VA, USA, 27–29 June 2005; Begel House Inc.: Williamsburg, VA, USA, 2005; pp. 935–940.

17. Tsukahara, T.; Iwamoto, K.; Kawamura, H.; Takeda, T. DNS of Heat Transfer in a Transitional Channel Flow Accompanied by a Turbulent Puff-Like Structure. In Proceedings of the Fifth International Symposium on Turbulence, Heat and Mass Transfer, Dubrovnik, Croatia, 25–29 September 2006; Begel House Inc.: Dubrovnik, Croatia, 2006; pp. 193–206.

18. Tsukahara, T.; Kawaguchi, Y.; Kawamura, H.; Tillmark, N.; Alfredsson, P.H. Turbulence stripe in transitional channel flow with/without system rotation. In *Seventh IUTAM Symposium on Laminar-Turbulent Transition*; Springer: Stockholm, Sweden, 2009; pp. 421–426.

19. Hashimoto, S.; Hasobe, A.; Tsukahara, T.; Kawaguchi, Y.; Kawamura, H. An experimental study on turbulent-stripe structure in transitional channel flow. In Proceedings of the Sixth International Symposium on Turbulence, Heat and Mass Transfer, Rome, Italy, 14–18 September 2009; Begel House Inc.: New York, NY, USA, 2009; pp. 193–196.

20. Tuckerman, L.S. Turbulent-laminar banded patterns in plane Poiseuille flow. In Proceedings of the 23rd International Congress of Theoretical and Applied Mechanics, Beijing, China, 19–24 August 2012; pp. 19–20.

21. Sano, M.; Tamai, K. A universal transition to turbulence in channel flow. *Nat. Phys.* **2016**, *12*, 249–253. [CrossRef]

22. Shimizu, M.; Manneville, P. Bifurcations to turbulence in transitional channel flow. *Phys. Rev. Fluids* **2019**, *4*, 113903. [CrossRef]

23. Tao, J.; Xiong, X. The unified transition stages in linearly stable shear flows. In Proceedings of the 14th Asian Congress of Fluid Mechanics, Hanoi and Halong, Vietnam, 15–19 October 2013; pp. 55–62.

24. Xiong, X.; Tao, J.; Chen, S.; Brandt, L. Turbulent bands in plane-Poiseuille flow at moderate Reynolds numbers. *Phys. Fluids* **2015**, *27*, 041702. [CrossRef]

25. Tao, J.; Eckhardt, B.; Xiong, X. Extended localized structures and the onset of turbulence in channel flow. *Phys. Rev. Fluids* **2018**, *3*, 011902(R). [CrossRef]

26. Paranjape, C.S.; Duguet, Y.; Hof, B. Bifurcation scenario for turbulent stripes in plane Poiseuille flow. In Proceedings of the Sixteenth European Turbulence Conference, Stockholm, Sweden, 21–24 August 2017.

27. Paranjape, C. Onset of Turbulence in Plane Poiseuille Flow. Ph.D. Thesis, Institute of Science and Technology Austria, Klosterneuburg, Austria, 2019.

28. Paranjape, C.S.; Duguet, Y.; Hof, B. Oblique stripe solutions of channel flow. *J. Fluid Mech.* **2020**, *897*, A7. [CrossRef]

29. Seki, D.; Matsubara, M. Experimental investigation of relaminarizing and transitional channel flows. *Phys. Fluids* **2012**, *24*, 124102. [CrossRef]

30. Kushwaha, A.; Park, J.S.; Graham, M.D. Temporal and spatial intermittencies within channel flow turbulence near transition. *Phys. Rev. Fluids* **2017**, *2*, 024603. [CrossRef]

31. Whalley, R.D.; Park, J.S.; Kushwaha, A.; Dennis, D.J.; Graham, M.D.; Poole, R.J. Low-drag events in transitional wall-bounded turbulence. *Phys. Rev. Fluids* **2017**, *2*, 034602. [CrossRef]

32. Whalley, R.; Dennis, D.; Graham, M.; Poole, R. An experimental investigation into spatiotemporal intermittencies in turbulent channel flow close to transition. *Exp. Fluids* **2019**, *60*, 102. [CrossRef]

33. Agrawal, R.; Ng, H.C.-H.; Davis, E.A.; Park, J.S.; Graham, M.D.; Dennis, D.J.; Poole, R.J. Low-and high-drag intermittencies in turbulent channel flows. *Entropy* **2020**, *22*, 1126. [CrossRef]

34. Kashyap, P.V.; Duguet, Y.; Dauchot, O. Flow statistics in the transitional regime of plane channel flow. *Entropy* **2020**, *22*, 1001. [CrossRef]

35. Meneveau, C. Lagrangian dynamics and models of the velocity gradient tensor in turbulent flows. *Annu. Rev. Fluid Mech.* **2011**, *43*, 219–245. [CrossRef]

36. Wang, L.; Fu, S. Development of an intermittency equation for the modeling of the supersonic/hypersonic boundary layer flow transition. *Flow Turbul. Combust.* **2011**, *87*, 165–187. [CrossRef]

37. Dean, R.B. Reynolds number dependence of skin friction and other bulk flow variables in two-dimensional rectangular duct flow. *J. Fluids Eng.* **1978**, *100*, 215–223. [CrossRef]

38. Gomé, S.; Tuckerman, L.S.; Barkley, D. Statistical transition to turbulence in plane channel flow. *Phys. Rev. Fluids* **2020**, *5*, 083905. [CrossRef]

39. Liu, J.; Xiao, Y.; Zhang, L.; Li, M.; Tao, J.J.; Xu, S. Extension at the downstream end of turbulent band in channel flow. *Phys. Fluids* **2020**, *32*, 121703. [CrossRef]

40. Rodriguez, I.; Borell, R.; Lehmkuhl, O.; Segarra, C.D.P.; Oliva, A. Direct numerical simulation of the flow over a sphere at *Re* = 3700. *J. Fluid Mech.* **2011**, *679*, 263–287. [CrossRef]

41. Johansson, A.; Alfredsson, H. On the structure of turbulent channle flow. *J. Fluid Mech.* **1982**, *122*, 295–314. [CrossRef]

42. Tuckerman, L.S.; Chantry, M.; Barkley, D. Patterns in Wall-Bounded Shear Flows. *Annu. Rev. Fluid Mech.* **2020**, *52*, 343–367. [CrossRef]

Publisher's Note: MDPI stays neutral with regard to jurisdictional claims in published maps and institutional affiliations.

Article

Kinematics and Dynamics of Turbulent Bands at Low Reynolds Numbers in Channel Flow

Xiangkai Xiao and Baofang Song *

Center for Applied Mathematics, Tianjin University, Tianjin 300072, China; xiaoxk2018@tju.edu.cn
* Correspondence: baofang_song@tju.edu.cn

Received: 19 September 2020; Accepted: 13 October 2020; Published: 16 October 2020

Abstract: Channel flow turbulence exhibits interesting spatiotemporal complexities at transitional Reynolds numbers. In this paper, we investigated some aspects of the kinematics and dynamics of fully localized turbulent bands in large flow domains. We discussed the recent advancement in the understanding of the wave-generation at the downstream end of fully localized bands. Based on the discussion, we proposed a possible mechanism for the tilt direction selection. We measured the propagation speed of the downstream end and the advection speed of the low-speed streaks in the bulk of turbulent bands at various Reynolds numbers. Instead of measuring the tilt angle by treating an entire band as a tilted object as in prior studies, we proposed that, from the point of view of the formation and growth of turbulent bands, the tilt angle should be determined by the relative speed between the downstream end and the streaks in the bulk. We obtained a good agreement between our calculation of the tilt angle and the reported results in the literature at relatively low Reynolds numbers.

Keywords: turbulent bands; obliqueness; advection speed; wave generation; inflectional instability

1. Introduction

Much below the linear critical Reynolds number of the parabolic channel flow, transition to turbulence can occur under finite-amplitude perturbations, i.e., via a subcritical transition. Numerous studies have established that turbulence takes the form of discrete turbulent bands that are oblique to the streamwise direction, interspersed with laminar flow, at transitional Reynolds numbers [1–10]. Similar banded turbulent structures have also been observed in other quasi-two-dimensional flows, i.e., systems with one confined dimension and two extended dimensions, such as plane Couette [11–13], Taylor Couette [14,15], annular pipe [16] and Wallefe flows [17]. Therefore, the coexistence of laminar and turbulent states in the form of banded turbulent structures is a common feature of turbulence at transitional Reynolds numbers of a broad variety of shear flows. Recent investigations into these structures have greatly advanced the understanding of the subcritical transition in these flows [10,18]. In the following discussion, for channel flow, the streamwise, wall-normal and spanwise directions are denoted as x, y and z, respectively, time is denoted as t and the half-channel-height as h. The flow is assumed driven by a constant volume flux and the Reynolds number is defined as $Re = \frac{U_c h}{\nu}$, where U_c is the centerline velocity of the unperturbed parabolic flow and ν the kinematic viscosity of the fluid.

The first observation and many numerical studies of turbulent bands in channel flow were performed by numerical simulations in relatively small computational domains, either normal or tilted, in which the structure, kinematics and dynamics of turbulent bands are rather restrained [1,2,4,19,20]. Particularly, narrow tilted domains force turbulent bands to be parallel to the narrow edge, which practically assumes infinitely long bands in combination with periodic boundary conditions. Nevertheless, this greatly reduces the computational cost and allows studying the kinematics and dynamics of bands over large time scales [9,20] and offers conveniences for studying the mean flow and wavelength of the band pattern [4,10]. In a domain tilted by 24°, Tuckerman et al. [4] reported that turbulent bands propagate approximately at the bulk speed of the flow, with a slight decreasing trend with the Reynolds number (the speed crosses the bulk speed at $Re \simeq 1100$). In a similar approach as Avila et al. [21] for pipe flow and Shi et al. [22] for plane Couette flow, Gomé et al. [20] also showed finite lifetime and splitting nature of bands and determined the on-set of sustained turbulence in channel flow to be at $Re \approx 950$ by balancing the super-exponential decay and splitting processes, in a domain also tilted by 24°. The subcritical transition to turbulence in plane Couette flow in tilted domains has been concluded to fall in the universality class of directed percolation [23] and the work of Gomé et al. [20] seems to suggest the same transition scenario in channel flow. However, the imposed tilt angle of the domain seems to affect the statistical results. For example, the simulations in a domain tilted by 45° [9] showed very different lifetimes of bands from the results of Gomé et al. [20]. Specifically, the former reported that turbulent bands are sustained at $Re > 620$, whereas the latter suggested that in fact the lifetime stays finite and is below 200 time units at $Re < 700$. The effect of the imposed tilt angle has not been thoroughly investigated. Besides, the usual narrow tilted domain only allows multiple bands to form parallel band pattern, i.e., bands are forced to take the same orientation.

Large domains pose a lesser restriction on turbulent bands. In recent years, a few studies have been dedicated to turbulent bands in large normal domains in experiments [3,9] and simulations [5–8,24,25]. If the domain is large enough, given a proper localized perturbation, turbulence elongates obliquely with respect to the streamwise direction and forms a fully localized band (localized both in its length direction and in its width direction). The existence of the two ends of the band adds further complexity to the flow. Paranjape [9] reported in experiment that at $Re < 660$, a turbulent band shrinks and will decay so that the flow will relaminarize in the end, because the growth at the downstream end (referred to as the head hereafter) is slower than the decay at the upstream end (referred to as the tail hereafter). At higher Reynolds numbers, a turbulent band becomes sustained because the growth at the head outperforms the decay at the tail and will grow in length. Numerical studies [6,7,24] agree with the experiments. Therefore, it has been confirmed that the growth of a band is unidirectional, driven by the head [7–9,24]. Because streaks decay at the tail and are generated at the head, an individual band undergoes a spanwise shift as a whole, aside from being advected in the streamwise direction. Shimizu and Manneville [8] mentioned that the spanwise drift speed is 0.1 and Xiao and Song [24] reported a close value of 0.08. Noticing the periodic streak generation at the head, Kanazawa [7] and Xiao and Song [24], respectively, proposed mechanisms behind the wave generation at the head, which are discussed in more detail below.

In fact, it was found that the length of a band does not grow infinitely. The length 'at equilibrium' of a band at $Re = 660$ was shown to be about 300 h and the length seems to increase with Re [7]. As the length is sufficiently large, the fast decay of the tail limits the growth, and splitting may occur with a daughter band nucleated. At relatively low Reynolds numbers, the splitting is longitudinal, i.e., the daughter band is parallel to the mother band. As Reynolds number increases ($Re \gtrsim 800$), transverse splitting (or branching) can also occur, nucleating daughter bands with the opposite tilt direction such that the flow pattern becomes two-sided (the criss-cross pattern) [8,9]. However, the study of the splitting of bands and the underlying mechanism is still rare.

In the presence of multiple bands, given that bands have a spanwise shift speed as a whole and can grow in length, close bands with opposite orientations may collide. Even parallel bands, when located sufficiently close to each other, were shown to interact also [6,8]. The dynamics of individual bands and the interaction between bands determine the pattern that bands can form and therefore, determine the statistical aspect of the transition to turbulence [8]. Using unprecedented large domain and simulating up to very large times (up to $\mathcal{O}(10^5)$ time units), Shimizu and Manneville [8] showed that turbulent bands can only form one-sided (parallel) pattern at low Reynolds numbers ($Re \lesssim 924$), breaking the spanwise symmetry, which is restored only at higher Reynolds numbers. Directed percolation was found to reasonably well describe the transition process toward featureless turbulence at further higher Reynolds numbers, as also proposed by Sano and Tamai [26] in experiments. Interestingly, the one-sided pattern of turbulent bands at lowest Reynolds numbers seems to justify the use of tilted domain in which bands are forced to be parallel, although the tilt angle was shown about 40°–45° below $Re \simeq 900$ [6–9] rather than 24° as used in some studies [4,20].

Although a great advancement in the understanding of turbulent bands has been made in recent studies, many problems even for individual turbulent bands have not been well understood, for example the mechanisms underlying the growth of bands at the head and the decay at the tail, the tilt angle selection and the self-sustaining mechanism of the bulk of turbulent bands. We discuss some of these problems in this paper.

2. The Head

2.1. Propagation Speed of the Head

Firstly, we investigated the advection speed of the head. It has been reported that the head of a turbulent band, which is always located at the downstream end, propagates in both streamwise and spanwise directions [6,8,24]. The spanwise motion can be in either positive or negative spanwise direction and the specific direction is correlated with the orientation of the band (see Figure 1). The head of the upper band moves downward (in negative spanwise direction) while that of the lower band moves upward, given their opposite orientations. Bands with similar orientation as the upper one are referred to as right-going bands, and those with the opposite orientation are referred to as left-going bands. This correlation can be intuitively understood because the head continually generates turbulence by invading laminar flow region on one side. We revisit this point in Section 2.2. Xiao and Song [24] measured the speeds at $Re = 750$ by tracking the head and reported a streamwise speed of $c_x = 0.85$ and a spanwise speed of $c_z = 0.1$ (absolute value).

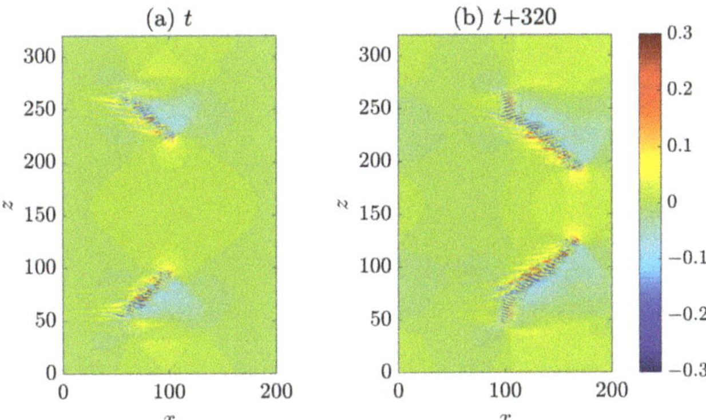

Figure 1. Turbulent bands with different orientations at $Re = 750$. (**a,b**) The streamwise direction is in the positive x direction and z denotes the spanwise direction. Streamwise velocity fluctuations in the x-z cut plane at $y = -0.5$ are plotted as the colormap with blue representing low speeds and red representing high speeds compared to the basic flow. The two panels are separated by 320 time units.

To investigate the *Re*-dependence of the speeds and also for calculating the tilt angle of turbulent bands in Section 4, we measured the speeds in the low Reynolds number regime ranging from $Re = 670$, which is nearly the lowest Reynolds number for sustained bands, to $Re = 1050$ at which frequent splitting and branching of bands were reported to occur [8,9]. For this study, the Reynolds numbers, domain sizes and resolutions are listed in Table 1. It has been shown that, at $Re = 660$, a band can continuously grow up to the length of approximately 300 h [7]. The length can be much larger at higher Reynolds numbers [7,8]. The domain sizes used in our study are not large enough for the band to reach the length 'at equilibrium', rather we only require the domain size to offer sufficiently long time for the head to reach its characteristic propagation speed. The simulation was stopped when the head and tail were too close to each other and started to interact due to the periodic boundary conditions. Xiao and Song [24] already showed that the speed of the head of turbulent bands at $Re = 750$ is not affected by the domain size by comparing the speeds measured in domains with $L_x = L_z = 120\ h$ and $L_x = L_z = 320\ h$.

Table 1. The Reynolds number Re, domain size L_x and L_z, number of wall-normal grid point N and the ratio between h and the grid spacing in x and z directions, Δx and Δz, respectively.

Re	$L_x \times L_z$	N	$h/\Delta x$	$h/\Delta z$
670	$120\ h \times 120\ h$	72	4.3	6.4
750	$120\ h \times 120\ h$	72	4.3	6.4
850	$160\ h \times 160\ h$	72	4.8	6.4
950	$160\ h \times 160\ h$	72	4.8	6.4
1050	$240\ h \times 240\ h$	72	4.8	6.4

At each Reynolds number, we generated a fully localized turbulent band directly at low Reynolds numbers using the method proposed by Song and Xiao [25]. After the band has sufficiently developed, the head was tracked over a time window of $\mathcal{O}(500)$ time units and the average speed was calculated based on the position and time separation. The results in Figure 2 show that both the streamwise and spanwise speeds stay nearly constant for all Reynolds numbers investigated, at 0.85 and 0.1, respectively. Besides, the speeds were shown to be rather stable, i.e., only fluctuate slightly in time around the respective averaged values for $Re = 750$ [24], which is also the case for other Reynolds numbers in this study.

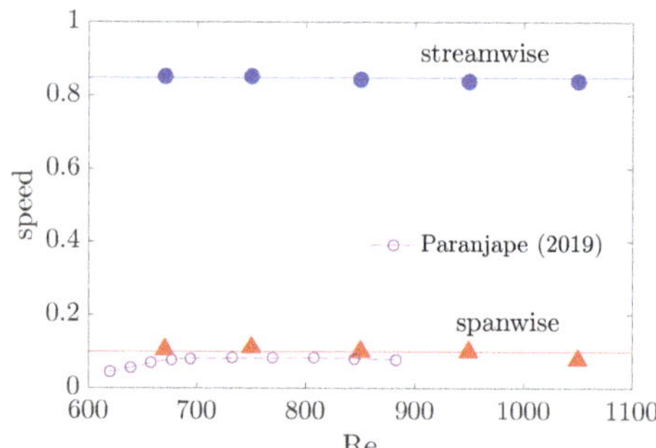

Figure 2. The streamwise (circles) and spanwise (triangles) speed of the head of turbulent bands at various Reynolds numbers. Note that it is the absolute value of the spanwise speed plotted given that the speed can take either positive or negative values. The two solid lines at 0.85 and 0.1 are plotted to guide the eyes. The experimental measurement of the spanwise speed [9] is plotted as the dashed-circle line for comparison.

In experiments, Paranjape [9] showed that the spanwise speed of the head slowly decreases from 0.085 at $Re \simeq 700$ to 0.08 when the Reynolds number is increased to $Re \simeq 850$ (see the dashed-circle line in Figure 2). Besides, Paranjape [9] reported a streamwise speed of the entire band of about 0.75 between $Re = 670$ and 900, but did not report the streamwise speed of the head. They also reported the speeds between $Re = 600$ and 670, in which regime we could not obtain a sustained turbulent band in our DNS. It can be seen that our spanwise speed is systematically larger than the experimental measurement [9] (see Figure 2). The difference could possibly be attributed to the periodic boundary condition used in our numerical simulations, although Xiao and Song [24] mentioned that the $L_x = L_z = 120\,h$ box gives the same speed as that given by the $L_x = L_z = 320\,h$ box at $Re = 750$. It may equally be attributed to the side-wall effect in experiments. Simulations in much larger periodic boxes or in a channel with side walls are needed to confirm about this point. Nevertheless, the two sets of speeds are close to each other.

2.2. Wave Generation at the Head and the Tilt Direction of the Band

In this section, firstly we discuss about some recent studies on the dynamics of the head. Therefore, a part of the results shown below is not original. It has been noticed that the head drives the growth of turbulent bands by continually generating waves, in the form of alternating high- and low-speed streaks and arrays of vortices, while moving into the adjacent laminar region [7–9,24]. Figure 3 shows the wave-like structure of the head. Contours of streamwise velocity fluctuation are plotted in the x-z plane at $y = -0.8$ (close to the wall, see Figure 3a), at $y = -0.5$ (Figure 3b) and in the mid-plane $y = 0$ (see Figure 3c). It can be seen that the flow is characterized by high-speed streaks close to the wall. In the mid-plane, the flow is characterized by low-speed streaks in the bulk, which almost merge and form a connected low-speed region, and is characterized by a high speed region at the head (see the yellow spot in Figure 3c). At $y = -0.5$, the flow exhibits wave-like alternating low and high-speed streaks. The large-scale (compared with the wave-like streaky structures) flow in the neighborhood of the head manifests a circulation (see [6,24]), which is counter-clockwise for a right-going band as shown. Duguet and Schlatter [27] proposed a mechanism for the formation of large-scale flow around turbulent bands in plane-shear flows. Their theory applies to the large-scale flow associated with the bulk region of the band and describes the band as the advection of small-scale structures (streaks) by the large-scale flow. However, they did not explicitly study the large-scale flow at the head.

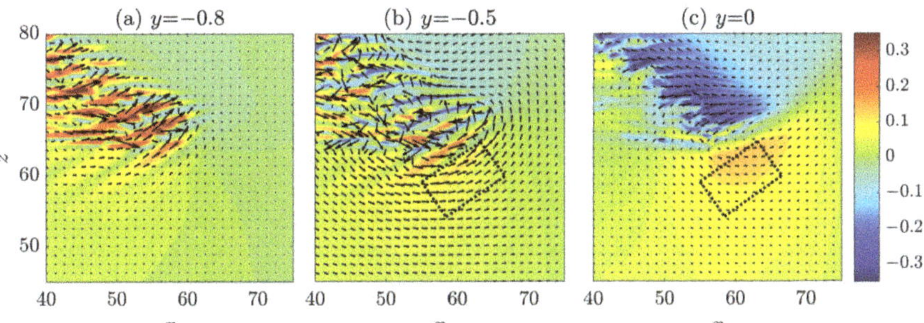

Figure 3. Large-scale circulation flow in the neighborhood of the head. Streamwise velocity is plotted as the colormap in the x-z cut plane at: $y = -0.8$ (**a**); $y = -0.5$ (**b**); and $y = 0$ (**c**). In each panel, the in-plane velocities are plotted as vectors. The dotted rectangle (size $6\,h \times 10\,h$) marks approximately the area where the first visible wave that is continually generated at the head in the frame of reference co-moving with the head.

The dotted rectangle in Figure 3 marks the approximate region in which the first visible high speed streak is periodically generated. The vector plot of the in-plane velocities shows that, at $y = -0.5$ (see Figure 3b), the vectors in the rectangle overall point to the positive z direction, and, at $y = 0$ (see Figure 3c), the vectors overall point to the negative z direction. This hints that there should be an inflection in the spanwise velocity profile in this region, which may be inflectionally unstable. Based on this observation, Xiao and Song [24] investigated the local mean flow at the head and attributed the wave generation at the head to an inflectional instability associated with the modified local mean flow. For the ease of discussion, we measured the averaged velocity profiles at the head again in a different region and for a different turbulent band compared to those reported in [24] (see Figure 4). Both streamwise and spanwise velocity profile (the parabolic base flow is not included) show inflection. These profiles are measured at a right-going band similar to the upper one in Figure 1 and the one shown in Figure 3. Figure 4b shows the unstable region in the wavenumer plane (the region enclosed by the bold line) and Figure 4c shows the streaky flow pattern of the most unstable disturbance (see also [24,25]). It can be seen that these streaks are tilted about the streamwise direction and the tilt direction is the same as the waves that can be seen at the head of right-going bands in Figures 1 and 3. Besides, the most unstable wave move downward, i.e., in the negative spanwise direction (see the arrow), just as the head of the right-going band. By the symmetry of channel flow about the x-y plane, it can be inferred that the velocity profiles at the head of a left-going band will be similar to those shown in Figure 4d, with the sign of the spanwise velocity changed. We performed a similar linear analysis here and show the unstable region in the wavenumber plane in Figure 4e and the most unstable disturbance in Figure 4f. Clearly, we can see a spanwise symmetry in the distribution of eigenvalues and in the flow pattern by comparing to Figure 4b,c. The waves shown in Figure 4f are tilted in the opposite direction compared with the waves in Figure 4c and move in the positive spanwise direction, which is consistent with the structure and kinematics of the head of a left-going band. In a word, linear stability analysis gives qualitatively similar flow structures and kinematics as that of the head. The nonlinear development of disturbances was shown to give similar flow structures as those at the head [24]. Therefore, Xiao and Song [24] proposed that the growth of turbulent bands is driven by the inflectional instability locally at the head. Further, Song and Xiao [25] performed a non-modal analysis of the inflectional velocity profiles and showed an Orr-mechanism via which disturbances can achieve a fast growth in energy at the early stage (by a factor of 100 within about 15 time units for $Re = 750$). Subsequently, the modal instability takes part and starts to dominate the growth at later points of time. The linear instability together with the fast non-normal growth at the early stage are able to result in a fast growth of the unstable waves at the head. Reaching a certain amplitude, the waves become turbulent when nonlinearity sets in and subsequently evolve inside the bulk of the band in the form of streaks and vortices.

Based on these discussions, here we propose that the moving direction and the tilt direction of a band are probably determined by what type of local flow is formed when a localized perturbation is introduced: One similar to that shown in Figure 4a generates a right-going band and one similar to that shown in Figure 4d generates a left-going band. In fact, the technique proposed by Song and Xiao [25], with which we generated the bands in Figure 1, is based on this mechanism. The key of the technique is to impose a localized body force that moves with the speed of the head and induces a locally inflectional flow. It can offer a control on the tilt direction of the generated bands because it offers a control on the spanwise velocity profile (to be similar to either the one in Figure 4a or the one in Figure 4c) and on the moving direction of the force. The efficacy of the technique in turn supports that some key characteristics of a band are determined by the local inflectional mean flow at the head.

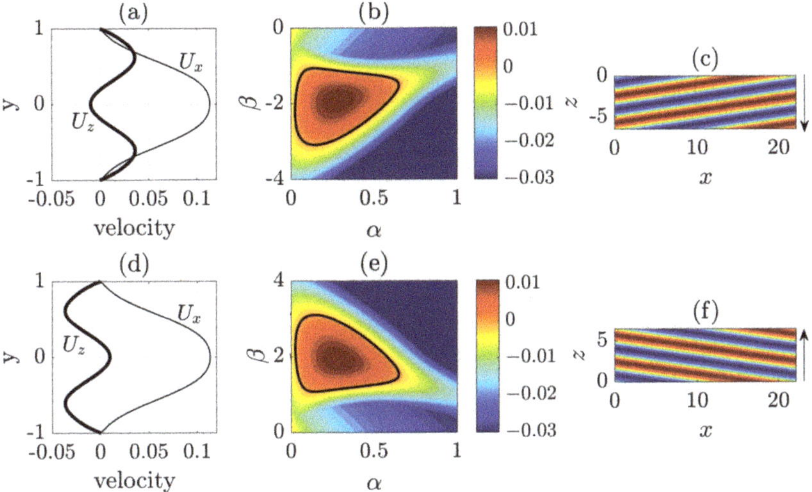

Figure 4. Linear instability of the modified velocity profile at the head that is spatially and temporally averaged in the rectangle shown in Figure 3b: left-going bands (**a**–**c**); and right-going bands (**d**,**e**). (**a**,**d**) Velocity profiles. Wall-normal component is very small and neglected; (**b**,**d**) The maximum eigenvalue in the wave number plane, in which α is the streamwise wave number and β the spanwise wavenumber. The bold line marks the neutral stability curve; (**c**,**f**) Contours of streamwise velocity of the most unstable disturbances at the cut plane $y = -0.5$. Red and blue colors represent high speed and low speed regions, respectively. The arrows show the direction of the spanwise wave speed. Similar analysis for the modified velocity profiles averaged in different regions were reported in [24,25].

Although the linear instability, as well as the non-normality, associated with the local mean flow seem to be the mechanism underlying the wave generation and growth of turbulent bands, how this inflectional local mean flow is formed and sustained is still not sufficiently understood. Tao et al. [6] observed that, when the computational domain is too small, a band may interact with its periodic image and decay. Based on this observation, they proposed that the sustainment of a turbulent band relies on the secondary large-scale flow surrounding the band, and a close neighbor may affect this large-scale flow and eliminate the band. Given that a turbulent band is driven by the head, this observation seems to imply that the head of a band is sustained by the large-scale flow, see Figure 3. However, Kanazawa [7] proposed a completely different scenario. They added a damping term to the Navier–Stokes equations, using which they suppressed the formation of the body of a band and isolated the head, and observed that the head can be self-sustained as a nonlinear periodic orbit. This periodic orbit is characterized by an array of streaks and vortex tubes that resemble the flow structure at the head. Because the band does not form under the damping, the large-scale flow is also absent, although there is still a local circulation flow associated with the localized periodic orbit itself. This seems to contradict the conclusion of Tao et al. [6] that a band relies on the large-scale flow surrounding the band. Further, Kanazawa [7] studied the bifurcation of the periodic orbit in the damped system and reported a saddle-node bifurcation that gives rise to the periodic orbit. Below the saddle-node bifurcation point, no such exact coherent structures exist. Therefore, the authors proposed that this self-sustained periodic orbit and the subsequent bifurcations to torus and chaos is responsible for the formation and sustaining mechanism of turbulent bands. However, they failed to obtain a periodic orbit and reproduce the bifurcations as the damping parameter vanishes, i.e., in the

Navier–Stokes equations without an artificial damping. Obtaining such a periodic solution may finally elucidate the appearance and self-sustaining mechanism of fully localized turbulent bands [7].

Kanazawa [7] did not show why and how exactly this periodic orbit generates wave-like streaks or vortices, rather, only described them as the characteristics of the periodic orbit. In fact, the inflectional instability proposed by Xiao and Song [24] may be related to this periodic orbit. The possible connection is that the circulation associated with the periodic orbit may be locally inflectional and responsible for the wave generation. The inflectional profiles of Xiao and Song [24] are just temporal-spatial averages at the head and only depend on y. The averaging leaves out the streamwise and spanwise dependence of the real local flow at the head; therefore, Xiao and Song [24] pointed out that this may be why their stability analysis cannot quantitatively capture some characteristics of the waves at the head, such as the value of the tilt angle of the waves with respect to the streamwise direction. The analysis of this three-dimensional periodic orbit may be needed to more quantitatively understand the dynamics of the head.

3. The Bulk

The bulk of a turbulent band is defined as the elongated part that is sufficiently far from the head and tail, which does not significantly vary on large-scale and can be considered to be at an 'equilibrium state'.

3.1. The Flow Structure

Many studies have noticed the wave-like form of the bulk of turbulent bands [6,7,9,24], i.e., regularly aligned and distributed streaks along the band. In Figure 5a, the streamwise velocity fluctuations are plotted as the colormap in the x-z plane at $y = -0.5$ (blue color shows low speed and red shows high speed region). Low-speed streaks (blue) are nearly parallel to the streamwise direction and show nearly a periodic pattern. On the upstream edge, high-speed streaks (red) can be observed but do not show a strong periodic pattern as the low-speed streaks. It should be noted that the tilt angle, with respect to the streamwise direction, of the steaks in the bulk is significantly lower than that at the head. Both low speed and high-speed streaks are nearly parallel to the streamwise direction. However, it still can be noticed that these two groups of streaks exhibit opposite tilt directions. The four dashed lines mark the positions of four cut planes perpendicular to the wall, in which streamwise velocity fluctuations are plotted to visualize the structure of the band in the wall-normal direction (see Figure 5b–e). A two-layer structure can be observed, which can be expected from the symmetry of the base flow about the channel center-plane. Each layer consists of staggered high- and low-speed streaks, and, in each layer, high-speed streaks are located near the wall and low-speed streaks near the channel center-plane. Figure 5e and the part between $s = 20$ and 50 in Figure 5d show that, on the upstream, high-speed streaks are the dominate structures. On the downstream, low-speed streaks dominate (see Figure 5b and the part between $s = 0$ and 20 in Figure 5c). In between, high-speed and low-speed streaks are comparable (see the part between $s = 20$ and 50 in Figure 5c and between $s = 0$ and 20 in Figure 5d), and this is the most energetic and turbulent region.

Xiao and Song [24] showed that the generated streaks move away from the head in the frame of reference co-moving with the head, and that streaks decay at the tail of the band. To show this process explicitly, we selected a low-speed streak and tracked it (see Figure 6). The tracking lasted for hundreds of time units until the streak reaches the tail of the band, without a significant change in the shape of the streak.

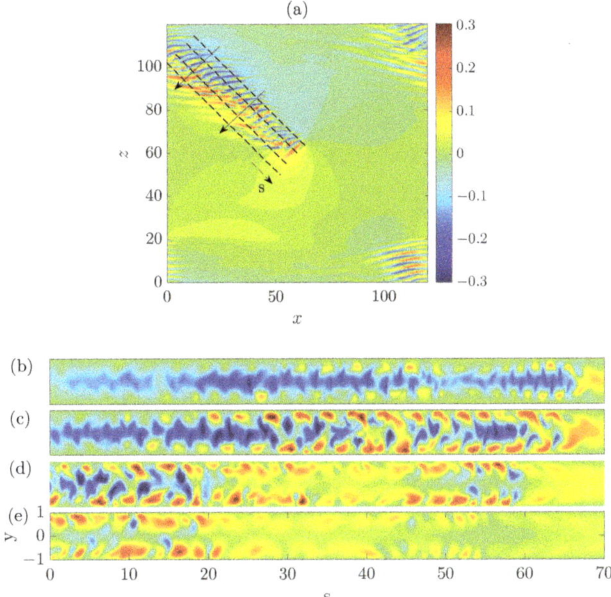

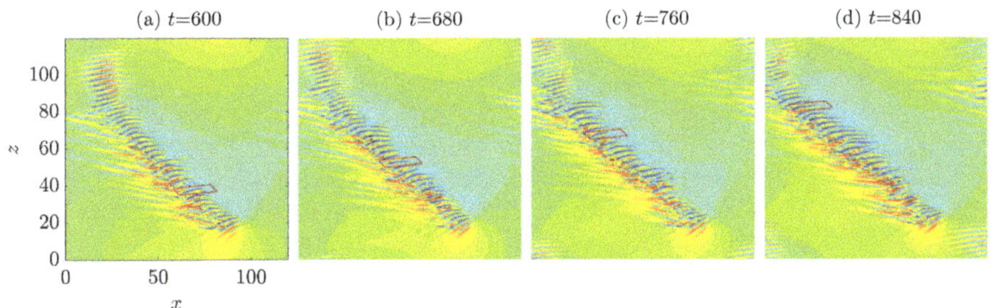

Figure 5. (**a**) Streamwise velocity fluctuations of a band at $Re = 750$ plotted in the x-z plane at $y = -0.5$. The four dashed lines mark the positions of four cut planes perpendicular to the wall, in which streamwise velocity fluctuations are plotted in (**b**–**e**). The s-axis goes from top-left to bottom-right along the lines. The two arrows, at $s = 10$ and 40, respectively, show the sequence of (**b**–**e**). The length in y direction is stretched by a factor of 3.

Figure 6. Tracking of a low-speed streak (enclosed by a parallelogram) at $Re = 750$ in the frame of reference co-moving with the head. (**a**) $t = 600$; (**b**) $t = 680$; (**c**) $t = 760$; (**d**) $t = 840$.

3.2. Advection Speed of the Streaks inside the Bulk

Next, we quantitatively studied the advection speed of the streaks. The advection speed can be estimated by tracking an individual low-speed streak, as shown in Figure 6. Alternatively, it is possible to measure the speed of an array of streaks as a whole. We adopted the latter approach. We used the velocity data on the cut plane of $y = -0.5$, which well cuts through the streaks and offers a nearly optimal visualization of the flow pattern (see Figure 5). Nevertheless, a cut plane close to the wall, which would cut through high-speed streaks that are located close to the wall (see Figure 5), is equally applicable. We used the Structural Similarity Index Measure (SSIM) method [28] from image processing, which accesses the similarity between two images based on luminance, contrast and structure of the images. The method is detailed in Section 6.

The advection speed of low-speed streaks for a few Reynolds numbers are shown in Figure 7. The results show that, in the low Reynolds number regime between 670 and 1050, the streamwise advection speed slowly decreases from 0.68 to 0.63, whereas the spanwise speed seems to stay nearly constant at around 0.07. Note that the streamwise speed is very close to the bulk speed of the flow, which is 0.67. In fact, in Figure 6, the parallelogram was moved at the speeds we measured in this way and very well tracked the streak over hundreds of time units. Paranjape [9] reported that the phase speeds of the exact nonlinear traveling wave solution they obtained at $Re = 720$ are $c_x = 0.77$ and $c_z = 0.06$, which are close to our results, suggesting a strong connection between their traveling wave solution and turbulent bands.

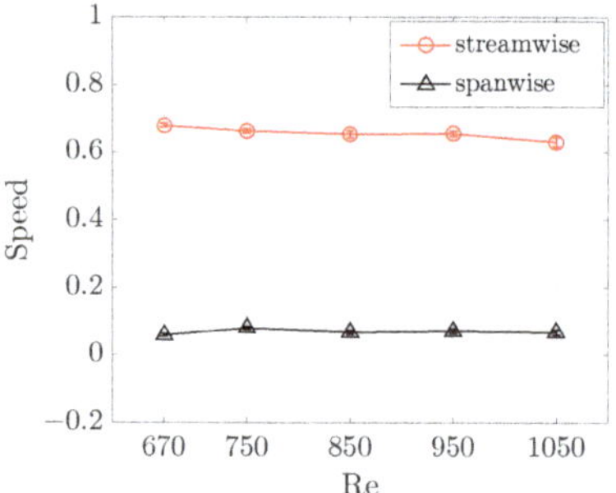

Figure 7. Advection speed of the low-speed streaks at a few Reynolds numbers. The spanwise speeds are $c_z = 0.068, 0.076, 0.069, 0.071$ and 0.068, and the streamwise speeds are $c_x = 0.68, 0.66, 0.65, 0.65$ and 0.63, for $Re = 670, 750, 850, 950$ and 1050, respectively.

4. Tilt Angle of Turbulent Bands

The tilt angle of turbulent bands at $Re < 1000$ was reported in experiments by Paranjape [9]. Their measurements showed that the angle stays nearly constant close to $45°$ below $Re \simeq 900$ and decreases to approximately $30°$ above $Re = 950$. The decreasing trend was also reported by Shimizu and Manneville [8]. A few numerical studies also reported the tilt angle at some Reynolds numbers; for example, Kanazawa [7] reported $41°$ at $Re = 660$, Tao et al. [6] reported approximately $40°$ at $Re = 700$ and Xiao and Song [24] reported an angle of about $39°$ at $Re = 750$, which are lower than but close to the experimental results of Paranjape [9]. The small difference may be attributed to the periodic boundary condition used in simulations and to the specific methods of quantifying the tilt angle.

However, the mechanism underlying the tilt angle selection is still not well-understood. Prior studies simply measured the tilt angle by considering the entire band as a tilted object based on image processing or in similar manners [6,9]. Differently, here we propose that the tilt angle should be more fundamentally determined by the propagation speed of the head and the advection speed of the streaks inside the bulk. More specifically, the speed of the streaks inside the bulk relative to the head should determine the tilt angle of the band. Based on our measurements shown in Figures 2 and 7, we calculated the tilt angle of the band as

$$\theta = \arctan \frac{|c_{z,\text{streak}} - c_{z,\text{head}}|}{|c_{x,\text{streak}} - c_{x,\text{head}}|}. \tag{1}$$

The result is shown in Figure 8. Our calculations agree well with the experimental result of Paranjape [9] below $Re \simeq 900$. However, at $Re = 1050$, our calculation appears to be much higher than their measurement: our calculation gives 37° for $Re = 1050$, whereas it was estimated to be around 30° in experiments. Nevertheless, our calculation gives the decreasing trend in the tilt angle as Re is increased to around $Re = 1000$ and above.

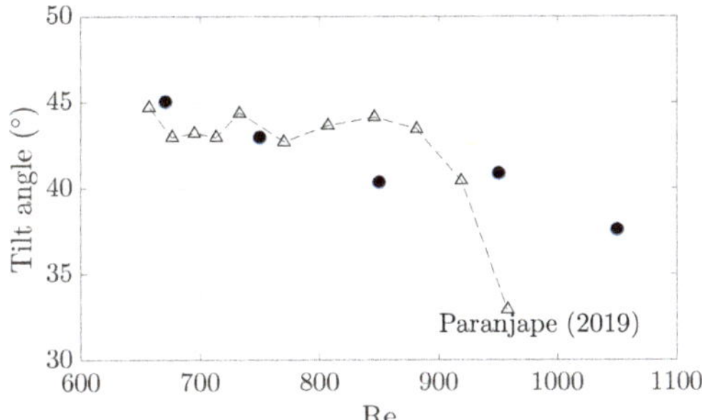

Figure 8. The tilt angle of turbulent bands, calculated with Equation (1), at a few Reynolds numbers. The experimental measurements of Paranjape [9] are plotted as the dashed-triangle line for comparison.

The possible reason for the significant difference between our calculation and the experimental measurements at $Re = 1050$ can possibly be understood by inspecting the structure of the band as Re increases (see Figure 9). We can see that, at $Re = 670$, the band has a well-defined banded structure, i.e., the width (e.g., the streamwise extension) of the band does not significantly change along the band (see Figure 9a). At $Re = 950$, the tail of the band seems to broaden and the width of the band may not be constant along the length direction any more (see Figure 9b). Further at $Re = 1050$, the band significantly delocalizes: The bulk broadens gradually towards the tail and part of the band turns into an extended turbulent area (see Figure 9c). By image processing the entire band, as in the measurements of Paranjape [9] and Tao et al. [6], the calculated tilt angle at $Re = 1050$ will certainly be smaller than our calculation that is only based on the information of the low-speed streaks and the head. This disagreement will be small at low Reynolds numbers when turbulence is well-banded.

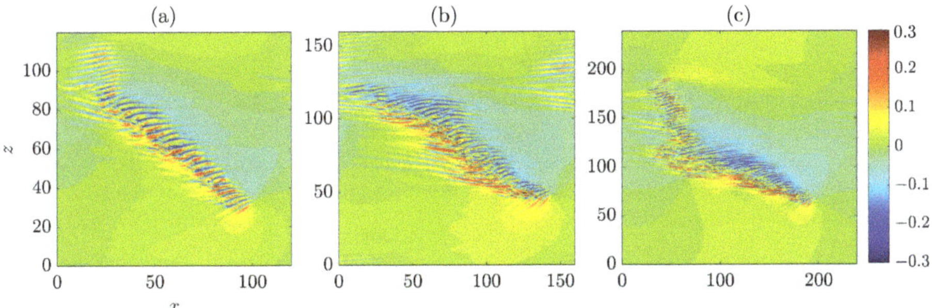

Figure 9. Turbulent bands at: $Re = 670$ (**a**); $Re = 950$ (**b**); and $Re = 1050$ (**c**).

The agreement between our calculation and the reported speeds in the literature supports our speculation that the tilt angle of the band is determined jointly by the propagation speed of the head and the advection speed of the streaks inside the bulk. However, what mechanism determines the advection speed of the streaks is still to be investigated. A quantitative study of the large-scale flow may give a hint to the advection of the streaks [9,27,29,30].

It should be noted that the two ends of turbulent bands may not exist in relatively small normal periodic domains or narrow tilted domains, therefore, seemingly our formulation of the tilt angle (Equation (1)) does not apply. In those cases, it is not clear what mechanism determines the tilt angle of turbulent bands. Our speculation is that the tilt angle may be indefinite and is strongly affected by the specific domain selection if the head does not exist. This might explain, for the same Reynolds number, why turbulent bands can exist in tilted domains with very different tilt angles [4,9,20] and why the nonlinear traveling wave solutions that Paranjape et al. [19] obtained can exist in a broad tilt-angle range from $20°$ to $70°$.

5. Discussion

The wave generation at the head, the tilt direction, the advection of the head, the streaks inside turbulent bands and the tilt angle of the band are discussed and investigated in this paper. The inflectional-instability argument of Xiao and Song [24] for the wave generation at the head and its potential relationship with the localized periodic-orbit theory of Kanazawa [7] are discussed. Based on the discussion, we propose that the tilt direction should probably be determined by the local inflectional spanwise velocity profile generated/introduced by the initial perturbation. The opposite tilt directions are rooted in the mirror symmetry of the spanwise velocity component. Besides, we measured the propagation speed of the head and the advection speed of the low-speed streaks in the bulk of turbulent bands at low Reynolds numbers up to $Re = 1050$. We found that the head propagates at constant speeds of $c_x = 0.85$ and $c_z = 0.1$ (absolute value) at all Reynolds numbers investigated. The low-speed streaks are advected roughly at the speed of the bulk speed in the streamwise direction with a slight decreasing trend as the Reynolds number increases, and the spanwise advection speed is nearly constant at approximately 0.07. Prior studies measured the tilt angle by treating the band as a tilted object [6,9]; alternatively, we here propose that the tilt angle of turbulent bands should be determined by the kinematics of the head and the streaks generated at the head. Specifically, the tilt angle can be calculated using the relative speed between the streaks in the bulk and the head, and, at least for $Re \lesssim 900$, we obtained a good agreement with the experimental measurements of Paranjape [9]. We also speculate that the tilt angle of a band may be indefinite and system-dependent if the head does not exist as in narrow tilted domains and relatively small normal domains.

A few problems remain poorly understood and should be investigated in order to further understand the transition in channel flow.

- The sustaining mechanism of the wave-generating head. The formation and sustainment of the locally inflectional flow at the head, whether or not the head is locally self-sustained and the relationship between the head and the large-scale flow are still not clear. If the head is indeed locally self-sustained and independent of the bulk, as proposed by Kanazawa [7], how the flow can be locally excited to this periodic orbit is also not clear. This problem is relevant to the generation and control of turbulent bands at low Reynolds numbers.
- The mechanism underlying the advection speed of the head. Xiao and Song [24] speculated that the speeds are possibly determined by the speeds of the unstable waves resulting from the local inflectional instability. They reported a close spanwise speed of the most unstable wave for $Re = 750$, which is about 0.1 and is close to the actual spanwise of the head (see Figure 2). However, the streamwise speed of the most unstable wave is roughly 0.55 (can be calculated from the eigenvalues and wavenumbers associated with the most unstable wave reported by them) and is significantly

lower than the values shown in Figure 2, which is about 0.85. This discrepancy may be attributed to the over-simplification of the local mean flow at the head by temporal and spatial averaging in their linear stability analysis, as well as by the region selection for the averaging. A possibility to elucidate the mechanism underlying the advection speed is to investigate the speed of the periodic orbit of Kanazawa [7].

- The mechanism underlying the self-sustainment and advection speed of the streaks. Paranjape et al. [19] obtained exact traveling wave solutions that have some key characteristics of turbulent bands and identified the solutions as the precursors of turbulent bands. Further, for these solutions, they speculated that the streaks are sustained by the tilting effect of the large-scale flow, instead of the self-sustaining process of wall turbulence at high Reynolds numbers in which sinuous streaks break down, generating streamwise vortices, and are regenerated by streamwise vortices [31,32]. The same mechanism may also apply to turbulent bands. In our simulations, we indeed observed that streaks in the bulk are long-lived and move with a characteristic speed without a clear breakdown and regeneration. Duguet and Schlatter [27] described turbulent bands in plane shear flows as the advection of small-scale structures (streaks and vortices) by the large-scale flow, which also seems to suggest the important role of the advection by the large-scale flow in the sustainment of the streaks.

- The mechanism underlying the decay of streaks at the tail as well as the splitting and branching of turbulent bands. At relatively higher Reynolds numbers, a band may also nucleate a band with the opposite tilt direction [8,9]. The splitting scenario, at least partially, determines the flow pattern.

6. Materials and Methods

For solving the incompressible Navier–Stokes equations in channel geometry, we used our in-house code as described in [24,25], which adopts a high-order finite-difference method with a centered nine-point stencil in the wall-normal direction and Fourier-spectral method in the periodic streamwise and spanwise directions. Readers are referred to OPENPIPEFLOW [33] for details about the finite-difference scheme and the parallelization of the code. The Navier–Stokes equations were integrated using the method of Hugues and Randriamampianina [34], which adopts a second-order-accurate backward-differentiation scheme, combined with the Adamas–Bashforth scheme for the nonlinear term, for the temporal discretization and a projection method to impose the incompressibility condition. The time-step size was fixed at $\Delta t = 0.01$ for the simulations presented in this paper, which was shown to be sufficiently small for the Reynolds number regime considered [4,6].

We adopted the method proposed by Song and Xiao [25] to generate turbulent bands in large domains. The method firstly derives a body force that is needed to maintain an inflectional velocity profile that bears a sufficiently strong instability. Given a target velocity profile $U(y)$, the body force is derived as

$$f = -\frac{1}{Re}\nabla^2 U(y). \tag{2}$$

Then, the body force is multiplied by a localization factor such that the force is localized in the x-z plane. The size of the localization region should be comparable with the size of the head of a turbulent band and the forcing region is moved at the speed of $c_z = 0.1$ (absolute value) and $c_x = 0.85$ (see Figure 2). If the profile $U(y)$ is sufficiently inflectional, the instability can generate sufficiently strong tilted waves (streaks and vortices) and trigger turbulent bands. Once triggered, the length of the band increases, and the force can be switched off after the band has sufficiently developed. The tilt direction of the band can be controlled by the signs of the spanwise component of $U(y)$ and the moving speed c_z.

We measured the advection speed of the low-speed streaks inside the bulk using the Structural Similarity Index Measure (SSIM) method proposed by Wang et al. [28], which is commonly used in image processing to measure the similarity between two images. The SSIM index is defined as:

$$SSIM(x,y) = l(x,y)^{\alpha} \cdot c(x,y)^{\beta} \cdot s(x,y)^{\gamma}, \tag{3}$$

where x and y are one-dimensional vectors containing all the pixel values of the two images to be compared, respectively, and

$$l(x,y) = \frac{2\mu_x \mu_y + c1}{\mu_x^2 + \mu_y^2 + c1}, \tag{4}$$

$$c(x,y) = \frac{2\sigma_x \sigma_y + c2}{\sigma_x^2 + \sigma_y^2 + c2}, \tag{5}$$

and

$$s(x,y) = \frac{\sigma_{xy} + c3}{\sigma_x \sigma_y + c3}, \tag{6}$$

measure the luminance, contrast and structural similarity, respectively. The exponents $\alpha > 0$, $\beta > 0$ and $\gamma > 0$ are used to tune the relative weight of respective factor, and here we set all of them to 1 according to the suggestion of Wang et al. [28]. In Equation (4), μ_x and μ_y denote the mean of x and y, respectively. In Equation (5), σ_x and σ_y denote the standard deviation of x and y, respectively. In Equation (6), σ_{xy} is the covariance of x and y. Parameters $c1 = (k_1 L)^2$ and $c2 = (k_2 L)^2$, where $k1$ and $k2$ are set to 0.01 and 0.03, respectively, and L is the maximum of the pixel value, which is set to 255 for unit8 data and 1 for floating point data. In our calculation, the flow velocities, which are floating point data, were taken as the pixel value x and y. The parameter c_3 is set such that $c_3 = c_2/2$ in practice according to the suggestion of Wang et al. [28]. Thus, we have

$$SSIM(x,y) = \frac{(2u_x u_y + c_1)(2\sigma_{xy} + c_2)}{(u_x^2 + u_y^2 + c_1)(\sigma_x^2 + \sigma_y^2 + c_2)}. \tag{7}$$

The result is a value between -1 and 1, and the larger is the result, the higher is the similarity.

Firstly, we take the streamwise velocities in the cut plane $y = -0.5$ from two different snapshots s_1 and s_2 that are separated in time by δt, after the tilt angle of the band has stopped changing considerably due to the initial transients. In the frame of reference co-moving with the head, i.e., moving with a streamwise speed of 0.85 and a spanwise speed of -0.1 (we considered a right-going band), the bulk of the band is located in a nearly fixed area (see Figure 10). Therefore, we set a rectangular area in which the data inside were considered for calculating the SSIM index. We set the data outside this area to zero so that we eliminated the influence of the data outside this area. Further, to highlight the low-speed streaks, only the streamwise velocities in this area that satisfies $u_x < 0$ and $u_x^2 > 0.002$ were retained. Secondly, we shifted the data from s_2 inside the rectangular over the time separation δt with a streamwise speed c_x and a spanwise speed c_z. The original data from s_1 and the shifted data from s_2 were used to calculate the SSIM index. Thus, for a given speed pair (c_x, c_z), there is a corresponding SSIM index. By varying the speed pair, the SSIM index will maximize with certain speeds, which we considered as the mean advection speeds of the streaks. The contours of the SSIM index in the c_x and c_z for $Re = 750$ are shown in Figure 11.

Note that, in practice, we set $-c_x$ and c_z to be between 0.1 and 0.4 (the band we considered is a right-going one; therefore, $c_x < 0$ and $c_z > 0$) because the actual speeds were estimated by eye to exist in this range, and note that the shift speeds are relative to the propagation of the head. Obtaining the contours of the SSIM index, we could estimate the advection speed of the streaks to be $c_x = -0.185$ and $c_z = 0.18$, i.e., the location of the local peak at the left-bottom corner in Figure 11. It can be seen that there is another local peak at the right-top corner, which shows a lower SSIM index. That peak was reached when the s_2 data were shifted by more than one wave-length associated with the pattern of the low-speed streaks. The lower SSIM index of the top-right peak, i.e., lower similarity, indicates that the streaky pattern slowly change as it is advected in the bulk.

Note that the time separation δt between s_1 and s_2 cannot be too small, otherwise the streaks would have moved too little over the time separation and the speed measurement would be inaccurate. Likewise, it cannot be too large in which case the streaks would have moved by multiple wavelengths, which would also affect the speed calculation. In practice, estimated by eyes, a value between $\delta t = 10$ and 15 is a good choice, and $\delta t = 10$ in Figures 10 and 11. In the end, by varying the time instant of s_1, we can obtain the average advection speed as a function of time and calculate the temporal average, which is plotted in Figure 7 (the speed of the head is added back in that figure).

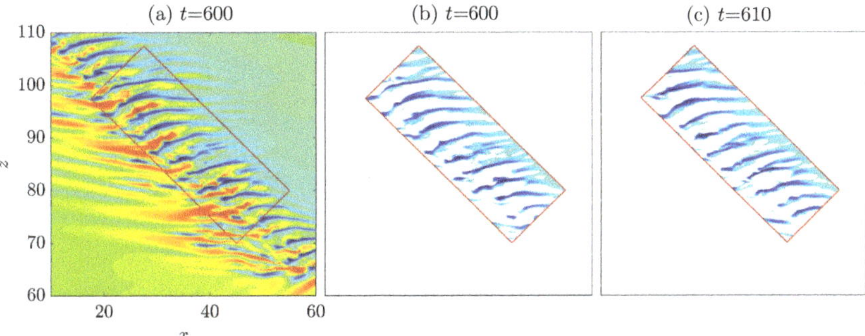

Figure 10. The selection of low-speed streaks for the advection speed calculation. The tilt angle of this rectangle is $45°$, which is close to the tilt angle of the band. The area of the rectangle should be large enough to contain sufficient streaks and meanwhile reduce the influence of the tilt angle of the rectangle. (**a**) The contours of streamwise speedat $y = -0.5$ at $t = 600$. The advection speed of the streaks in the region enclosed by the rectangle is calculated; (**b**,**c**) The filtered low-speed streaks at $t = 600$ and $t = 610$.

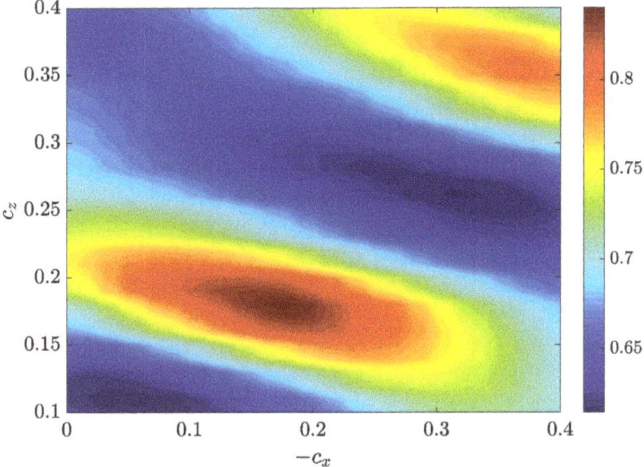

Figure 11. The contours of the SSIM index in the c_x-c_z plane for the case shown in Figure 10.

Author Contributions: B.S. designed the research and X.X. performed the simulations. X.X. and B.S. analyzed the data and wrote the manuscript. All authors have read and agreed to the published version of the manuscript.

Funding: This research was funded by the National Natural Science Foundation of China under grant numbers 91852105 and 91752113 and by Tianjin University under grant number 2018XRX-0027.

Acknowledgments: The simulations were performed on Tianhe-2 supercomputer at the National Supercomputer Center in Guangzhou. Discussions with Dwight Barkley and Jianjun Tao are highly appreciated.

Conflicts of Interest: The authors declare no conflict of interest.

References

1. Tsukahara, T.; Seki, Y.; Kawamura, H.; Tochio, D. DNS of turbulent channel flow at very low Reynolds numbers. In Proceedings of the Fourth International Symposium on Turbulence and Shear Flow Phenomena, Williamsburg, VA, USA, 27–29 June 2005; pp. 935–940.
2. Tsukahara, T.; Iwamoto, K.; Kawamura, H.; Takeda, T. DNS of Heat Transfer in a Transitional Channel Flow Accompanied by a Turbulent Puff-like Structure. *arXiv* **2014**, arXiv:1406.0586v2.
3. Tsukahara, T.; Kawaguchi, Y.; Kawamura, H. An experimental study on turbulent-stripe structure in transitional channel flow. *arXiv* **2014**, arXiv:1406.1378.
4. Tuckerman, L.S.; Kreilos, T.; Shrobsdorff, H.; Schneider, T.M.; Gibson, J.F. Turbulent-laminar patterns in plane Poiseuille flow. *Phys. Fluids* **2014**, *26*, 114103. [CrossRef]
5. Xiong, X.M.; Tao, J.; Chen, S.; Brandt, L. Turbulent bands in plane-Poiseuille flow at moderate Reynolds numbers. *Phys. Fluids* **2015**, *27*, 041702. [CrossRef]
6. Tao, J.J.; Eckhardt, B.; Xiong, X.M. Extended localized structures and the onset of turbulence in channel flow. *Phys. Rev. Fluids* **2018**, *3*, 011902. [CrossRef]
7. Kanazawa, T. Lifetime and Growing Process of Localized Turbulence in Plane Channel Flow. Ph.D. Thesis, Osaka University, Osaka, Japan, 2018.
8. Shimizu, M.; Manneville, P. Bifurcations to turbulence in transitional channel flow. *Phys. Rev. Fluids* **2019**, *4*, 113903. [CrossRef]
9. Paranjape, C.S. Onset of Turbulence in Plane Poiseuille flow. Ph.D. Thesis, IST Austria, Klosterneuburg, Austria, 2019.
10. Tuckerman, L.S.; Chantry, M.; Barkley, D. Patterns in Wall-Bounded Shear Flows. *Ann. Rev. Fluid Mech.* **2020**, *52*, 343–67. [CrossRef]
11. Prigent, A.; Gregoire, G.; Chate, H.; Dauchot, O.; van Saarloos, W. Large-Scale Finite-Wavelength Modulation within Turbulent Shear Flows. *Phys. Rev. Lett.* **2002**, *89*, 014501. [CrossRef]
12. Barkley, D.; Tuckerman, L.S. Computational study of turbulent laminar patterns in Couette Flow. *Phys. Rev. Lett.* **2005**, *94*, 014502. [CrossRef]
13. Duguet, Y.; Schlatter, P.; Henningson, D.S. Formation of turbulent patterns near the onset of transition in plane Couette flow. *J. Fluid Mech.* **2010**, *650*, 119–129. [CrossRef]
14. Coles, D. Transition in circular couette flow. *J. Fluid Mech.* **1965**, *3*, 385–425. [CrossRef]
15. Avila, K. Shear Flow Experiments: Characterizing the Onset of Turbulence as A Phase Transition. Ph.D. Thesis, Georg-August-Universiät Göttingen, Göttingen, Germany, 2013.
16. Ishida, T.; Duguet, Y.; Tsukahara, T. Transitional structures in annular Poiseuille flow depending on radius ratio. *J. Fluid Mech.* **2004**, *794*, R2. [CrossRef]
17. Chantry, M.; Tuckerman, L.S.; Barkley, D. Universal continuous transition to turbulence in a planar shear flow. *J. Fluid Mech.* **2017**, *824*, R1. [CrossRef]
18. Manneville, P. Laminar-turbulent patterning in transitional flows. *Entropy* **2017**, *19*, 316. [CrossRef]
19. Paranjape, C.S.; Duguet, Y.; Hof, B. Oblique stripe solutions of channel flow. *J. Fluid Mech.* **2020**, *897*, A7. [CrossRef]
20. Gomé, S.; Tuckerman, L.S.; Barkley, D. Statistical transition to turbulence in plane channel flow. *Phys. Rev. Fluids* **2020**, *5*, 083905. [CrossRef]
21. Avila, K.; Moxey, D.; De Lozar, A.; Avila, M.; Barkley, D.; Hof, B. The Onset of Turbulence in Pipe Flow. *Science* **2011**, *333*, 192–196. [CrossRef]

22. Shi, L.; Avila, M.; Hof, B. Scale Invariance at the Onset of Turbulence in Couette Flow. *Phys. Rev. Lett.* **2013**, *110*, 204502. [CrossRef]

23. Lemoult, G.; Shi, L.; Avila, K.; Jalikop, S.V.; Avila, M.; Hof, B. Directed percolation phase transition to sustained turbulence in Couette flow. *Nat. Phys.* **2016**, *12*, 254–258. [CrossRef]

24. Xiao, X.; Song, B. The growth mechanism of turbulent bands in channel flow at low Reynolds numbers. *J. Fluid Mech.* **2020**, *883*, R1. [CrossRef]

25. Song, B.; Xiao, X. Trigger turbulent bands directly at low Reynolds numbers in channel flow using a moving-force technique. *J. Fluid Mech.* **2020**, *903*, A43. [CrossRef]

26. Sano, M.; Tamai, K. A universal transition to turbulence in channel flow. *Nat. Phys.* **2016**, *12*, 249–253. [CrossRef]

27. Duguet, Y.; Schlatter, P. Oblique laminar-turbulent interfaces in plane shear flows. *Phys. Rev. Lett.* **2013**, *110*, 034502. [CrossRef]

28. Wang, Z.; Bovik, A.C.; Sheikh, H.R.; Simoncelli, E.P. Image quality assessment: from error visibility to structural similairty. *IEEE Trans. Image Process.* **2004**, *13*, 600–612. [CrossRef]

29. Rolland, J. Formation of spanwise vorticity in oblique turbulent bands of transitional plane Couette flow, part 1: Numerical experiments. *Eur. J. Mech. B-Fluids* **2015**, *50*, 52–59. [CrossRef]

30. Rolland, J. Formation of spanwise vorticity in oblique turbulent bands of transitional plane Couette flow, part 2: Modelling and stability analysis. *Eur. J. Mech. B-Fluids* **2016**, *56*, 13–27. [CrossRef]

31. Hamilton, J.; Kim, J.; Waleffe, F. Regeneration mechanisms of near-wall turbulence structures. *J. Fluid Mech.* **1995**, *287*, 317–348. [CrossRef]

32. Waleffe, F. On a self-sustaining process in shear flows. *Phys. Fluids* **1997**, *9*, 883–900. [CrossRef]

33. Willis, A.P. The Openpipeflow Navier–Stokes Solver. *SoftwareX* **2017**, *6*, 124–127. [CrossRef]

34. Hugues, S.; Randriamampianina, A. An improved projection scheme applied to pseudospectral methods for the incompressible Navier-Stokes equations. *Int. J. Num. Meth. Fluids* **1998**, *28*, 501–521. [CrossRef]

Publisher's Note: MDPI stays neutral with regard to jurisdictional claims in published maps and institutional affiliations.

Article

Intermittency and Critical Scaling in Annular Couette Flow

Kazuki Takeda [1], Yohann Duguet [2] and Takahiro Tsukahara [1,*]

[1]   Department of Mechanical Engineering, Tokyo University of Science, Chiba 278-8510, Japan; tairyuuuu88@gmail.com

[2]   LIMSI-CNRS, Université Paris-Saclay, F-91400 Orsay, France; duguet@limsi.fr

*   Correspondence: tsuka@rs.tus.ac.jp; Tel.: +81-4-7122-9352

Received: 14 August 2020; Accepted: 1 September 2020; Published: 4 September 2020

Abstract: The onset of turbulence in subcritical shear flows is one of the most puzzling manifestations of critical phenomena in fluid dynamics. The present study focuses on the Couette flow inside an infinitely long annular geometry where the inner rod moves with constant velocity and entrains fluid, by means of direct numerical simulation. Although for a radius ratio close to unity the system is similar to plane Couette flow, a qualitatively novel regime is identified for small radius ratio, featuring no oblique bands. An analysis of finite-size effects is carried out based on an artificial increase of the perimeter. Statistics of the turbulent fraction and of the laminar gap distributions are shown both with and without such confinement effects. For the wider domains, they display a cross-over from exponential to algebraic scaling. The data suggest that the onset of the original regime is consistent with the dynamics of one-dimensional directed percolation at onset, yet with additional frustration due to azimuthal confinement effects.

Keywords: subcritical phenomenon; transition to turbulence; direct numerical simulation

---

1. Introduction

The dynamics at the onset of turbulent fluid flow, as the parameters are varied, is one of the most puzzling issues of hydrodynamics. Subcritical flows are known to feature two regimes in competition, namely a laminar and a turbulent one. As the Reynolds number (their main control parameter) is varied, this competition takes the form of laminar-turbulent coexistence featuring some interesting analogies with phase transitions in thermodynamics. The onset of this coexistence in wall-bounded shear flows has been speculated to follow a statistical scenario called directed percolation (DP). It involves a critical point (a critical Reynolds number) in the vicinity of which fluctuations diverge algebraically [1,2]. The directed percolation scenario has gained theoretical importance because it appears as the usual rule for a one-dimensional systems obeying a set of specific properties, notably a unique absorbing state and short-range interactions [3,4]. However, it quickly proved difficult to isolate similar phenomena experimentally [5]. The main limitations happen to be finite-size effects, as well as the presence of defects [6–8] or issues revolving around nucleation rates [9,10]. The first experimental evidence for directed percolation in a two-dimensional physical system, with a complete set of critical exponents, occurred in electroconvection in nematic liquid crystals [11]. More recent experiments and numerical simulations with inert liquids were aimed at establishing the critical exponents relevant for the laminar-turbulent transition. The only meaningful experimental results are to be found in Ref. [12] for the flow inside an annulus driven by the revolutions of the outer wall, where all critical exponents match those of (1 + 1)-D DP. All other experimental attempts in effectively two-dimensional geometries have so far lead to ambiguous results [13,14]. A few numerical studies based on other geometries have also confirmed the DP hypothesis in one dimension, among them [15]. The most notorious system displaying one-dimensional spatiotemporal intermittency (STI) is cylindrical pipe flow. Although (1 + 1)-D DP has been widely speculated and is found in the most recent modelling approaches [16–18], clean experimental evidence seems to require facilities of a size beyond anything engineerable [19]. The only convincing two-dimensional study to date based on the (underesolved) Navier–Stokes equations and supporting the DP hypothesis is found in Ref. [20]. There again, a cost compromise was necessary between accuracy of the Navier–Stokes solutions and size effects. There the set of critical exponents differs from their unidimensional counterpart and corresponds to (2 + 1)-D DP. The status of the application of (2 + 1)-D DP to other planar flows is still open: for plane Couette flow (pCf), finite-size effects wrongly predict to discontinuous scenarios [21], whereas plane Poiseuille flow (pPf) seems to display a two-stage behavior so far poorly understood [22–24]. At a finite distance from the critical point, these two planar flows feature more structured arrays of turbulent stripes, all oblique to the mean flow direction (see, e.g., [21,25–30] for recent reviews).

Given the current status of DP affairs in shear flows, new flow candidates where to probe the DP hypothesis are encouraged, irrespective of the effective dimension considered (one or two). In the present article, we revisit transition in annular Couette flow (aCf) in the light of critical scaling. This flow has a geometry similar to cylindrical pipe flow, however, with a solid cylinder at its centerline. The geometry is determined by the radius ratio η between the radius of the outer pipe and that of the inner one. This flow supports both turbulence [31] as well as a linearly stable base flow for all Reynolds number of interest, hence transition has to be of the subcritical type. Unlike annular pipe flow [32–34], no pressure gradient is applied, instead the fluid is entrained by the translating motion of the inner cylinder [35]. Earlier work by some of us [36] on this flow have lead to surprising results: although the transitional flow reported for $\eta \geq 0.5$ consists of helical bands of turbulence wrapping around the inner rod, for lower values of η, a new regime of laminar-turbulent alternations was reported. This regime is characterized by slightly shorter streamwise correlations and non-oblique structures, explained by the azimuthal confinement and by the impossibility to host azimuthal large-scale flows [37]. The aim of the present article is to give a more detailed characterization of the novel low-η intermittent regime and of its onset. In particular, the azimuthal extension of aCf is investigated in a range of parameters beyond that used by Kunii et al. [36]. As will be seen, this new choice of geometrical parameters leads to new conclusions regarding the critical exponents. This new parametric study allows one to rationalize once and for all the quantitative comparison between original geometry and the extended one.

The plan of this article unfolds as follows: the geometry and the numerical methods are explained in Section 2, and the statistics of STI are reported in Section 3 and discussed in Section 4.

2. Set-Up and Methodology

2.1. Geometry of aCf

Annular Couette flow is the flow in the interstice between two coaxial cylinders of formally infinite length, driven by the motion at velocity $U_w > 0$ of the inner cylinder in the x-direction. The annular geometry of this flow is common to both Taylor–Couette flow and annular Pipe flow; however, the forcing is different and no spin of the walls is considered. A sketch of that geometry is displayed in Figure 1 with the usual notations for the cylindrical coordinates (x, r, θ). Assuming that the inner and outer cylinder have respective dimensional radii r_{in} and r_{out}, the main geometrical parameter of this study is the radius ratio $\eta = r_{in}/r_{out}$, which varies in the open interval $(0, 1)$. We also introduce the gap h between the two cylinders $h = r_{out} - r_{in}$.

Computationally, the pipes require to have either finite length or to be spatially periodic. The use of a spectral Fourier-based method to solve the pressure Poisson equation requires axial and azimuthal periodicity. This introduces the two wavelengths L_x and L_θ, respectively, as the domain length and the angular periodicity. While L_x is a free parameter, the natural value for L_θ is 2π because of the cylindrical geometry. However, there is no computational obstruction to choosing other values for L_θ, for instance $L_\theta = 8\pi$ or 16π as in Ref. [36]. In what follows, we keep the generic notation L_θ.

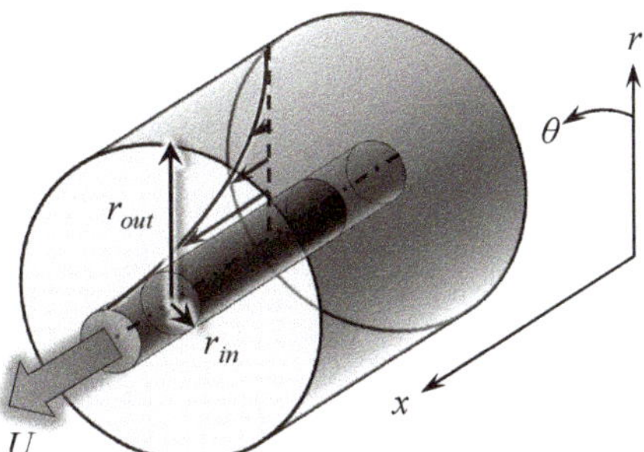

Figure 1. Sketch of annular Couette flow in the cylindrical coordinate system.

Like in other wall-bounded shear flows, the main lengthscale ruling out the transitional dynamics at onset is the gap h between the two solid walls, which here depends directly on the value η via the relation $h = r_{out}(1 - \eta)$. The perimeter on the internal cylinder, at mid-gap or on the external cylinder, now expressed in units of h, is shown in Figure 2 when the original dimensional value of L_θ is 2π (Figure 2a). The inner perimeter is also displayed when L_θ is a multiple of 2π (Figure 2b), with $L_\theta = 2\pi n$. The theory developed in Refs. [33,36] shows that azimuthal large-scale flows cannot be accommodated by the geometry unless $L_\theta r/h \gg 1$ everywhere in the domain. The data for the inner cylinder play the role of a lower bound. For $L_\theta = 2\pi$, it is clear from Figure 2a that, for the lowest values of η, no azimuthal large-scale flow is possible. However, increasing n leads to azimuthal large-scale flows being possible for smaller and smaller values of η. This leads to the possibility to artificially restore large-scale flows otherwise ruled out by geometrical confinement.

2.2. Governing Equations and Computational Methods

Whereas η is a geometrical parameter only, we also introduce the Reynolds number $Re_w = U_w h/4\nu$, based on the half velocity of the cylinder sliding $U_w/2$, the half gap width $h/2$, and the kinematic viscosity ν of the fluid. The reason why half-gap and half-velocities are considered to non-dimensionalize the equations is a simple way to reconnect with the standard conventions for pCf as η goes towards unity. By choosing this convention for all values of η, the non-dimensional incompressible equations ruling the flow dynamics without any turbulence model read

$$\nabla^* \cdot \boldsymbol{u}^* = 0, \tag{1}$$

$$\frac{\partial \boldsymbol{u}^*}{\partial t^*} + (\boldsymbol{u}^* \cdot \nabla^*)\,\boldsymbol{u}^* = -\nabla^* p^* + \frac{1}{4Re_w}\Delta^* \boldsymbol{u}^*, \tag{2}$$

where superscripts $*$ indicate quantities non-dimensionalized with U_w and h, and where $\boldsymbol{u} = (u_x, u_r, u_\theta)$ and p represent the velocity field and the pressure field, respectively.

Equation (2) is discretized in space using finite differences and with fine enough grid resolutions according to the standard criteria of direct numerical simulation (DNS) [26]. The time discretization is carried out using a second-order Crank–Nicolson scheme, and an Adams–Bashforth scheme for the wall-normal viscous term and the other terms, respectively. Further details about the numerical methods used here can be found in Ref. [38]. Table 1 lists the parameters used in this computational study.

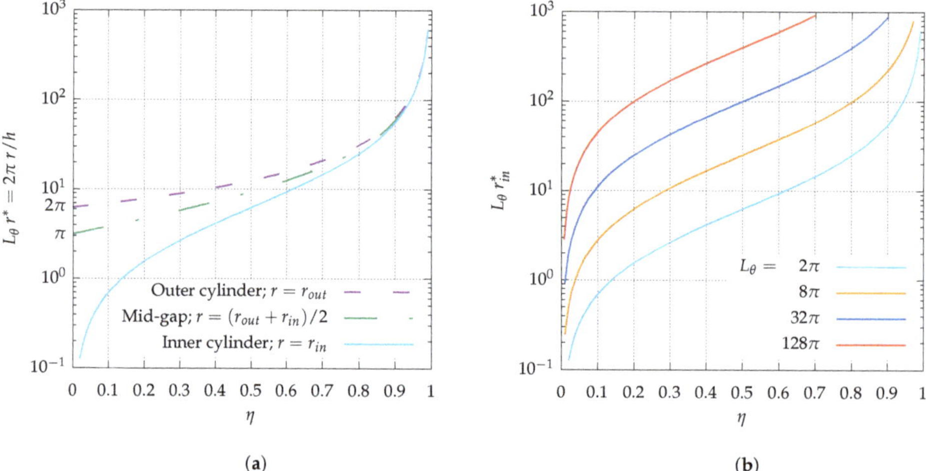

(a) (b)

Figure 2. (a) circumference of original annular pipe system at the outer cylinder, at mid-gap, and at the inner cylinder; (b) circumference at the inner cylinder for $L_\theta \geq 2\pi$.

Table 1. Computational conditions. L_i^* : length of the computational domain in the direction i, non-dimensionalized by the gap width $h = (r_{out} - r_{in})$; L_{out}^* (resp. L_{in}^*) the circumference of the outer (resp. inner) cylinder surfaces, normalized by h; N_i : the number of grids.

$\eta = r_{in}/r_{out}$	0.1			0.15	0.2		0.3
$L_x^* \times L_r^*$	512 × 1				409.6 × 1		
L_θ	2π	32π	128π	128π	112π	2π	96π
$L_{out}^* \ (= L_\theta\, r_{out}^*)$	7.0	111.7	446.8	473.1	439.8	9.0	430.8
$L_{in}^* \ (= L_\theta\, r_{in}^*)$	0.7	11.2	44.7	71.0	88.0	2.7	129.2
$N_x \times N_r$				2048 × 64			
N_θ	32	512	2048	2048	2048	64	2048

3. Statistics at the Onset of Transition

3.1. Global Stability and Coherent Structures Close to Onset

In the present subsection, we recall some key results of Ref. [36] together with some updated predictions. The investigation of the onset of turbulence starts with the determination of the global Reynolds number Re_g, defined as the highest Reynolds number below which no turbulence can survive (at least in the thermodynamic limit, i.e., over infinite observation times in unbounded domains). Since the flow is subcritical, using a given type of initial condition for this task can lead to overestimates of Re_g. The commonly adopted strategy, both in experiments and numerics, is that of an adiabatic descent [39] initiated from a turbulent state at sufficiently high Reynolds number. In the limit where the waiting time between successive diminutions of Re is sufficient long, the value at which turbulence gets extinct is a good approximation of Re_g. Figure 3 displays information about Re_g depending on the radius ratio η. For $L_\theta = 2\pi$ ($n = 1$), Re_g increases monotonically with decreasing η. For larger L_θ, Re_g is always smaller than for the case with $L_\theta = 2\pi$ and the same value of η, with a now decreasing trend for $Re_g(\eta)$ which is even more marked once $\eta \leq 0.3$. The values of L_θ needed to obtain this curve robustly are all listed in Table 1. As for the case of artificially extended aCf at $\eta = 0.1$, the result for $L_\theta = 128\pi$ is plotted in the figure. The parameter range strictly below $\eta = 0.1$ has not been investigated.

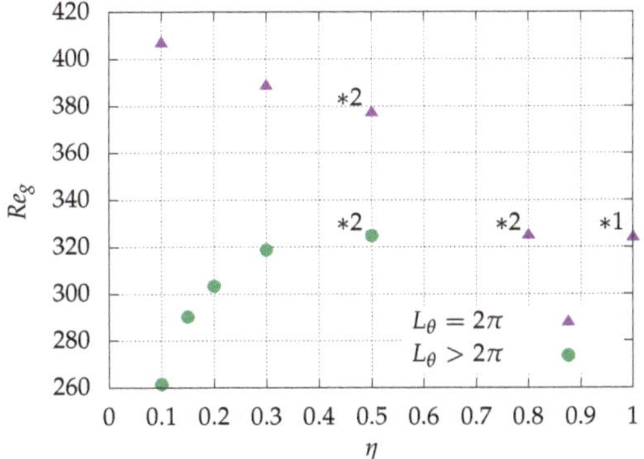

Figure 3. Radius ratio η dependency of the global critical Reynolds number Re_g. The plot includes the pCf limit $\eta \to 1$ from Ref. [21] (labeled "*1"), as well as DNS data from Ref. [36] for $\eta = 0.5$ and 0.8 (labeled "*2"). Triangles: original aCf with $L_\theta = 2\pi$ is plotted using triangles; circles: artificially extended aCf ($L_\theta > 2\pi$).

The fact that artificially extended systems display a lower threshold in Re indicates that some specific spatiotemporal regimes, specific to large L_θ and not allowed for in narrow domains, are able to maintain themselves against relaminarization. As in Ref. [36], we can compare typical snapshots of the velocity fields in the corresponding regime in order to highlight the qualitative differences. Figures 4 and 5 display instantaneous snapshots of the radial velocity at mid-gap (i.e., $r = (r_{in} + r_{out})/2$) at respectively $\eta = 0.3$ and 0.1, one very close to Re_g (left column) and the other slightly above it (right column). Each row corresponds to a different value of the integer n ($n = 1$, 16, 48, and 64), i.e., another value of L_θ. When $n = 1$, the one-dimensional intermittency is reminiscent of the dynamics in cylindrical pipe flow [40]. The differences between different values of η emerge only for higher n. For $\eta = 0.3$, the stripe patterns exhibit an obliqueness typical of most laminar-turbulent patterns [25,26,37,41]. However, it is visually clear that the situation is different for $\eta = 0.1$, with shorter structures and less pronounced obliqueness. It is not immediately clear whether the effective dimensionality of the proliferation process is rather one or two. These issues can be addressed using the determination of critical exponents, as will be done in the next subsection.

3.2. Data Binarization

Velocity fluctuations with respect to the mean flow are defined as $u' = u - \bar{u}$, where $\bar{u}$ is the space-averaged time-dependent velocity averaged along x and θ, as defined in Equation (3). Here, y denotes the (dimensional) distance from the inner cylinder to the outer cylinder as $y = r - r_{in}$, instead of using r.

$$\bar{u}(y,t) = \frac{1}{L_x L_\theta} \int_0^{L_x} \int_0^{L_\theta} u(x,y,\theta,t)\mathrm{d}x\mathrm{d}\theta. \tag{3}$$

The flow is separated into its laminar and turbulent components by postulating a threshold independently of the Reynolds number. The local criterion chosen is $|u'_r/U_w| \geq 0.01$ for turbulence and $|u'_r/U_w| < 0.01$ for laminar flow, with u'_r the radial velocity component, which vanishes everywhere for strictly laminar flow. As in Figures 4 and 5, localized turbulent regions are visualized by contours of u'^*_r in steps of ± 0.01. The turbulent fraction F_t is evaluated at mid-gap ($y = h/2$) by estimating the percentage of grid points for which the turbulent criterion above is fulfilled.

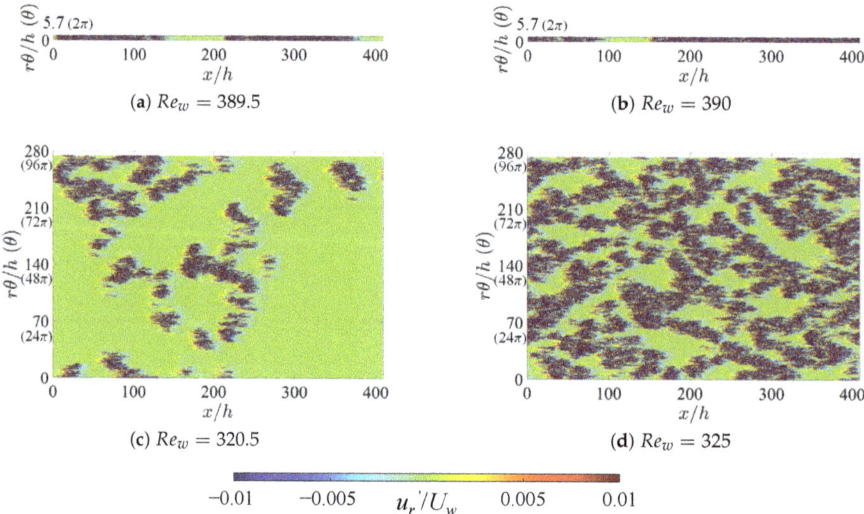

Figure 4. Contours of radial velocity fluctuations u_r^* at mid-gap for $\eta = 0.3$ around $Re_w = Re_g$. Typical snapshots of instantaneous flow fields obtained after reaching each equilibrium state are shown here. The main flow is from left to right. (**a,b**) original aCf with $L_\theta = 2\pi$, and (**c,d**) artificially extended with $L_\theta = 96\pi$.

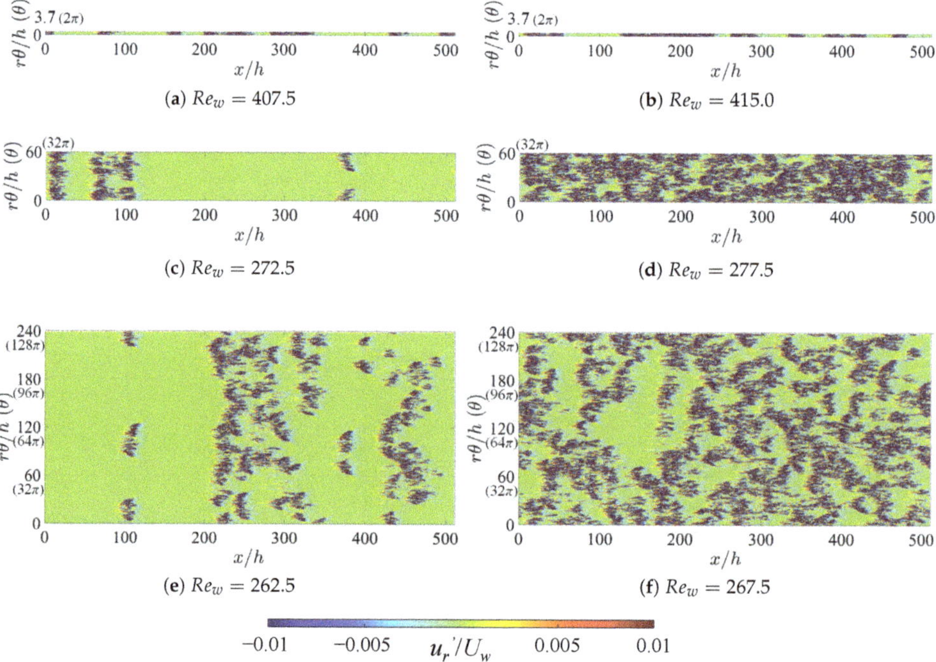

Figure 5. The same as Figure 4, but for $\eta = 0.1$. (**a,b**) $L_\theta = 2\pi$; (**c,d**) $L_\theta = 32\pi$; and (**e,f**) $L_\theta = 128\pi$.

The dynamics of the proliferation process for $\eta = 0.1$ and 0.3 is illustrated in Figure 6 using space-time diagrams and compared one to another in the case $n = 1$. The spatial variable is $x - U_f t$, i.e., the streamwise coordinate in a frame moving with constant velocity U_f, which is close to the average velocity $\overline{u_m}$. The space-time diagram is based on the binarized radial velocity u'_r. The absolute value of the radial velocity evaluated at mid-gap is first averaged azimuthally according to

$$\langle u'_{rms}\rangle_\theta(x,t) = \sqrt{\frac{1}{2\pi}\int_0^{2\pi} u'^2_r(x,h/2,\theta,t)\,\mathrm{d}\theta}.\tag{4}$$

and the binarization criterion is $\langle u'_{rms}\rangle_\theta/U_w \geq 0.01$. The frame velocity U_f for $\eta = 0.3$ is chosen to be same with $\overline{u_m}$, which is estimated in two steps. First, a spatially average velocity is evaluated at every time t

$$u_m(t) = \frac{1}{L_x(r^2_{out} - r^2_{in})L_\theta}\int_0^{L_x}\int_{r_{in}}^{r_{out}}\int_0^{L_\theta} u_x(x,y,\theta,t)\,r\mathrm{d}x\mathrm{d}r\mathrm{d}\theta,\tag{5}$$

then it is time-averaged using a classical moving average technique over a time interval ΔT (with $\Delta T > 10^4 h/U_w$ after reaching equilibrium).

$$\overline{u_m} = \frac{1}{\Delta T}\int_T^{T+\Delta T} u_m(t)\mathrm{d}t.\tag{6}$$

We found that, for $\eta = 0.1$, an optimal value of U_f for the frame to move with puffs was slightly slower than $\overline{u_m}$. For each value of η, three space-time diagrams are displayed, respectively below, close to and above the corresponding critical point $Re_g(\eta)$. The shorter aspect of the coherent structures for $\eta = 0.1$ is striking compared to $\eta = 0.3$. Many more splitting and decay events, qualitatively similar to the pipe flow case [40,42,43], occur for $\eta = 0.1$ despite equal pipe lengths. This suggests that the status of the present simulations for $\eta = 0.1$ is qualitatively much closer to the thermodynamic limit than it is for $\eta = 0.3$. As a by-product, the critical scaling is expected to converge at a lower price than at higher η. Given the cost obstacles induced by the diverging lengthscales/timescales in most critical phenomena, the above conclusion is positive news.

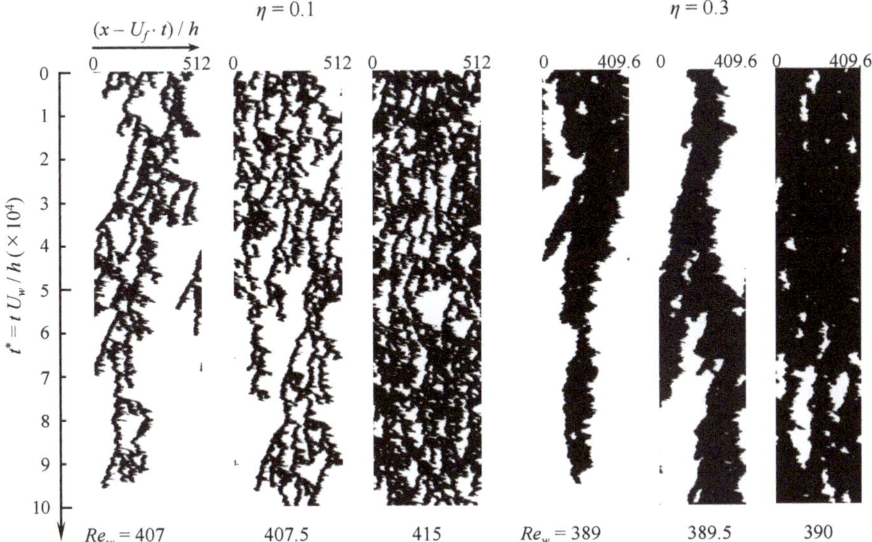

Figure 6. Space-time ($x - U_f t$) diagram of original aCf ($L_\theta = 2\pi$) for $\eta = 0.1$ (three leftmost columns) and 0.3 (three rightmost columns). Black: turbulence according to the criterion $\langle u'_{rms}\rangle_\theta/U_w \geq 0.01$. The values of the frame velocity U_f for $\eta = 0.1$ are $0.288U_w$ at $Re_w = 407$, $0.2875U_w$ at $Re_w = 407.5$, $0.2815U_w$ at $Re_w = 415$, and those for $\eta = 0.3$ are approximately equal to $\overline{u_m}$.

3.3. Intermittency Statistics

The statistical post-processing protocol for STI is vastly similar to that used by other authors: the first step is to monitor the decay in the time of the turbulence fraction $F_t(t)$ when the system is initiated with turbulence everywhere. By dichotomy, this yields a good approximation of Re_g and allows one to define the reduced control parameter $\varepsilon = (Re_w - Re_g)/Re_g$. This decay is expected to be algebraic exactly at onset, i.e., of the form $F_t(t) = O(t^{-\alpha})$. This yields as well the so-called dynamic exponent α. In a second phase, the equilibrium turbulent fraction (i.e., its time average) is monitored as a function of ε. For $\varepsilon > 0$, the data *versus* the expected scaling $F_t(t) = O(\varepsilon^\beta)$ yield the exponent β. Eventually, the mean correlation length $\xi(Re_w)$ (either ξ_x in the streamwise direction or ξ_θ in the azimuthal one) can be estimated at equilibrium by monitoring the cumulative distribution function (CDF) of the laminar gaps $P_{lam}(l_x > L)$, where l_x stands for the length of a laminar trough and L is a dummy variable. A critical exponent $\mu_\perp$ can be evaluated from fits as the algebraic decay exponent of the CDF.

We begin by describing the results from the critical quench experiments of Figure 7 for $\eta = 0.1$ and $n = 64$. The initial condition corresponds to a turbulent velocity field from a long simulation well above Re_g, here taken as $Re_w = 280$. The same initial condition is used for new simulations at another target value of Re_w, in principle such that Re_w is "close" to Re_g. As expected, the flow relaminarizes (attested by the monotonic decrease of $F_t(t)$) for sufficiently low values of Re_w, whereas it stays turbulent for the higher values. In the latter case, the turbulent fraction reaches a non-zero mean value $\overline{F_t}$, which will be reported in the next figure. The set of colored curves in Figure 7a straddle the decay curve corresponding to the critical value $Re_w = Re_g$, whose best approximation in the figure is the red curve associated with $Re_w = 262.5$. For continuous phase transitions, the corresponding decay is expected to be of power-law type, i.e., $F_t = O(t^{-\alpha})$. This fact of $260 < Re_g < 262.5$ yields an approximation of $Re_g = 261.7$, which allows for defining ε as before. The present approach rests on the hypothesis of a critical scaling in the vicinity of the critical point. If that hypothesis is correct then, by rescaling time and turbulent fraction, the curves of Figure 7a should collapse onto two master curves, one for the relaminarization process and the other for the saturation process. This is tested in Figure 7b by plotting $t^\alpha F_t(t)$ as a function of the rescaled time $t|\varepsilon|^{\nu_\parallel}$. As for α and $\nu_\parallel$, the approximate values from $(1+1)$-D DP theory, respectively 0.451 and 1.733, have been used for the rescaling. The match is satisfying, which confirms that a critical range has been identified in this system.

As a by-product of Figure 7, the values of the mean turbulent fraction $\overline{F_t}$, obtained after reaching equilibrium, are reported in Figure 8 as functions of Re_w. Critical theories all predict a scaling $\overline{F_t} = O(\varepsilon^\beta)$ close enough to the critical point. The algebraic scaling revealed in the previous plots of critical quench suggests that, for instance, $Re = 262.5$ belongs to the range where algebraic fits apply for $\eta = 0.1$ and $L_\theta = 128\pi$. Consequently, if, for these parameters, ε is defined using the approximated $Re_g = 261.7$, the dependence of $\overline{F_t}$ *versus* ε is also expected to be algebraic in the same range of values of Re. In that case, the power-law exponents can be classically estimated using log-log plots and compared to those from DP theories. Algebraic fits of $\overline{F_t}$ are shown in Figure 8 both for $\eta = 0.1$ (left) and 0.3 (right). For each case, the main plot of $\overline{F_t}$ *versus* Re_w is displayed in linear coordinates, while the inset displays $\overline{F_t}$ *versus* ε in log-log coordinates, in order to highlight the quality of the estimation of the power-law exponent.

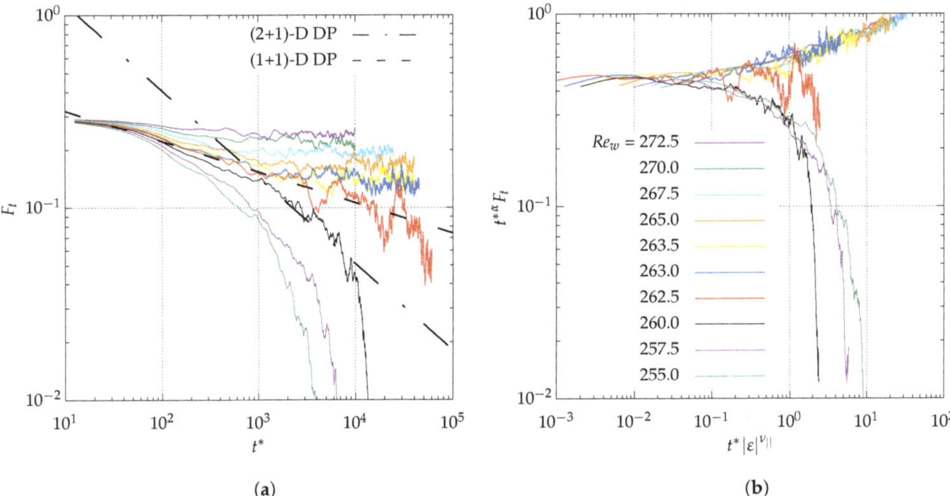

(a)

(b)

Figure 7. Critical quenches from $Re_w = 280$ to each Reynolds number. Temporal variation of turbulent fraction F_t for $\eta = 0.1$ and $L_\theta = 128\pi$ (log-log scale). In (**a**), the black dashed-dotted line and dashed line each indicate possible algebraic fits with the dynamic exponent α from $(2+1)$-D and $(1+1)$-D directed percolation (respectively $\alpha = 0.451$ and 0.159). See also Supplementary; (**b**) test of the 1D scaling hypothesis by plotting $t^\alpha F_t$ vs. $t\varepsilon^{\nu_{||}}$ (log-log scale), with $\nu_{||}$=1.733 for $(1+1)$-D DP.

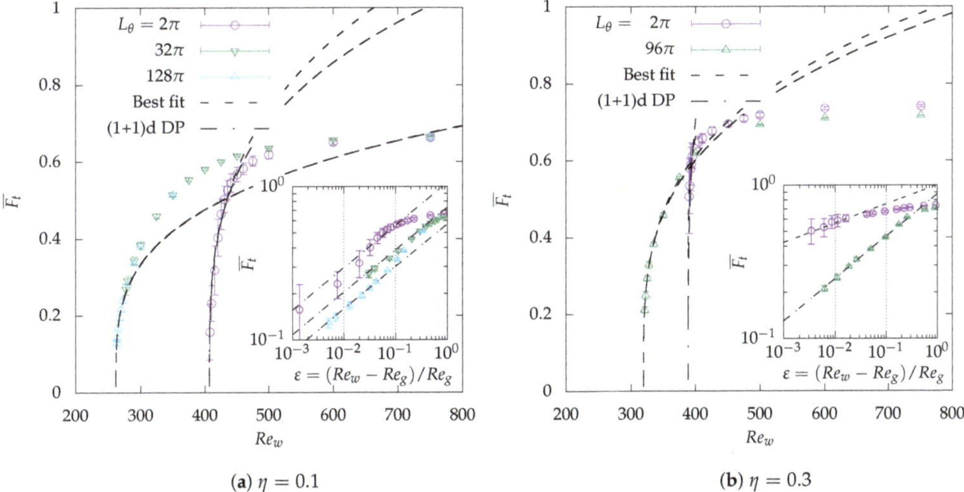

(**a**) $\eta = 0.1$

(**b**) $\eta = 0.3$

Figure 8. Reynolds-number dependence of the time-averaged turbulent fraction $\overline{F_t}$ vs Re_w for the different radius ratios in the original domain ($L_\theta = 2\pi$) and in artificially extended domains ($L_\theta \gg 2\pi$). Vertical error bars: standard deviations of F_t during the averaging period. Dashed/dashed-dotted line: algebraic fits $\overline{F_t} = O(\varepsilon^\beta)$, with exponent β obtained either as best fit β_{fit} or from the $(1+1)$-D DP universality class $\beta_{1D} = 0.276$. In each figure, the insets are plotted in log-log coordinates *versus* ε that is determined with Re_g presented in Table 2.

The details of the fitting procedure for the various parameters used are given in Table 2. It includes the values of the best fitted exponents as well as the approximate fitting range. As could already be deduced graphically from the insets in Figure 8a, for $\eta = 0.1$, the compatibility of the exponent β with the theoretical value of $\beta_{1D} = 0.276$ from $(1+1)$-D DP is good (to the second digit). This is confirmed for both $\eta = 32\pi$ and $\eta = 128\pi$, which suggests that the thermodynamic limit is already reached, at least as far as the determination of the exponent β is concerned. For $L_\theta = 2\pi$ the approximated exponent is 0.31 which constitutes a less accurate, but still consistent approximation of the theoretical exponent. For $L_\theta = 2\pi$, the range of validity of the algebraic fits extends up to $\approx 5\%$, whereas it exceeds 10% for $L_\theta \geq 32\pi$. For $\eta = 0.3$, the situation is slightly different: for a large azimuthal extent $L_\theta = 96\pi$, there is a very good match with the 1D theoretical exponent all the way up to $\varepsilon \approx 20\%$. For $L_\theta = 2\pi$, however, although an algebraic fit seems consistent with the data below $\varepsilon < 1\%$ the measured exponent is closer to 0.12 than to 0.276: none of these values matches any of the percolation theories.

Table 2. Critical Reynolds number Re_g and critical exponent β depending on geometrical parameters η (radius ratio) and L_θ (azimuthal extension). In addition, shown is the fitting range to estimate Re_g and β. †: not measured.

$\eta = r_{in}/r_{out}$	L_θ	Fitting Range	Re_g	β
0.10	2π	407.5–460.0	406.9	0.31(3)
0.10	32π	277.5–300.0	269.0	0.26(2)
0.10	128π	263.0–270.0	261.7	0.28(2)
0.15	128π	— †	290.5	— †
0.20	112π	— †	303.5	— †
0.30	2π	389.0–395.0	388.7	0.12(2)
0.30	96π	320.5–375.0	319.0	0.28(1)
(1 + 1)-D DP model	—	—	—	0.276

The interpretation is delicate. On one hand, algebraic fits seem always verified as soon as ε is small enough; on the other hand, (1 + 1)-D percolation exponents are well approximated only for sufficient azimuthal extension of the order of 100π or more. The original system with $L_\theta = 2\pi$ hence needs to be interpreted as a system with the DP property that experiences a *geometrical frustration* due to lateral confinement. The present data support the hypothesis that the frustration effect is stronger for $\eta = 0.3$ than for $\eta = 0.1$, and thus that the quality of the DP fit will be correspondingly worse. Conversely, the convergence towards the thermodynamic limit seems slower for larger η.

Importantly, we emphasize the main difference between the present conclusion and that by Kunii et al. [36], where the azimuthal extension for $\eta = 0.1$ was limited to $L_\theta = 16\pi$ (to be compared to the present values of 32π and 128π). The fits reported in Figure 16 of that article suggested a fit compatible with the (2 + 1)-D exponent $\beta_{2D} = 0.583$. This former result, in the light of the present computations, is re-interpreted now as a finite-size effect.

A power-law dependence of $\overline{F_t}$ alone does not warrant the proximity to the critical point, as pointed out by Shimizu and Manneville [23] for pPf. Although the critical quenches reported earlier also suggest power-law statistics near the picked up values for Re_g, the classical determination relies on, at least, three independent algebraic exponents. In order to lift this ambiguity, we chose to report in Figure 9 statistics of laminar gap size for different values of Re_w near the suspected critical point. Expecting possible anisotropy when the domain is artificially extended in θ, two kinds of statistics have been monitored, similarly to the study of Chantry et al. [20]. The axial extent of the gaps for $\eta = 0.1$ and $L_\theta = 128\pi$ is shown in Figure 9a in log-log coordinates (and Figure 9c in lin-log representation). The azimuthal extent of the laminar gaps is shown in Figure 9b in log-log coordinates (and Figure 9d in lin-log representation). All four figures support a cross-over from exponential to power-law statistics as Re_w approaches the value of 262.5, with a decay exponent graphically compatible with the decay exponent $\mu_\perp$ of (1 + 1)-D DP. The cross-over appears, however, more clearly in the azimuthal where the match with the theoretical value of $\mu_\perp$ is valid over a full decade. In the streamwise direction, the trend is not clear enough to extract a critical exponent with full accuracy. This confirms, however, that the present statistics are indeed gathered in a relevant neighborhood of the critical point and that, for these parameters, $Re_w = 262.5$ is a decent working approximation of Re_g.

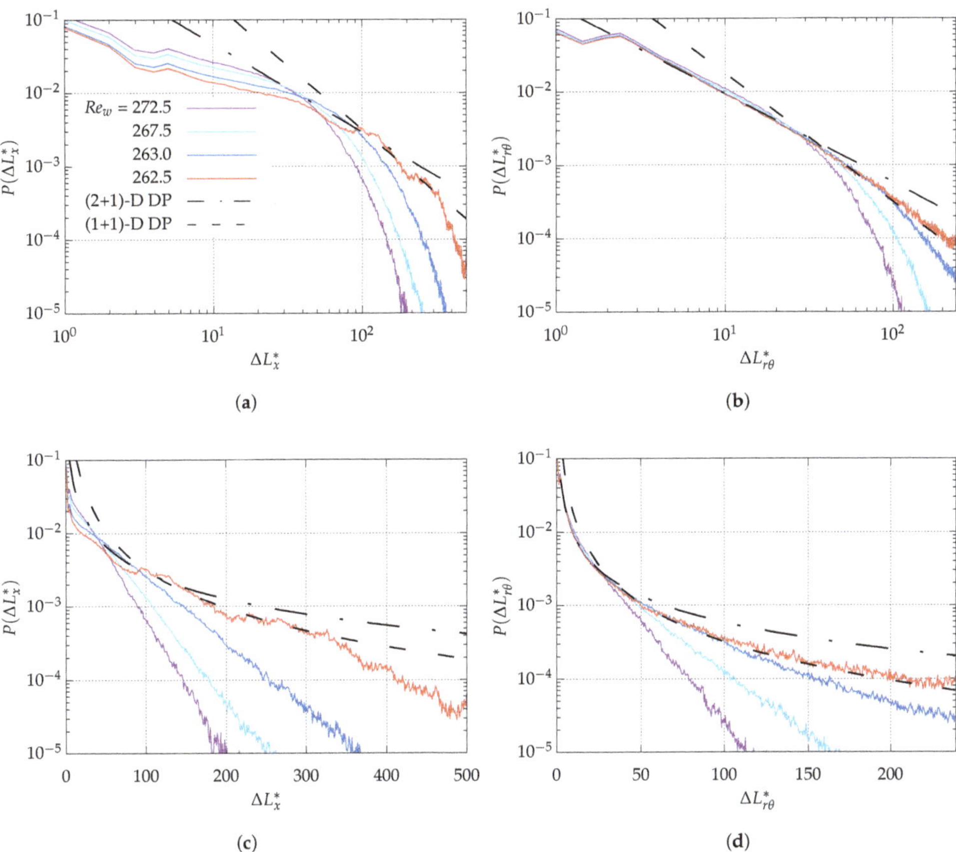

Figure 9. Time-averaged distributions of laminar gap in (**a**,**c**) the streamwise direction and (**b**,**d**) the azimuthal direction, evaluated at mid-gap. (**a**,**b**) log-log plots vs. (**c**,**d**) lin-log plots. $L_\theta = 128\pi$, $\eta = 0.1$ as in Figure 5e,f. In both figures, black dashed-dotted line (- · -) and dashed lien (- - -) indicate theoretical distributions $P(\Delta L^*) \sim \Delta L^{*-\mu_\perp}$ with exponents $\mu_\perp$ from the universality classes of $(2+1)$-D DP and $(1+1)$-D DP, respectively, i.e., $\mu_{\perp 2D} = 1.84$, and $\mu_{\perp 1D} = 1.748$.

3.4. Dynamics of Localized Turbulent Patches

In this last subsection, we address the issue of the influence of azimuthal confinement/extension on the lower transition threshold Re_g, as the estimations from Figure 3 suggest. In Ref. [36], a similar trend was noted (from measurements in shorter and narrower domains). The mechanism suggested in this former work addressed the presence of oblique stripes rather than their influence on the value of Re_g. It was thereafter realized that the phenomenon governing the value of Re_g, and by extension all statistics of the turbulent fraction, is the way different coherent structures interact together dynamically rather than the shape of such individual structures (although that shape certainly influences the interactions). In analogy with pipe [16,42,43] and channel [44,45], the finite turbulent fraction is the result of a dynamical competition between the proliferation of coherent structures and their tendency to decay in number. The transitional range where $\overline{F_t} > 0$ is dominated by the splitting of coherent structures, whereas instantaneous relaminarizations become rare. We hence focus on the dynamics of splitting events in two different computational domains, namely those with $L_\theta = 32\pi$ and 128π. Figure 10 contains zooms on the radial velocity plotted for different values of $y = cst$ surfaces (a different value for each row) and for different times (different columns). In Figure 10, the value of L_θ is fixed to 32π, but the circumference in terms of $r\theta/h$ varies according to r. The global dynamics of these flows can also been scrutinized in the videos made available as Supplementary Materials. The comparison of different values of y is useful to confirm that, for all parameters, the spots remain coherent over the gap even during splitting events.

Lateral splitting events are considered in each of these figures and videos. Because of the different advection velocities in the azimuthal direction, spanwise collisions can occur. During spanwise collisions, usually one of the two spots disappears (see also Ref. [21] for similar observations in pCf). This tends to reduce the turbulent fraction while the other surviving spot is still active. In the presence of a short enough spanwise periodicity, a spot collides with itself rather than with a different neighbor. In such periodic domains, the local relaminarization of one spot is equivalent to the extinction of an infinity of identical spots. Hence, the turbulent fraction decreases more than in large domains where individual spots behave more like independent entities. We thus expect more turbulence to proliferate more for larger L_θ. As a consequence, the critical Reynolds number Re_g, for which the rate of proliferation balances the probability to relaminarize locally, is lowered when L_θ is increased, consistently with the thresholds reported in Figure 3 and Table 2. This effect is more marked at lower η.

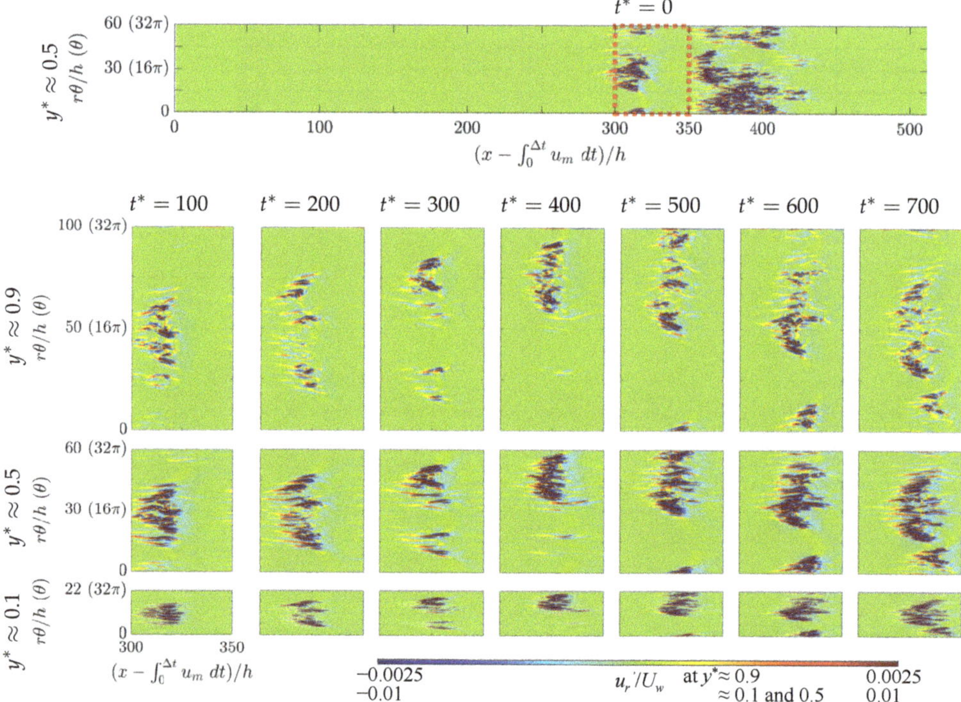

Figure 10. Snapshots of splitting and self-colliding events in aCf for $Re_w = 262.5$ with $L_\theta = 32\pi$ and $\eta = 0.1$. Radial velocity in a frame moving with bulk velocity $\overline{u_m}$. Here, $t = 0$ is an arbitrary time instant after reaching equilibrium. Top row, $y^* = y/h \approx 0.9$; center row, $y^* \approx 0.5$; lower row, $y^* \approx 0.1$.

4. Conclusions

The present DNS study deals with the statistical aspects of the intermittent transitional regime of aCf, with an emphasis on the low values of the radius ratio η close to 0.1. It is an extension of the simulations reported recently by Ref. [36]. The paper compares two computational situations, respectively the case of a realistic geometry and the one where the azimuthal extent is larger than the original value of 2π. In Ref. [36], this parametric trick was introduced in an explicit attempt to decouple the effects of wall curvature effects from the effects of azimuthal confinement induced by the geometry. The main conclusion for large η was that the reported absence of oblique laminar-turbulent patterns was due to azimuthal confinement, since they could re-appear for $L_\theta > 2\pi$. In the present article, the same trick is introduced for $\eta = 0.1$; however, larger values of L_θ have been tried up to 128π (i.e., 64 times the original value). The oblique patterns do not reappear and a new percolating regime takes place with shorter spatial correlations. The statistical analysis of the STI is convergent as L_θ grows, and is consistent with (1 + 1)-D DP. This updates the results of Ref. [36] where (2 + 1)-D DP was suggested from fits with $L_\theta = 16\pi$. The present results suggest now that the $L_\theta = 16\pi$ algebraic statistics was still far from the true thermodynamic limit, while $L_\theta = 128\pi$ seems to yield more decent results.

To our knowledge, there has been only poor evidence for the cross-over from exponential to algebraic scaling in the shear flow literature, as far as well-resolved simulations of the Navier–Stokes equations are concerned [2]. An exception is the work by Shi et al. [46] in a tilted periodic domain of pCf, which again is not a fully realistic numerical domain. It is interesting to speculate how much the present results can teach us something about a fully realistic system such as cylindrical pipe flow. Naive homotopy of the turbulent regimes is ruled out because of the singularity near the centerline. Instead, we can compare the rate at which these two effectively one-dimensional percolating systems tend towards their own thermodynamic limit. This issue was raised recently in the experimental study by Mukund and Hof [19]. There, despite pipes as long as 3000 diameters, no critical regime (with power-law statistics) was identified, only classical STI as reported in Refs. [47,48]. This issue was attributed to the narrowness of the critical range, and to a clustering property of puffs which delays the convergence to the thermodynamic limit. Here, in aCf with $\eta = 0.1$, the situation is different but depends on this artificial parameter L_θ. To our surprise, power-law statistics of the turbulent fraction as well as of the laminar gap distributions do appear in our simulations as Re_w is reduced. All cases shown in Figure 8 suggest a cross-over from turbulent to power-law behavior as Re_w is within $\approx 1\%$ of the critical point. For $L_\theta = 2\pi$ or around, the turbulent fraction curve still suggests an unconverged power-law. For $L_\theta = 32\pi$ or 128π, power-law statistics of $\overline{F_t}$ are fully consistent with one-dimensional DP appear. This occurs despite a value of L_x of only $512h$, i.e., much less than the pipe flow case and even less if one counts in outer pipe diameters. A possible interpretation is that azimuthal extension, by modifying the interaction with neighboring spots, can suppress the tendency to form clusters, and hence converge faster towards the thermodynamic limit. This is consistent with lower transition thresholds in Re_w as well. One is left wondering if a similar approach to cylindrical pipe flow could also easily yield the percolation exponents from simulation measurements.

We conclude by noting that artificially modifying both the shape of turbulent patches and their interaction, as done here using azimuthal extension, is more than an esoteric thought experiment or an exotic parameter study. It is used here as a legitimate strategy in order to untangle complex phenomena, e.g., to decouple confinement from curvature effects. As demonstrated in our recent work using a simple modeling approach [49,50], wall roughness can have similar effects on transitional flows and change the way turbulence invades laminar flows. We expect similar strategies of artificial domain extension to be relevant to such cases too.

Supplementary Materials: Video S1: Time evolution of turbulent fraction $F_t(t)$ and of fluctuating velocity fields visualized at mid-gap, for $\eta = 0.1$ with an artificially extended azimuthal domain size of $L_\theta = 128\pi$. On the right column, contours show x-θ distributions of the radial velocity fluctuation u_r' normalized by the inner-cylinder velocity U_w. Top (orange box and curve in the graph) : above the global critical Reynolds number Re_g. Middle (red) : near Re_g. Bottom (black) : below Re_g. A supporting video article is available at https://doi.org/10.5281/zenodo.3985963.

Author Contributions: Conceptualization and methodology, T.T. and Y.D.; simulations and acquisition, K.T.; post-processing, K.T. and T.T.; writing Y.D. and T.T.; funding acquisition, T.T. All authors have read and agreed to the published version of the manuscript.

Funding: This work was funded by Grant-in-Aid for JSPS (Japan Society for the Promotion of Science) Fellowship JP16H06066 and JP19H02071.

Acknowledgments: Numerical simulations were performed on SX-ACE supercomputers at the Cybermedia Center of Osaka University and the Cyberscience Center of Tohoku University.

Conflicts of Interest: The authors declare no conflict of interest. The funders had no role in the design of the study; in the collection, analyses, or interpretation of data; in the writing of the manuscript, or in the decision to publish the results.

Abbreviations

The following abbreviations are used in this manuscript:

aCf	annular Couette flow
CDF	cumulative distribution function
DNS	direct numerical simulation
DP	direct percolation
pCf	plane Couette flow
pPf	plane Poiseuille flow
rms	root-mean-square value
STI	spatiotemporal intermittency

References

1. Pomeau, Y. Front motion, metastability and subcritical bifurcations in hydrodynamics. *Phys. D Nonlinear Phenom.* **1986**, *23*, 3–11. [CrossRef]
2. Eckhardt, B. Transition to turbulence in shear flows. *Phys. A Stat. Mech. Appl.* **2018**, *504*, 121–129. [CrossRef]
3. Grassberger, P. On phase transitions in Schlögl's second model. *Zeitschrift für Physik B Condensed Matter* **1982**, *47*, 365–374. [CrossRef]
4. Janssen, H.K. On the nonequilibrium phase transition in reaction-diffusion systems with an absorbing stationary state. *Zeitschrift für Physik B Condensed Matter* **1981**, *42*, 151–154. [CrossRef]
5. Daviaud, F.; Dubois, M.; Bergé, P. Spatio-temporal intermittency in quasi one-dimensional Rayleigh-Bénard convection. *EPL (Europhys. Lett.)* **1989**, *9*, 441. [CrossRef]
6. Chaté, H.; Manneville, P. Spatio-temporal intermittency in coupled map lattices. *Phys. D Nonlinear Phenom.* **1988**, *32*, 409–422. [CrossRef]
7. Bohr, T.; van Hecke, M.; Mikkelsen, R.; Ipsen, M. Breakdown of universality in transitions to spatiotemporal chaos. *Phys. Rev. Lett.* **2001**, *86*, 5482. [CrossRef]
8. Duguet, Y.; Maistrenko, Y.L. Loss of coherence among coupled oscillators: From defect states to phase turbulence. *Chaos Interdiscip. J. Nonlinear Sci.* **2019**, *29*, 121103. [CrossRef]
9. Daviaud, F.; Bonetti, M.; Dubois, M. Transition to turbulence via spatiotemporal intermittency in one-dimensional Rayleigh-Bénard convection. *Phys. Rev. A* **1990**, *42*, 3388. [CrossRef]
10. Kreilos, T.; Khapko, T.; Schlatter, P.; Duguet, Y.; Henningson, D.S.; Eckhardt, B. Bypass transition and spot nucleation in boundary layers. *Phys. Rev. Fluids* **2016**, *1*, 043602. [CrossRef]
11. Takeuchi, K.A.; Kuroda, M.; Chaté, H.; Sano, M. Directed percolation criticality in turbulent liquid crystals. *Phys. Rev. Lett.* **2007**, *99*, 234503. [CrossRef] [PubMed]
12. Lemoult, G.; Shi, L.; Avila, K.; Jalikop, S.V.; Avila, M.; Hof, B. Directed percolation phase transition to sustained turbulence in Couette flow. *Nat. Phys.* **2016**, *12*, 254–258. [CrossRef]
13. Sano, M.; Tamai, K. A universal transition to turbulence in channel flow. *Nat. Phys.* **2016**, *12*, 249–253. [CrossRef]
14. Avila, K. Shear Flow Experiments: Characterizing the Onset of Turbulence as a Phase Transition. Ph.D. Thesis, Niedersächsische Staats-und Universitätsbibliothek Göttingen, Göttingen, Germany, 2013.
15. Hiruta, Y.; Toh, S. Subcritical laminar–turbulent transition as nonequilibrium phase transition in two-dimensional Kolmogorov flow. *J. Phys. Soc. Jpn.* **2020**, *89*, 044402. [CrossRef]

16. Barkley, D. Theoretical perspective on the route to turbulence in a pipe. *J. Fluid Mech.* **2016**, *803*, P1. [CrossRef]

17. Linga, G. Fluid Flows with Complex Interfaces: Modelling and Simulation from Pore to Pipe. Ph.D. Thesis, The Niels Bohr Institute, Faculty of Science, University of Copenhagen, København, Danmark, 2018.

18. Shih, H.-Y.; Lemoult, G.; Vasudevan, M.; Lopez, J.; Shih, H.Y.; Linga, G.; Hof, B.; Mathiesen, J.; Goldenfeld, N.; Hof, B. Statistical model and universality class for interacting puffs in transitional turbulence. *Bull. Am. Phys. Soc.* **2020**, *65*, W25.00007.

19. Mukund, V.; Hof, B. The critical point of the transition to turbulence in pipe flow. *J. Fluid Mech.* **2018**, *839*, 76–94. [CrossRef]

20. Chantry, M.; Tuckerman, L.S.; Barkley, D. Universal continuous transition to turbulence in a planar shear flow. *J. Fluid Mech.* **2017**, *824*, R1. [CrossRef]

21. Duguet, Y.; Schlatter, P.; Henningson, D.S. Formation of turbulent patterns near the onset of transition in plane Couette flow. *J. Fluid Mech.* **2010**, *650*, 119–129. [CrossRef]

22. Tao, J.; Eckhardt, B.; Xiong, X. Extended localized structures and the onset of turbulence in channel flow. *Phys. Rev. Fluids* **2018**, *3*, 011902. [CrossRef]

23. Shimizu, M.; Manneville, P. Bifurcations to turbulence in transitional channel flow. *Phys. Rev. Fluids* **2019**, *4*, 113903. [CrossRef]

24. Kashyap, P.V.; Duguet, Y.; Dauchot, O. Flow statistics in the patterning regime of plane channel flow. *Entropy* **2020**, submitted.

25. Prigent, A.; Grégoire, G.; Chaté, H.; Dauchot, O.; van Saarloos, W. Large-scale finite-wavelength modulation within turbulent shear flows. *Phys. Rev. Lett.* **2002**, *89*, 014501. [CrossRef]

26. Tsukahara, T.; Seki, Y.; Kawamura, H.; Tochio, D. DNS of turbulent channel flow at very low Reynolds numbers. In *TSFP Digital Library Online*; Begel House Inc.: Danbury, CT, USA, 2005; pp. 935–940,

27. Brethouwer, G.; Duguet, Y.; Schlatter, P. Turbulent-laminar coexistence in wall flows with Coriolis, buoyancy or Lorentz forces. *J. Fluid Mech.* **2012**, *704*, 137–172. [CrossRef]

28. Manneville, P. Transition to turbulence in wall-bounded flows: Where do we stand? *Mech. Eng. Rev.* **2016**, *3*. [CrossRef]

29. Manneville, P. Laminar-turbulent patterning in transitional flows. *Entropy* **2017**, *19*, 316. [CrossRef]

30. Tuckerman, L.S.; Chantry, M.; Barkley, D. Patterns in Wall-Bounded Shear Flows. *Annu. Rev. Fluid Mech.* **2020**, *52*, 343–367. [CrossRef]

31. Chung, S.Y.; Rhee, G.H.; Sung, H.J. Direct numerical simulation of turbulent concentric annular pipe flow: Part 1: Flow field. *Int. J. Heat Fluid Flow* **2002**, *23*, 426–440. [CrossRef]

32. Ishida, T.; Duguet, Y.; Tsukahara, T. Transitional structures in annular Poiseuille flow depending on radius ratio. *J. Fluid Mech.* **2016**, *794*, R2. [CrossRef]

33. Ishida, T.; Duguet, Y.; Tsukahara, T. Turbulent bifurcations in intermittent shear flows: From puffs to oblique stripes. *Phys. Rev. Fluids* **2017**, *2*, 073902. [CrossRef]

34. Fukuda, T.; Tsukahara, T. Heat transfer of transitional regime with helical turbulence in annular flow. *Int. J. Heat Fluid Flow* **2020**, *82*, 108555. [CrossRef]

35. Frei, C.; Lüscher, P.; Wintermantel, E. Thread-annular flow in vertical pipes. *J. Fluid Mech.* **2000**, *410*, 185–210. [CrossRef]

36. Kunii, K.; Ishida, T.; Duguet, Y.; Tsukahara, T. Laminar-turbulent coexistence in annular Couette flow. *J. Fluid Mech.* **2019**, *879*, 579–603. [CrossRef]

37. Duguet, Y.; Schlatter, P. Oblique laminar-turbulent interfaces in plane shear flows. *Phys. Rev. Lett.* **2013**, *110*, 034502. [CrossRef]

38. Abe, H.; Kawamura, H.; Matsuo, Y. Direct numerical simulation of a fully developed turbulent channel flow with respect to the Reynolds number dependence. *J. Fluids Eng.* **2001**, *123*, 382–393. [CrossRef]

39. Prigent, A.; Dauchot, O. Transition to versus from turbulence in subcritical Couette flows. In *IUTAM Symposium on Laminar-Turbulent Transition and Finite Amplitude Solutions*; Springer: Dordrecht, The Netherlands, 2005; pp. 195–219.

40. Moxey, D.; Barkley, D. Distinct large-scale turbulent-laminar states in transitional pipe flow. *Proc. Natl. Acad. Sci. USA* **2010**, *107*, 8091–8096. [CrossRef]

41. Fukudome, K.; Tsukahara, T.; Ogami, Y. Heat and momentum transfer of turbulent stripe in transitional-regime plane Couette flow. *Int. J. Adv. Eng. Sci. Appl. Math.* **2018**, *10*, 291–298. [CrossRef]

42. Avila, K.; Moxey, D.; de Lozar, A.; Avila, M.; Barkley, D.; Hof, B. The onset of turbulence in pipe flow. *Science* **2011**, *333*, 192–196. [CrossRef]

43. Shimizu, M.; Manneville, P.; Duguet, Y.; Kawahara, G. Splitting of a turbulent puff in pipe flow. *Fluid Dyn. Res.* **2014**, *46*, 061403. [CrossRef]

44. Shimizu, M.; Kanazawa, T.; Kawahara, G. Exponential growth of lifetime of localized turbulence with its extent in channel flow. *Fluid Dyn. Res.* **2019**, *51*, 011404. [CrossRef]

45. Paranjape, C.S.; Duguet, Y.; Hof, B. Oblique stripe solutions of channel flow. *J. Fluid Mech.* **2020**, *897*. [CrossRef]

46. Shi, L.; Avila, M.; Hof, B. Scale invariance at the onset of turbulence in Couette flow. *Phys. Rev. Lett.* **2013**, *110*, 204502. [CrossRef] [PubMed]

47. Sreenivasan, K.; Ramshankar, R. Transition intermittency in open flows, and intermittency routes to chaos. *Phys. D Nonlinear Phenom.* **1986**, *23*, 246–258. [CrossRef]

48. Avila, M.; Hof, B. Nature of laminar-turbulence intermittency in shear flows. *Phys. Rev. E* **2013**, *87*, 063012. [CrossRef] [PubMed]

49. Ishida, T.; Brethouwer, G.; Duguet, Y.; Tsukahara, T. Laminar-turbulent patterns with rough walls. *Phys. Rev. Fluids* **2017**, *2*, 073901. [CrossRef]

50. Tsukahara, T.; Tomioka, T.; Ishida, T.; Duguet, Y.; Brethouwer, G. Transverse turbulent bands in rough plane Couette flow. *J. Fluid Sci. Technol.* **2018**, *13*, JFST0019. [CrossRef]

Article

Laminar–Turbulent Intermittency in Annular Couette–Poiseuille Flow: Whether a Puff Splits or Not

Hirotaka Morimatsu and Takahiro Tsukahara *

Department of Mechanical Engineering, Tokyo University of Science, Chiba 278-8510, Japan;
7515117@alumni.tus.ac.jp
* Correspondence: tsuka@rs.tus.ac.jp; Tel.: +81-4-7122-9352

Received: 31 October 2020; Accepted: 27 November 2020; Published: 30 November 2020

Abstract: Direct numerical simulations were carried out with an emphasis on the intermittency and localized turbulence structure occurring within the subcritical transitional regime of a concentric annular Couette–Poiseuille flow. In the annular system, the ratio of the inner to outer cylinder radius is an important geometrical parameter affecting the large-scale nature of the intermittency. We chose a low radius ratio of 0.1 and imposed a constant pressure gradient providing practically zero shear on the inner cylinder such that the base flow was approximated to that of a circular pipe flow. Localized turbulent puffs, that is, axial uni-directional intermittencies similar to those observed in the transitional circular pipe flow, were observed in the annular Couette–Poiseuille flow. Puff splitting events were clearly observed rather far from the global critical Reynolds number, near which given puffs survived without a splitting event throughout the observation period, which was as long as 10^4 outer time units. The characterization as a directed-percolation universal class was also discussed.

Keywords: subcritical transition; spatiotemporal intermittency; direct numerical simulation

1. Introduction

The discontinuous reverse transition of wall-bounded turbulence into a laminar flow is a fundamental problem that has been studied for many years, while the laminar-to-turbulent transition is rather smooth, or its critical point is often well predicted by linear stability theory. Subcritical flows in the reverse transition are known to feature two regimes in competition, namely, laminar and turbulent, in which there occurs large-scale intermittency that coexists spatially with a laminar flow. The large-scale nature of localized turbulence often forms a regular pattern once established. The intermittent structure or formation pattern of localized turbulence varies depending on the flow system, and a number of studies have been conducted on canonical flows, such as a circular pipe flow (CPF) and planar flows. In the CPF, a so-called equilibrium turbulent puff, or simply a "puff," is localized in the streamwise direction, resulting in uni-directional intermittency. The puff turbulence is sustained within a Reynolds-number range based on the bulk velocity U and the pipe diameter D of $Re_D = 2000$–2700 [1]: Although there are some differences depending on the experimental conditions, such as the disturbance introduction method and the pipe length [2–4], studies have indicated that the puff's nature is deeply related to the determination of the lower-limit Reynolds number (the global critical Reynolds number, Re_g), above which turbulent motions can survive globally. Streamwise-localized solutions underlying the puff have been found, and Hopf bifurcations to new branches including unstable periodic orbits are expected to cover the turbulent attractor [5]. It is also known that puffs can split (or proliferate) more frequently than their decay and have a finite lifetime even at $Re > Re_g$ [6–9]. Avila et al. [10] identified $Re_g = 2040 \pm 10$ for the CPF by monitoring both the puff-splitting time and the decay time. Recent attempts have been made to elucidate the puff-driving mechanism [11] and one-dimensional modeling [12], and an understanding of the puff's nature is progressing compared to other intermittent structures.

In planar flows, intermittent structures with bi-directionality were discovered during the last two decades (excluding the spiral turbulence in a Taylor–Couette flow [13]), which are called oblique turbulent stripes/bands with a certain inclination with respect to the streamwise direction, and were found in a plane Couette flow (pCf) [14–18] and a plane Poiseuille flow (pPf) [19–21]. A wall turbulence that is stably stratified by body forces, such as the Coriolis force and buoyancy, also undergoes the stripe regime [22,23]. The stripe pattern has attracted recent interest, and some studies have found families of relevant localized solutions [24,25]. As the Reynolds number approaches the relevant Re_g, the stripe pattern becomes isolated oblique bands, which fall into a non-equilibrium state accompanied by band growth, a break (not the same as the splitting of the puff), and a mutual collision [26–28]. Because the laminar gap surrounding the isolated bands is large at near criticality, a large-scale channel setup or computational domain is required for precise tracking of the process toward a fully laminar state and for estimating Re_g. For this reason, research is still ongoing, such as elucidating the mechanism of an isolated oblique band [29] and the statistical characteristics [30]. For details, also see recent review papers [31–33].

The two kinds of intermittent structures mentioned above were observed in different canonical flows, that is, the CPF and the planar flow, and the direct relationship between the turbulent puff and stripe is unknown. Our research group therefore focused on an annular flow between concentric cylinders. Depending on the radius ratio $\eta \equiv r_{\mathrm{in}}/r_{\mathrm{out}}$ (where r_{in} and r_{out} are the inner and outer cylinder radii, respectively), the curvature and the circumferential length (relative to the gap width) should change and may affect the large-scale nature of the intermittency. With $\eta \approx 1$ or 0, the flow system can be regarded as a planar flow or a CPF system, respectively. Ishida et al. [34–36] conducted direct numerical simulations (DNSs) to study the subcritical transition process of the annular Poiseuille flow (aPf) using η as a parameter in addition to the Reynolds number. The authors observed both the turbulent puff and the stripe according to η, i.e., a helical turbulence (i.e., a turbulent stripe in the annular flow) at $\eta \geq 0.5$, puff turbulence similar to the transitional CPF at $\eta < 0.2$, and an intermediate state at $0.2 \leq \eta \leq 0.4$. At $\eta = 0.1$, the observed puff split and decayed over time. A similar tendency was also uncovered in an annular Couette flow (aCf); it was reported that puffs occur at $\eta = 0.1$, and they split and attenuate over time [37,38]. The authors found a speckled irregular intermittent structure that differs from turbulent stripes and puffs, which was shown to have characteristics of the (1+1)-dimensional directed-percolation (DP) universal class. Recent studies have focused on the relationship between the subcritical transition phenomenon and DP [39–42].

In this study, by employing an annular system as a platform, we aim to unify uni- and bi-directional intermittent structures observed in the CPF and planar flows, respectively. The key to achieving this aim is bridging between the two different systems in terms of the base flow. The base flows of the studied aPf and aCf are qualitatively different from that of the CPF. This mismatch motivated us to simulate the annular Couette–Poiseuille flow (aCPf) at a low η, which should be more similar to the CPF. However, the presence of the inner cylinder may affect both the onset and splitting of the puff. The main purpose of this study is to answer whether puff splitting would occur in a low η. Moreover, Re_g and the Reynolds-number dependence of puff splitting are investigated, and the DP feature is discussed.

Previous DNS studies on Couette–Poiseuille flow mainly focused on the planar turbulence. Kuroda et al. [43] compared the mean velocity profiles and various turbulence statistics for three patterns of imposed mean pressure gradients in the flow path. In particular, among the three patterns, the authors analyzed the shear stress near the moving wall surface in a turbulent field such that it approaches zero. A similar attempt was also conducted by other researchers [44–46]. As an experimental study, Nakabayashi et al. [47] also measured the turbulence statistics of a plane Couette–Poiseuille flow at high Reynolds numbers, and classified the flow field into a Couette- or Poiseuille-type depending on the base flow. In addition, Klotz et al. [48,49] eliminated the net flow in a plane Couette–Poiseuille flow (pCPf), allowing localized turbulence to be tracked for long periods of time while stationary in the observation window. A recent quench experiment on the decay of Couette–Poiseuille turbulence is likely to approach the crossover of the decay rate, that is, the quantitative identification of Re_g [50]. However, to the best of our knowledge, except for in limited studies [51,52], there is no DNS available for the subcritical transition process of the aCPf. The present DNS is the first to explore laminar–turbulent intermittency in a low-η aCPf.

The remainder of this paper is organized as follows. Section 2 presents the flow configuration, dimensionless parameters, and equations used in our simulations. In Section 3, which is dedicated to the preliminary results, we validated the current code and illustrated the parameter dependence of the base flow in terms of the mean friction on the inner cylinder. Section 4 begins with a puff characterization of the observed turbulent patches. Space-time diagrams of a turbulent quantity revealing the puff splitting and decay are then presented. All results are summarized and discussed in Section 5.

2. Problem Setup and Methods

The problem under consideration is the turbulent annular flow of an incompressible Newtonian fluid, for which the governing equations are the equation of continuity and the Navier–Stokes equation, as described in the classical cylindrical coordinate system of (x, r, θ):

$$\nabla^* \cdot \mathbf{u}^* = 0, \tag{1}$$

$$\frac{\partial \mathbf{u}^*}{\partial t^*} + (\mathbf{u}^* \cdot \nabla^*) \, \mathbf{u}^* = -\nabla^* p^* + \frac{1}{Re_w} \Delta^* \mathbf{u}^* - \frac{dP^*}{dx^*} \mathbf{e_x}. \tag{2}$$

Here, the velocity vector is represented by $\mathbf{u}$, or (u_x, u_r, u_θ), which are the respective components in (x, r, θ); p is the pressure, and t is the time. These quantities are non-dimensionalized and are marked by a * superscript: $\mathbf{u}^* = \mathbf{u}/u_w$, $p^* = p/\rho u_w^2$ (ρ, density), $t^* = t u_w/h$, $x^* = x/h$, and $r^* = r/h$, where u_w and h are the inner-cylinder axial velocity and the gap between the two cylinder radii, respectively, as illustrated in Figure 1. The Reynolds number Re_w is therefore based on u_w, h, ρ, and the fluid viscosity μ, whereas another definition using one-quarter of Re_w is more conventional for studies on the pCf [15,16,22]. In only the axial-direction component of Equation (2), a constant pressure gradient in x is added as an external force term, $-dP/dx$, with the axial unit vector $\mathbf{e_x}$. In addition to the imposed pressure gradient, the flow is driven by an axial translation of the inner rod with a constant velocity of $u_w > 0$. The x-axis corresponds to the central axis common to both cylinders, and the radius ratio of η is an important geometrical parameter and is set to 0.1 for the main analysis in this study. Periodic boundary conditions are imposed in both the x and θ directions, and no-slip boundary conditions are enforced at the wall surfaces of the cylinders. In the following sections, the imposed pressure gradient is re-defined as the pressure gradient function $F(p)$, which is normalized as

$$F(p) \equiv -\frac{dP^*}{dx^*} \cdot Re_w = \frac{-dP/d(x/h)}{\mu u_w/h} \tag{3}$$

and can be interpreted as the ratio of the imposed pressure gradient (i.e., the Poiseuille-like driving force) against the wall-bounded viscous shear stress (the Couette-like driving force).

As introduced above, there are two control parameters for the flow under consideration, i.e., Re_w and $F(p)$. Poiseuille-like flows are realized for a large $F(p)$, whereas Couette-like flows are obtained for a small $F(p)$, and a specifically pure Couette flow corresponds to $F(p) = 0$. As indicated in [44], the ratio of the shear stress at the two walls, which can be defined by $\gamma = \tau_{in}/\tau_{out}$ in an aCPf, is another candidate of the control parameter relevant to a Couette–Poiseuille flow. Flows with $\gamma \approx 0$, or a shear-less inner cylinder wall, are of special interest because they exhibit nearly zero mean shear at the moving rod, and can thus be a model for an understanding of the puff dynamics in a pure pipe. Under such conditions, the inner cylinder practically affects the core flow only as an impermeable thin rod, and the coherent turbulent structures and turbulent production that dominantly occur near the static outer cylinder wall mimic those found in a canonical system of the CPF. Although the system chosen here is closer to a CPF than to an aPf or an aCf, it should be noted that the different boundary conditions regarding the inner rod preclude a mathematical homotopy continuation with the CPF. Except for a fully laminar flow state, $F(p)$ providing $\gamma = 0$ is not explicitly obvious, and thus, a parametric survey must be conducted for each given Re_w. In this study, we conducted a preliminary survey of the $F(p)$ dependence of τ_{in} for several Re_w values using a DNS with a medium-scale computational domain, as reported in Section 3. Based on these results, we selected $F(p)$, which will provide $\tau_{\mathrm{in}} \approx 0$ ($\gamma \approx 0$) at each tested Re_w, and accordingly applied the main DNS using a large-scale domain to reduce the spatial limitation on the laminar–turbulent coexistence.

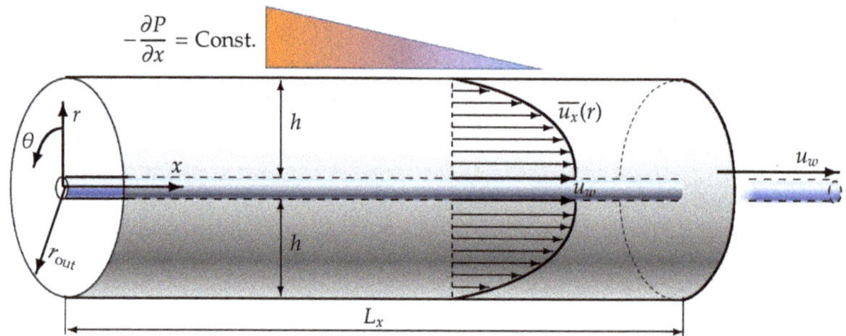

Figure 1. Couette–Poiseuille flow in an annular channel between two concentric cylinders with a radius ratio of $\eta \equiv r_{\mathrm{in}}/r_{\mathrm{out}} = 0.1$, driven by a constant pressure gradient and an inner-cylinder axial movement. In this study, the pressure gradient is adjusted such that the mean velocity gradient on the inner-cylinder surface is approximately zero; that is, $\tau_{\mathrm{in}} = \mu \left[\partial \overline{u_x}(r)/\partial r\right]_{r=r_{\mathrm{in}}} \approx 0$.

The numerical conditions of the preliminary and main simulations for $\eta = 0.1$ are summarized in Tables 1 and 2, respectively. Long domain sizes of $51.2h$ and $409.6h$ were employed in the axial direction to capture a single turbulent puff and expected multiple puffs, whereas the radial and azimuthal domain lengths were of geometrically determined values of h and 2π, respectively. The grid resolutions have been confirmed to be fine such that fine-scale eddies in turbulent patches are well resolved, at least for the particularly interesting transitional regime of $Re_w \leq 1600$.

Equations (1) and (2) were discretized using a staggered central finite-difference method, where the fourth-order central difference scheme was used in both x and θ, along with the second-order scheme in r on a non-uniform radial grid. A time advancement was performed using a fractional-step second-order Adams–Bashforth scheme in combination with a Crank–Nicolson scheme for the radial viscous term. The Courant–Friedrichs–Lewy (CFL) condition was continuously monitored in all directions, and accordingly, the time-step Δt constraint for the nonlinear terms was enforced to ensure stability. The details of the numerical method were reported in the literature [34,53]. The code validation carried out is discussed in the next section.

3. Preliminary Simulations

The reliability of the current simulation code may be demonstrated through a comparison with the existing pCPf DNS database at a comparable Reynolds number. Kasagi and coworkers [43,54] applied a DNS of several pCPfs and released their database obtained, from which a condition of $(Re_w, (p)) = (6000, 15.96)$ was chosen for the code validation during this test. At this Reynolds number and the mean pressure gradient, the pCPf is under a fully turbulent state throughout the channel, and no large-scale intermittency occurs. Its friction Reynolds number Re_τ, normalized by the friction velocity on the fixed wall and the half width of the gap, is 154. Kuroda et al. [43] adopted a spectral method with a 128×128 Fourier series in the horizontal directions and Chebyshev polynomials up to order 96 in the wall-normal direction. Their domain size was $2.5\pi h \times h \times \pi h$, whereas our counterpart simulation on an aCPf of $\eta = 0.9$ employed a nearly equal domain size of $8h \times h \times \pi/8$ ($\approx 3h$ at the gap center) in (x, r, θ). Figure 2 shows comparisons of the present mean and second-order statistics. An overbar, such as $\overline{u_x}(r)$, denotes an ensemble-averaged quantity with respect to t, x, and θ, and subscript 'rms' indicates a root-mean-square value. The present control parameters of Re_w and $F(p)$ for our aCPf are 6000 and 16.0, respectively, and the resulting friction velocity and the friction Reynolds number on the fixed outer cylinder wall are $0.050u_w$ and 151. The present results shown in Figure 2 are in reasonable agreement with the reference study, despite the wall curvature of the aCPf. A noticeable difference is detected only near the fixed wall ($y/h \approx 0.9$), where the profile of $\overline{u_x}$ exhibits a steep gradient, and thus, those of the streamwise turbulent intensity $u'_{x\,\mathrm{rms}}$ and Reynolds shear stress $\overline{u'_x u'_r}$ have peaks. The rather coarse grid resolution and the low-order spatial discretization (our finite difference code versus the previous spectral code) might affect the accuracy of the present simulation. In addition to the peak values of $u'_{x\,\mathrm{rms}}$ and $\overline{u'_x u'_r}$, the second-order statistics from the present DNS and those of Kuroda et al. [43] agree well, particularly considering the differences in the flow geometry. The current Fortran code has been employed in different studies for several different boundary conditions [34,35,37,38,55], and thus, no further validation will be shown here.

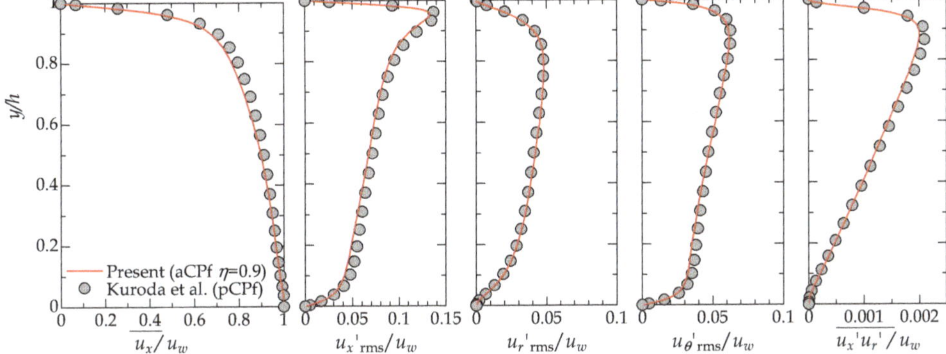

Figure 2. Code validation by comparison with a previous direct numerical simulation (DNS) study on Couette–Poiseuille flow at $Re_w = 6000$. (Left) Mean streamwise velocity profile; (middle three panels) root-mean-square values of velocity fluctuations in the streamwise, wall-normal (radial), and spanwise (azimuthal) directions; and (right) Reynolds shear stress. Lines and symbols represent the results obtained by this study for the annular Couette–Poiseuille flow (aCPf) with $\eta = 0.9$, and the result by Kuroda et al. [43] for plane Couette–Poiseuille flow (pCPf), respectively. Here, the wall-normal coordinate y represents the distance from the inner (bottom) wall, that is, $y = r - r_{\mathrm{in}}$.

In this study, we simulated a low-η annular flow that mimics a CPF by approximating the base flow, or the mean velocity profile, to that in the CPF. In the CPF, the velocity profile reaches its maximum at the pipe center; the velocity gradient becomes zero at the pipe center, and is at maximum on the surface of the pipe (i.e., the outer-cylinder surface). To match the base-flow characteristics of a CPF in an annular system, it is necessary to conduct a parametric investigation on the appropriate magnitude of the pressure gradient applied in the annular channel. As a preliminary analysis, we employed a medium-scale computational domain to reduce the computational cost of the parametric study. The computational domain size is smaller in the x direction than the present main analysis shown in Section 4. The streamwise length of the domain, L_x, was sufficient to capture one turbulent puff. The purpose of the preliminary analysis is to identify the value of the pressure gradient function

$F(p)$ at each Reynolds number such that the friction coefficient on the inner cylindrical wall, $C_{f,in}$, is practically zero, where $C_{f,in}$ is defined by the following:

$$C_{f,in} = \frac{\tau_{in}}{\frac{1}{2}\rho U^2},$$

(4)

where U is the bulk mean velocity obtained through a simulation. The positive/negative sign of $C_{f,in}$ corresponds to the positive/negative velocity gradient on the wall surface of the inner cylinder. Given $du/dr < 0$, $C_{f,in} < 0$, and vice versa. Table 1 shows the calculation conditions and the ranges of Re_w and $F(p)$ in the preliminary analysis. In the preliminary analysis, the calculation area in the mainstream direction was set to a smaller calculation area than that for the main analysis, but can capture one turbulent puff. Figure 3 shows the $F(p)$ dependence of $C_{f,in}$ at several values of Re_w near the global critical value. In each analysis plotted in the figure, a turbulent field with a high Reynolds number at equilibrium for each given $F(p)$ was set as the initial flow field, and the ensemble-averaged $C_{f,in}$ value was acquired after reaching a statistically steady state. Note that laminarization did not occur in any of the cases shown here. In general, as $F(p)$ increases, $C_{f,in}$ increases monotonically while changing from a negative to a positive value. This is consistent with the transition of the mean velocity profile from Couette-like to Poiseuille-like, and it can be confirmed that "the turning point" of $F(p)$ indicating $C_{f,in} = 0$ increases with Re_w. According to this Reynolds-number dependence, an extrapolation predicts a value of $F(p)$ that brings $C_{f,in} = 0$ at a lower Re_w, by which the main DNS analysis in the next section was executed.

Table 1. Numerical conditions for preliminary DNS with a moderate computational domain.

η	0.1
Domain size	$L_x \times L_r \times L_\theta = 51.2h \times h \times 2\pi$
Number of grids	$N_x \times N_r \times N_\theta = 512 \times 64 \times 128$
Re_w	1600–3000
$F(p)$	4.0–16.0

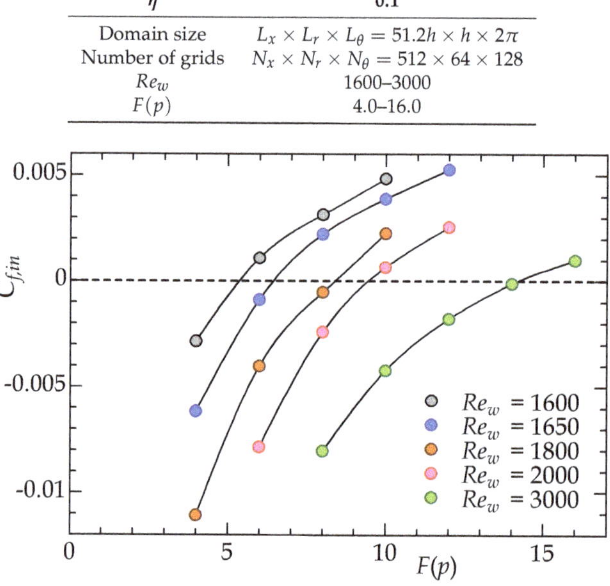

Figure 3. Friction coefficient on the inner cylinder surface as a function of the pressure function $F(p)$ for different Reynolds numbers, obtained through the preliminary DNS study on an aCPf.

4. Results and Discussion

The main DNS at $Re_w \leq 1600$ for which a laminar–turbulent intermittency was clearly confirmed through the preliminary analysis is presented in this section, and the characteristics of the localized turbulence are discussed. Table 2 summarizes the numerical conditions, including the friction Reynolds number Re_τ that was obtained. As in the preliminary analysis, the radius ratio was $\eta = 0.1$, and the computational domain was extended only in x; however, the grid resolution was not changed. Because Re_τ is lower than that of the preliminary analysis, the grid spacing in terms of the wall units was finer. The table also shows the grid resolutions based on the friction velocity u_τ under each condition:

$$u_\tau = \frac{\eta u_{\tau,in} + u_{\tau,out}}{\eta + 1},$$ (5)

where $u_{\tau,in}$ and $u_{\tau,out}$ are defined by the corresponding wall shear stress, τ_{in}, and τ_{out}, as well as by the relation $\tau = \rho u_\tau^2$, from which inner units can be defined. For a low Reynolds-number regime of $Re_w < 1600$, which is of interest in this study, the grid spacings of $\Delta x^+ < 8$, $\Delta r^+_{min} < 0.2$, $\Delta r^+_{max} < 3$, and $\Delta z^+ < 4$ are comparable to or higher in resolution than those in previous studies [35,37,53]. The initial conditions during each analysis adopted a turbulent field with a one-step-higher Reynolds number, but reduced the Reynolds number adiabatically. In other words, the study was carried out carefully such that the sudden drop in the Reynolds number will not be a proximate cause of laminarization.

Table 2. Numerical conditions for the main DNS with a long domain. The grid resolutions of $(\Delta x, \Delta r, \Delta\theta)$ are described in their dimensionless form based on u_τ and μ/ρ. The minimum and maximum Δr of the radial direction, in which we used non-uniform grids, are shown. [†] Laminar values from a laminarized case.

η	0.1						
Domain Size	$L_x \times L_r \times L_\theta = 409.6h \times h \times 2\pi$						
Number of Grids	$N_x \times N_r \times N_\theta = 4096 \times 64 \times 128$						
Re_w	3000	1600	1575	1550	1540	1530	1525	1500 [†]
$Re_\tau(= h^+)$	70.2	36.0	34.1	33.4	33.1	32.7	32.4	31.5
$F(p)$	14.2	6.5	6.0	5.8	5.7	5.6	5.5	5.3
Δx^+	14.1	7.14	6.82	6.68	6.61	6.53	6.47	6.30
Δr^+_{min}	0.37	0.19	0.18	0.18	0.17	0.17	0.17	0.17
Δr^+_{max}	4.44	2.26	2.16	2.11	2.09	2.07	2.05	1.99
$r^+_{in}\Delta\theta$	0.77	0.39	0.37	0.36	0.36	0.36	0.35	0.34
$r^+_{out}\Delta\theta$	7.66	3.89	3.72	3.64	3.61	3.56	3.53	3.44

4.1. Puffs in Annular-Pipe Flow

Figure 4 presents a three-dimensional visualization of localized turbulence in the form of puffs, which is observed as an equilibrium state reached after a lengthy simulation under the condition of $Re_w = 1600$ and $F(p) = 6.5$. The turbulent region can be clearly detected by showing the radial velocity fluctuations or the wall-normal velocity component. The threshold value of $\pm 0.03u_w$ for the iso-surface was arbitrarily chosen to extract its typical arrowhead shape similar to that of a puff. A slight change in this threshold value does not significantly affect the interpretation of the present results. In the snapshot, multiple turbulent patches, called 'puffs' hereafter, can be confirmed to be distributed intermittently with respect to the streamwise direction. The blank regions between neighboring puffs can be regarded as being in a laminar flow because of an insignificant fluctuating velocity, implying the well-established coexistence of laminar and turbulent regions in the aCPf. As is clear from the enlarged figure, the puff has an arrowhead shape, and the puff extends downstream in the center of the outer pipe. Although the average velocity gradient on the inner cylindrical wall surface is almost zero, this situation is considered to be due to the similar driving mechanism of the puff of the CPF. For the CPF, Shimizu et al. [11] reported that turbulence in the puff originates from low-speed streaks, as well as from streamwise vortices along the (outer-)pipe wall and across the trailing edge of the puff through the Kelvin–Helmholtz instability, which induces velocity fluctuations that propagate downstream faster than the puff itself in the core region. Such a driving mechanism of the puff is also common to the present aCPf with nearly zero $C_{f,in}$. The streamwise size of each puff is approximately 30 times the gap width h, which corresponds to 15 times the hydraulic diameter, and is consistent with that of the puff observed in the CPF [1,7,8,12]. The array of puffs seems variable in intervals, but is likely not less than $30h$. The wavelength and periodicity of the puffs are examined using two-point correlation functions of the turbulence quantities. In the x and θ directions, the auto-correlation coefficients are defined as follows:

$$R_{ii}(\Delta x) = \frac{\overline{u_i'(x, r_{\text{ref}}, \theta) u_i'(x + \Delta x, r_{\text{ref}}, \theta)}}{u_{i\,\text{rms}}'(r_{\text{ref}}) \cdot u_{i\,\text{rms}}'(r_{\text{ref}})} \tag{6}$$

and

$$R_{ii}(\Delta \theta) = \frac{\overline{u_i'(x, r_{\text{ref}}, \theta) u_i'(x, r_{\text{ref}}, \theta + \Delta \theta)}}{u_{i\,\text{rms}}'(r_{\text{ref}}) \cdot u_{i\,\text{rms}}'(r_{\text{ref}})}, \tag{7}$$

where $i \in (x, r, \theta)$. Figure 5 shows the two-point correlation coefficients of each velocity component for the case visualized in Figure 4. The statistical dataset was accumulated over the time of $5000h/u_w$ after achieving a pseudo-equilibrium state of multiple puffs.

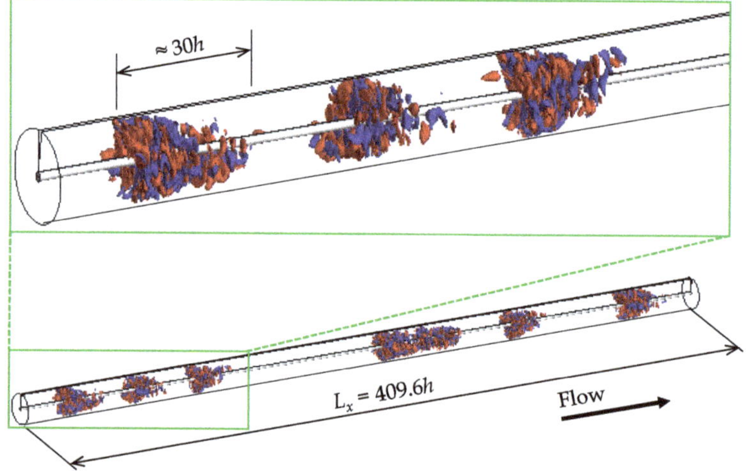

Figure 4. Instantaneous flow field for $Re_w = 1600$ and $F(p) = 6.5$. Iso-surfaces of radial velocity fluctuation are shown: red, $u_r' = 0.03u_w$; blue, $u_r' = -0.03u_w$. The left-to-right direction corresponds to the direction of the main flow, by which the observed puffs propagate. Not to scale.

From Figure 5a, the axial periodicity and interval of the puff can be estimated. First, we note that the three curves at different y^* exhibit consistency, implying that flow state and patterning are only weakly dependent on y or r. As also plotted in (b) and (c) for the other directional components, fine-scale turbulent structures inside a puff should have a rather short streamwise extent, and indeed, the profiles of R_{rr} and $R_{\theta\theta}$ fall to almost zero at $\Delta x < 5h$. The profile of R_{xx} also decreases drastically for a small Δx, although its significant oscillation for a long axial extent suggests a spatial coexistence of laminar and turbulent regions rather than turbulent structures, since these two flow states have different mean velocity profiles, particularly near the walls. The oscillations observed in Figure 5a are somewhat strong at both the inner and outer walls, relative to the gap center. The profile of R_{xx} takes the first negative local minimum at $\Delta \approx 30h$ and shows regular spikes at intervals of approximately $60h$. The correlation is not zero even at half the computational domain length ($L_x/2 = 204.8h$). Peaks at $60h$, $120h$, and $180h$ manifest the presence of seven distinct puffs in L_x on average. This suggests that the puffs at this Reynolds number tend to be arranged regularly throughout the axial extent. If the puff spacing is irregular, the correlation coefficient distribution should not show periodic fluctuations and should asymptotically approach zero. This regularity of the puff arrangement may differ from the characteristic of the DP universal class, which should exhibit a wide-scale invariant pattern close to the critical point [39,41]. Mukund and Hof [3] reported a similar aspect on multiple puffs in a CPF, where they referred to the wave-like fashion as 'puff clustering'; that is, the resultant pattern of clustering puffs was observed to propagate like waves. They also pointed out that interactions between puffs were responsible for the approach to the statistical steady state and strongly affected the percolation threshold. This may predict a difference in the global stability between a single puff (i.e., isolated puffs) and multiple puffs (puff clustering), as discussed in Section 4.2.

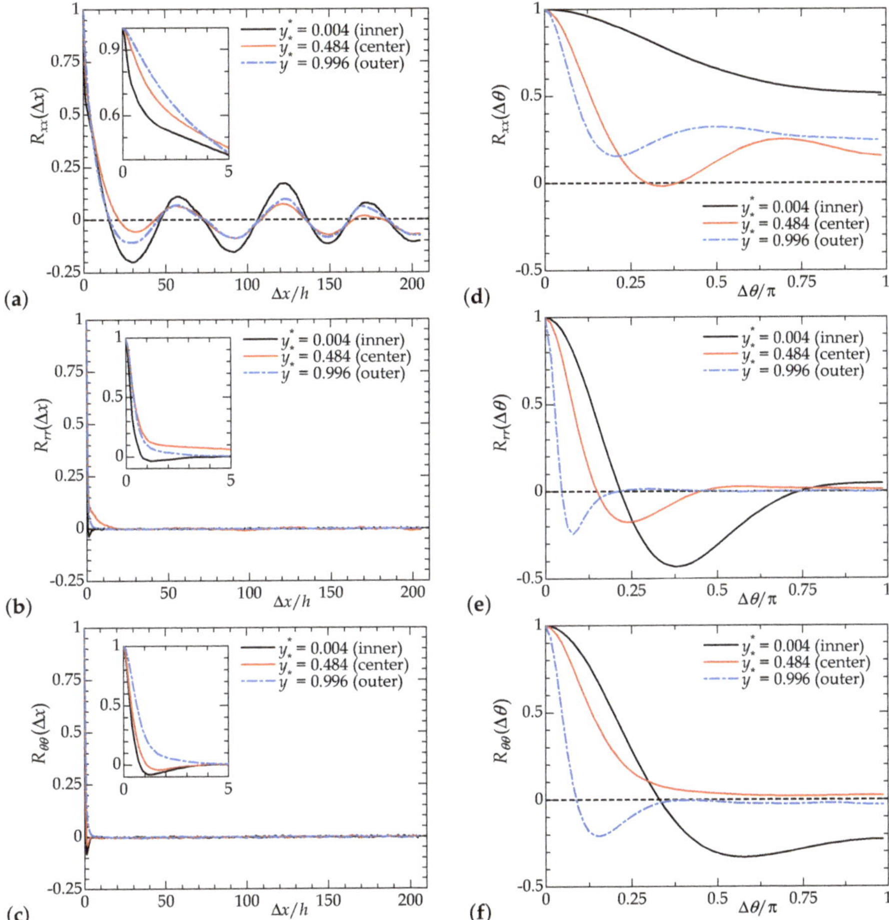

Figure 5. Two-point correlation coefficient of velocity fluctuation for $Re_w = 1600$ and $F(p) = 6.5$. (**a–c**) Streamwise spatial correlation as a function of Δx, and (**d–f**) azimuthal correlation as a function of $\Delta \theta$. (**a,d**) Auto-correlation of streamwise velocity component u'_x, (**b,e**) that of u'_r, and (**b,e**) that of u'_θ. Here, the reference radial position r_{ref} is translated as the inner-wall-normal height $y^* = r_{\text{ref}} - r/h$.

The azimuthal two-point correlation functions shown in Figure 5d–f indicate the azimuthal intervals between fine-scale turbulent structures, such as low-speed streaks inside the puff. There exists no large-scale pattern in the azimuthal direction, unlike those of the helically shaped turbulent patches in high-η aPf [35] and aCf [37]. The blue curve in Figure 5d, measured near the outer cylinder wall, only has a peak at $\Delta \theta = \pi/2$. The cross-sectional flow pattern observed here consists of four low-speed streaks close to the outer wall spaced at $\pi/2$. This azimuthal configuration regarding turbulence inside the puff is in agreement with those found in the CPF [6].

The presence of turbulent equilibrium puffs was observed even at $Re_w < 1600$, and the flow field finally reached the fully laminar state at $Re_w = 1500$. Although the space-time diagram (STD) and the turbulent fraction, $Ft(t)$ (the plot of which is shown later), reveal a tendency toward laminarization at $Re_w = 1525$, one turbulent puff was maintained in the present computational domain at least during the present observation time of $>1.3 \times 10^4$, and the laminarization was not completed. If normalized by the hydraulic equivalent diameter $2h$ and bulk velocity U, the Reynolds numbers of $Re_w = 1600$ and 1500 correspond to $Re_D = 2190$ and 2045, respectively. This range of $Re_D = 2045$–2190 is close to or slightly narrower than that for the counterpart of the CPF ($Re_D = 2000$–2700 [1], 2040–2400 [10], 2300–3000 [2], and 2000–2200 [4]). In particular, a discrepancy in the lower bound value of the subcritical transition regime, that is, the global critical Reynolds number, is of interest, although the similarity with the results by Avila et al. [10] is rather surprising. A cause of this discrepancy remains unclear: One of the main causes may be the presence of the inner cylinder, which suppresses turbulent motions across the central axis in the case of an aCPf. Another cause may be the non-slip inner-cylinder surface, which prevents a puff from splitting into two puffs in the case of an aCPf. Shimizu et al. [8] proposed a model process of puff splitting in the CPF, which starts with an azimuthally isolated streak propagating downstream through the laminar–turbulent interface of the puff. An emitted streaky disturbance can be a seed of a "daughter puff," which spreads again in the azimuthal direction and grows into a turbulent puff after leaving the parent puff sufficiently far away. As a system even closer to the CPF, an ideal aPf with a stress-free boundary condition at the inner wall can be analyzed, although such an unpractical situation will be considered as a future task. In terms of the conjecture that puff splitting is unlikely in the aCPf relative to the CPF, we traced puffs with lengthy simulations, and their STDs are shown in Section 4.2.

4.2. Space-Time Diagrams

As for the puff turbulence in CPF, it is well known that a turbulent puff can split into two puffs over time, the turbulence between puffs should attenuate and become a laminar pocket, and one or both puff(s) should decay quasi-stochastically because of their finite lifetime [3,6,8,9]. Avila et al. [10] observed the puff-turbulence sustainment only due to puff-splitting events that have time scale shorter than the puff-decay time scale. These features may be identified from the temporal development of the puff spatial distribution. The STDs of the present aCPf are shown in Figures 6–8, where the horizontal axis is the streamwise coordinate in a frame of reference moving at a certain velocity, and the vertical axis represents the dimensionless time at each Reynolds number. The frame-moving velocity is nearly the mean gap-center velocity, which also corresponds to the propagation velocity of an observed single puff. The color contour shows the azimuthal average of the radial velocity at mid-gap, $\langle u_r \rangle_\theta = \int_0^{2\pi} u_r(x, h/2, \theta, t)\,d\theta/2\pi$, such that the laminar and turbulent regions can be clearly distinguished. Although the apparent length of each turbulent puff depends on the criterion used to discriminate it from the surrounding laminar flow, a different choice does not change the qualitative conclusions obtained.

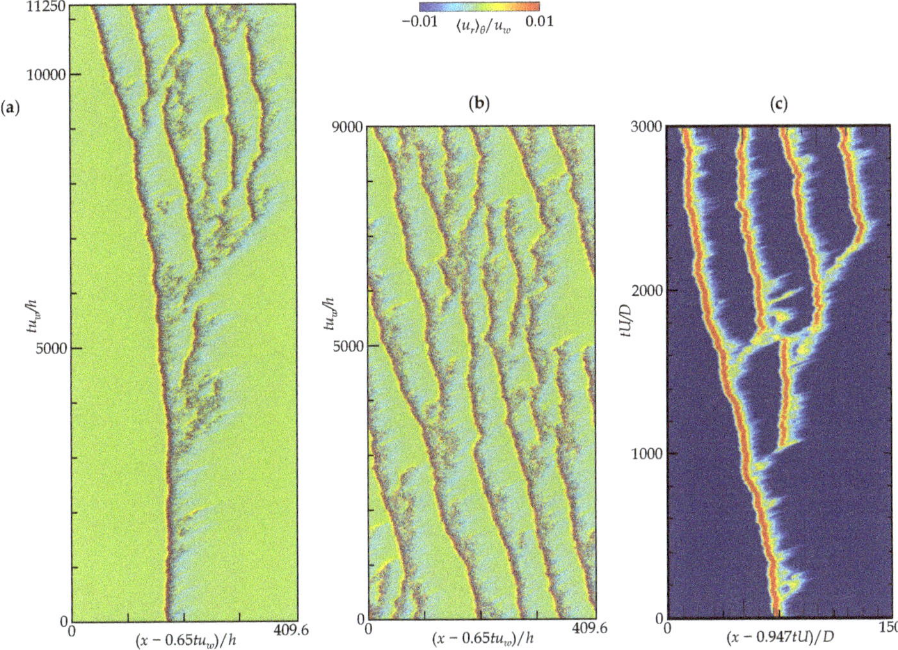

Figure 6. Space-time diagram for (**a**) Case 1 at $Re_w = 1600$ and $F(p) = 6.5$, (**b**) Case 2 at $Re_w = 1600$ and $F(p) = 6.5$ with a different initial condition from that in Case 1, and (**c**) a typical discrete expansion process of turbulence in a subcritical transitional pipe flow at $Re = 2300$, cited from Avila et al. [10]. In (**c**), the contour color indicates the cross-sectional average of the streamwise vorticity squared, where red and blue correspond to turbulent puff and laminar regions, respectively, and the Reynolds number Re is based on the mean velocity U and the pipe diameter D. In (**a**,**b**), the contour shows the azimuthally averaged radial velocity $\langle u_r \rangle_\theta$ at the gap center. The axial distribution is monitored from a moving frame of reference with a speed close to the puff propagation. The temporal development is monitored from $t = 0$, that is, the beginning of each DNS with a higher-Re_w field with more puffs; therefore, some initial puffs decayed immediately after the start of the simulation. Not to scale (aspect ratio $x{:}y = 10{:}1$).

We first present the results for $Re_w = 1600$ and $F(p) = 6.5$, as discussed in Section 4.1. The flow field visualized in Figure 4 was first achieved through an adiabatic decrease in Re_w (with a change in $F(p)$, accordingly) from a fully turbulent regime, and was then used as the initial condition for the following simulation to trace the behavior of the puff in the phase diagram of the L_x-space and time for as long as possible. The STD obtained is shown in Figure 6b, which monitors the pattern starting from an initial state with several puffs—the isolated turbulent patch featured as a red and blue segment at given time t. The overall puff pattern remains intrinsically spatiotemporally intermittent and exhibits both puff decay and splitting very frequently. These individual puffs have statistically well-defined lengths, similar to those in a CPF [8]. The number of puffs captured in the present domain is roughly constant between 5 and 7, and it is again confirmed that the puff intervals tend to be constant even if splitting or attenuation occurs in each individual puff. For this reason, Figure 5a reveals regular oscillations in the correlation function $R_{xx}(\Delta x)$, while a snapshot visualized in Figure 4 happens to have no periodicity in the puffs when considering the complete pipe length. Figure 6b may invoke an STD obtained from experimental and numerical observations of a DP-like feature in other flow systems [39,42]. Another DNS labeled as Case 1 was repeated for the same parameter set of $(Re_w, F(p))$, but with a different initial condition with a single puff, which was prepared from a lower-Re_w DNS (see Figure 6a). The initial single puff is sustained for a long period of $>5000h/u_w$, during which it splits irregularly, the first time at $tu_w/h \approx 3000$ and the second time at $tu_w/h \approx 4000$, but both newborn puffs decay after they are separated from their parents. A newly emitted daughter puff by the third splitting $tu_w/h \approx 5500$ grows and successively produces grandchild puffs. In addition, there are many signs of puff splitting. The puff turbulence eventually covers the entire domain, yet is intrinsically patchy, as in Case 2. It can be concluded that, in an aCPf similar to a CPF, the turbulent puff can split, regardless of the initial field, in qualitative agreement with a typical STD sample of a CPF [10], as displayed in Figure 6c.

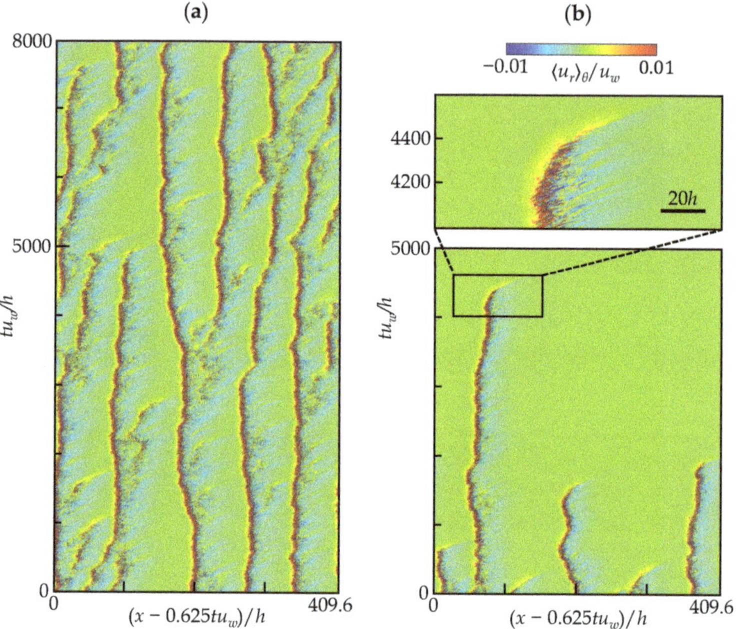

Figure 7. Space-time diagram for (a) $Re_w = 1600$ ($Re_D \approx 2190$) and $F(p) = 6.5$, as well as (b) $Re_w = 1500$ and $F(p) = 5.3$ ($Re_D \approx 2045$). The contour shows $\langle u_r \rangle_\theta$ at the gap center. The axial distribution was monitored from a moving frame of reference. The same initial condition was applied for all cases presented here. Not to scale (aspect ratio $x{:}y = 10{:}1$).

Figure 7a shows the STD of $Re_w = 1600$ and $F(p) = 6.5$ (Case 2), but the speed of the moving frame of reference is modified such that puffs appear to be stationary with respect to space. With this adjustment, the propagation speed of the puff can be estimated as approximately $0.625u_w$. According to Figure 6a, when tracking a single puff in Case 1, the propagation speed is slightly faster and $\approx 0.65u_w$. The result is reasonable because the bulk velocity generally decreases with the expanding turbulent region. Figure 7b is an STD at $Re_w = 1500$ with the same horizontal coordinate of $(x - 0.625u_w)/h$, showing the eventual return to laminar flow. Once a puff starts to decay, its turbulent patch seems to accelerate slightly and takes approximately $300u_w/h$ to attenuate completely. Before that, it took more than $4200h/u_w$ before the system settled to the fully laminar state. While the flow at $Re_D = 2190$ of Figure 7a exhibits frequent puff splitting or those signs during a period of $tu_w/h \approx 5000$, the flow at $Re_D = 2045$ in Figure 7b undergoes only the puff decay with no puff splitting, and the flow field simply reached a laminar flow.

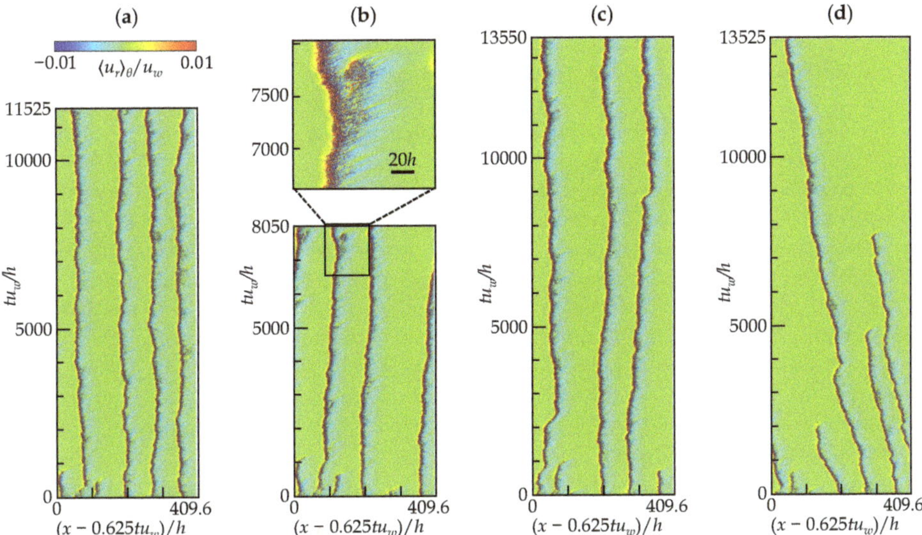

Figure 8. Space-time diagram for (**a**) $Re_w = 1550$, (**b**) $Re_w = 1540$, (**c**) $Re_w = 1530$, and (**d**) $Re_w = 1525$: see Table 1 for each given $F(p)$ value. The contour shows $\langle u_r \rangle_\theta$ at the gap center. The axial distribution is monitored from a moving frame of reference. The same initial condition was applied for all cases presented here. Not to scale (aspect ratio $x{:}y = 10{:}1$).

We further investigated the intermediate range between the two above-discussed cases ($2045 < Re_D < 2190$) to elucidate the trends in the frequency or time of the puff-splitting events. Figure 8 presents an STD at each control-parameter set. In all DNSs presented in the figure, the initial conditions are exactly the same. In the figure, six puffs can be seen initially, but two or three of them decay immediately, particularly in the lower-Reynolds-number cases. At the lowest Re_w shown in Figure 8d, puffs disappear one after another on a time scale of $O(1000h/u_w)$, and finally, one puff remains. There is no sign of decay in the surviving puff even after $13{,}000h/u_w$, but it is likely that the puffs will stochastically disappear and laminarize if a much longer simulation is available. This might also be true for the other cases presented here. At $Re_w < 1600$, no puff splitting was observed, resulting in only puff damping. In only Figure 8b, a sign of puff splitting is detected at $tu_w/h \approx 7500$, although the "daughter puff" is not perfectly formed, and is finally attenuated before leaving the parent puff. Note here that a further DNS indicates no qualitative change in the flow pattern at least until $tu_w/h = 11{,}500$ also for $Re_w = 1540$, although not shown in the figure. According to a similar type of study [10], the puff splitting in the CPF was observed both numerically and experimentally for $Re_D > 2200$, whereas clear splitting was measured in their experiments down to $Re_D = 2025 < Re_g\ (=2040)$. If our observations were continued as long as 10^7 outer time units, as Avila et al. [10] experimentally did, the current system of the aCPf could exhibit a puff-splitting event even below the true Re_g, which is not exactly determined as of now. At least, it can be said that the puff decay and splitting rates at this stage differ strongly from those observed at $Re_w = 1600\ (Re_D = 2190)$. As for this regime, a conclusion similar to an experimental study on a CPF [3] can be drawn, i.e., the cluster of puffs in a wave-like fashion results in fewer puff-splitting events in the STD, whose visual appearance differs from the STD for a DP universality class. Such well-organized distances between active sites (corresponding to the puffs) and the absence of splitting events are different features from those of the DP.

Figure 9 shows the temporal change in the turbulent fraction, $F_t(t)$, which is the spatial ratio of the turbulent region to the entire calculated region, including both the turbulent and laminar regions. Here, $F_t(t) \approx 1$ indicates a fully turbulent state, and $F_t(t) = 0$ is a fully laminar state. We set a threshold v_{th} to distinguish between laminar and turbulent regions such that $F_t(t) \approx 0.5$ in the

Reynolds number region where turbulent puffs densely appear in an axial extent, as in the case of $Re_w = 1600$ and $F(p) = 6.5$ visualized in Figure 7a. Figure 9a shows the temporal change of $F_t(t)$ at $Re_w = 1500$ and $F(p) = 5.3$, that is, the case diagnosed as a laminar regime by a visualization in Figure 7b, employing three different threshold values (v_{th}, v_{th2}, and v_{th3}). It can be confirmed that the time change of $F_t(t)$, particularly the gradient of the curve, does not depend on the threshold value. When $v_{th} = 0.005$, the temporal changes in $F_t(t)$ at several Reynolds numbers below $Re_w = 1600$ are plotted in Figure 9b. In the vicinity of the critical point, a (1+1)-D DP universality class should obey a power law of $F_t(t) \propto t^{-0.159}$ over time. From the figure, the current data at $Re_w = 1575$–1550 seem to be consistent with (1+1)-DP, although more data and more exponents will be needed to properly confirm this trend. However, it should be noted that, for $Re_w \leq 1550$, none of the puffs split and turbulent puffs were only attenuated, as shown in Figure 8a. This result suggests that a value close to the critical exponent of DP can be obtained even under a non-DP phenomenon of a simple decaying process without splitting. We should regard this result as a 'spurious' DP feature because the puff splitting (or an active site that creates offspring) is a requisite for the critical point and, hence, DP behavior. In other words, this reminds us to take caution regarding the judgment of a DP within the laminar–turbulent intermittency.

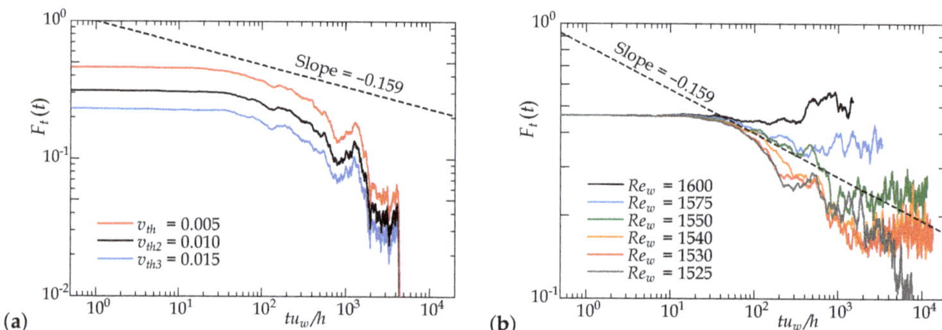

Figure 9. Time series of turbulent fractions: (**a**) for $Re = 1500$ with different threshold values and (**b**) for $Re = 1525$–1600 with a threshold value of v_{th}.

5. Conclusions

We performed direct numerical simulations (DNSs) of the concentric annular Couette–Poiseuille flow (aCPf) and investigated the laminar–turbulent intermittent field of the so-called puff turbulence, particularly during its subcritical transition. From previous studies, the laminar–turbulent intermittency in annular flows (a pure Couette flow [37] or Poiseuille flow [34]) exhibits the helically shaped turbulent pattern with bi-directional spatial intermittency and puff turbulence with uni-directional intermittency, depending on the radius ratio. This fact leads to a unified understanding of the formation of localized turbulence patterns of different systems, including planar and circular pipe flows (CPFs); however, these analyses were conducted under conditions in which the basic velocity profiles do not qualitatively match those of the CPF. In this study, the radius ratio (of the inner/outer radii) was as low as 0.1, and the mean pressure gradient was imposed such that the inner-cylinder surface had a zero velocity gradient on average, so that the CPF was alternatively simulated by an annular system. Multiple puffs were demonstrated using a long computational domain in the axial direction, and the presence or absence of the puff-splitting event and its onset Reynolds number were investigated using a long-term DNS. The Reynolds number was reduced adiabatically from the fully turbulent field, and the following results were obtained.

- At $Re_w = 1600$, puff-splitting events occur along with stochastic puff decay, resulting in wave-like fashion of multiple puffs with constant intervals.

- At $Re_w < 1600$, no puff-splitting event occurs, but initially given individual puffs survive over a present observation time of at least $10^4 h/u_w$, maintaining the intervals among the puffs.
- At $Re_w \approx 1550$, a 'spurious' feature of (1+1)D-DP was detected during the quenching process even without puff splitting, and a lower Re_w deviates from the DP critical exponent.
- At $Re_w = 1500$, the flow becomes fully laminar after the non-trivial finite lifetime of the puff.
- The range of $Re_w = 1500$–1600 (with the accordingly changed $F(p)$) corresponds to the bulk Reynolds-number range of $Re_D = 2045$–2190 based on the hydraulic diameter and bulk velocity.

The question considered in this study was whether puff splitting can occur in an aCPf, which essentially has a non-slip inner cylinder. In fact, puff splitting was clearly observed at $Re_w = 1600$, and a sign of splitting was detected at $Re_w = 1540$, which may be close to the global critical point, Re_g. This result guarantees that the planar system and the in-pipe system can be linked via the annular system. Near the criticality, oblique turbulent stripes grow or split in the longitudinal direction of the band, but the mainstream directional splitting, as seen in the CPF, is less pronounced in the planar flows. Our results suggest that the localized structures seen in both the planar and pipe flows can cause mainstream directional splitting. However, we should note that no completed puff splitting was detected near Re_g. The puff splitting could be observed for $Re_w < 1600$ and even below Re_g by increasing both the observation time and domain by orders of magnitude. Such a task to explore the exact Re_g value as well as the Reynolds-number dependence of the puff-splitting time scale near Re_g is a challenging one that is almost impossible at present. Another possible approach is to study lifetimes of single puffs [56] and time scales of splitting [10] at conditions away from Re_g, as in earlier studies on the CPF. This may allow us to discuss whether the current system behaves more like quasi-1D Couette flow or like pipe flow, as done by Shi et al. [57] for a pCf. We would like to report on this issue in another paper. Moreover, the characterization of a DP universal class remains skeptical. Similarly to the critical phenomena of the DP universal class, the region of the absorbing state (laminar-flow gap among puffs) should increase as the criticality approaches. From these facts, it is important to verify the DP feature after further expanding the axial computational domain. In addition, since the transition process of an aCPf has a dependence on the radius ratio and $F(p)$, a parametric study will also be addressed in the future.

Author Contributions: Conceptualization, T.T.; methodology, H.M. and T.T.; simulation and formal analysis, H.M.; visualization, H.M.; writing—original draft preparation, H.M. and T.T.; project administration, T.T.; funding acquisition, T.T.; all authors discussed the results and commented on the manuscript. All authors have read and agreed to the published version of the manuscript.

Funding: This work was funded by Grant-in-Aid for JSPS (Japan Society for the Promotion of Science) Fellowship 16H06066 and 19H02071.

Acknowledgments: Numerical simulations were performed on SX-ACE supercomputers at the Cybermedia Centre of Osaka University and the Cyberscience Centre of Tohoku University.

Conflicts of Interest: The authors declare no conflict of interest. The funders had no role in the design of the study; in the collection, analyses, or interpretation of data; in the writing of the manuscript, or in the decision to publish the results.

Abbreviations

The following abbreviations are used in this manuscript:

aCf	annular Couette flow
aCPf	annular Couette–Poiseuille flow
aPf	annular Poiseuille flow
CPF	Circular pipe flow
DNS	direct numerical simulation
DP	direct percolation
pCPf	plane Couette–Poiseuille flow

pPf   plane Poiseuille flow
rms   root-mean-square value

References

1. Wygnanski, I.J.; Champagne, F.H. On transition in a pipe. Part 1. The origin of puffs and slugs and the flow in a turbulent slug. *J. Fluid Mech.* **1973**, *59*, 281–335. [CrossRef]
2. Nishi, M.; Ünsal, B.; Durst, F. Laminar-to-turbulent transition of pipe flows through puffs and slugs. *J. Fluid Mech.* **2008**, *614*, 425–446. [CrossRef]
3. Mukund, V.; Hof, B. The critical point of the transition to turbulence in pipe flow. *J. Fluid Mech.* **2018**, *839*, 76–94. [CrossRef]
4. Priymak, V. Direct numerical simulation of quasi-equilibrium turbulent puffs in pipe flow. *Phys. Fluids* **2018**, *30*, 064102. [CrossRef]
5. Avila, M.; Mellibovsky, F.; Roland, N.; Hof, B. Streamwise-localized solutions at the onset of turbulence in pipe flow. *Phys. Rev. Lett.* **2013**, *110*, 224502. [CrossRef] [PubMed]
6. Hof, B.; Van Doorne, C.W.; Westerweel, J.; Nieuwstadt, F.T.; Faisst, H.; Eckhardt, B.; Wedin, H.; Kerswell, R.R.; Waleffe, F. Experimental observation of nonlinear traveling waves in turbulent pipe flow. *Science* **2004**, *305*, 1594–1598. [CrossRef]
7. Moxey, D.; Barkley, D. Distinct large-scale turbulent-laminar states in transitional pipe flow. *Proc. Natl. Acad. Sci. USA* **2010**, *107*, 8091–8096. [CrossRef]
8. Shimizu, M.; Manneville, P.; Duguet, Y.; Kawahara, G. Splitting of a turbulent puff in pipe flow. *Fluid Dyn. Res.* **2014**, *46*, 061403. [CrossRef]
9. Barkley, D. Theoretical perspective on the route to turbulence in a pipe. *J. Fluid Mech.* **2016**, *803*. [CrossRef]
10. Avila, K.; Moxey, D.; de Lozar, A.; Avila, M.; Barkley, D.; Hof, B. The onset of turbulence in pipe flow. *Science* **2011**, *333*, 192–196. [CrossRef]
11. Shimizu, M.; Kida, S. A driving mechanism of a turbulent puff in pipe flow. *Fluid Dyn. Res.* **2009**, *41*, 045501. [CrossRef]
12. Barkley, D.; Song, B.; Mukund, V.; Lemoult, G.; Avila, M.; Hof, B. The rise of fully turbulent flow. *Nature* **2015**, *526*, 550–553. [CrossRef] [PubMed]
13. Coles, D. Transition in circular Couette flow. *J. Fluid Mech.* **1965**, *21*, 385–425. [CrossRef]
14. Prigent, A.; Grégoire, G.; Chaté, H.; Dauchot, O.; van Saarloos, W. Large-scale finite-wavelength modulation within turbulent shear flows. *Phys. Rev. Lett.* **2002**, *89*, 014501. [CrossRef] [PubMed]
15. Barkley, D.; Tuckerman, L.S. Computational study of turbulent laminar patterns in Couette flow. *Phys. Rev. Lett.* **2005**, *94*, 014502. [CrossRef]
16. Duguet, Y.; Schlatter, P.; Henningson, D.S. Formation of turbulent patterns near the onset of transition in plane Couette flow. *J. Fluid Mech.* **2010**, *650*, 119–129. [CrossRef]
17. Couliou, M.; Monchaux, R. Large-scale flows in transitional plane Couette flow: a key ingredient of the spot growth mechanism. *Phys. Fluids* **2015**, *27*, 034101. [CrossRef]
18. De Souza, D.; Bergier, T.; Monchaux, R. Transient states in plane Couette flow. *J. Fluid Mech.* **2020**, *903*, A33. [CrossRef]
19. Tsukahara, T.; Seki, Y.; Kawamura, H.; Tochio, D. *DNS of Turbulent Channel Flow at Very Low Reynolds Numbers*; TSFP Digital Library Online; Begel House Inc.: Danbury, CT, USA, 2005; pp. 935–940.
20. Hashimoto, S.; Hasobe, A.; Tsukahara, T.; Kawaguchi, Y.; Kawamura, H. *An Experimental Study on Turbulent-Stripe Structure in Transitional Channel Flow*; ICHMT Digital Library Online; Begel House Inc.: Danbury, CT, USA, 2009.
21. Xiong, X.; Tao, J.; Chen, S.; Brandt, L. Turbulent bands in plane-Poiseuille flow at moderate Reynolds numbers. *Phys. Fluids* **2015**, *27*, 041702. [CrossRef]
22. Tsukahara, T.; Tillmark, N.; Alfredsson, P. Flow regimes in a plane Couette flow with system rotation. *J. Fluid Mech.* **2010**, *648*, 5–33. [CrossRef]
23. Brethouwer, G.; Duguet, Y.; Schlatter, P. Turbulent-laminar coexistence in wall flows with Coriolis, buoyancy or Lorentz forces. *J. Fluid Mech.* **2012**, *704*, 137–172. [CrossRef]
24. Reetz, F.; Kreilos, T.; Schneider, T.M. Exact invariant solution reveals the origin of self-organized oblique turbulent-laminar stripes. *Nat. Commun.* **2019**, *10*, 2277. [CrossRef] [PubMed]

25. Paranjape, C.S.; Duguet, Y.; Hof, B. Oblique stripe solutions of channel flow. *J. Fluid Mech.* **2020**, *897*, A7. [CrossRef]

26. Shimizu, M.; Manneville, P. Bifurcations to turbulence in transitional channel flow. *Phys. Rev. Fluids* **2019**, *4*, 113903. [CrossRef]

27. Tsukahara, T.; Ishida, T. Lower bound of sub-critical transition in plane Poiseuille flow. *Nagare* **2014**, *34*, 383–386. (In Japanese)

28. Takeda, K.; Tsukahara, T. Subcritical transition of plane Poiseuille flow as (2+1)d and (1+1)d DP universality classes. In Proceedings of the 8th Symposium on Bifurcations and Instabilities in Fluid Dynamics (BIFD2019), Limerick, Ireland, 16–19 July 2019.

29. Xiao, X.; Song, B. Kinematics and dynamics of turbulent bands at low Reynolds numbers in channel flow. *Entropy* **2020**, *22*, 1167. [CrossRef]

30. Kashyap, P.V.; Duguet, Y.; Dauchot, O. Flow statistics in the transitional regime of plane channel flow. *Entropy* **2020**, *22*, 1001. [CrossRef]

31. Manneville, P. Transition to turbulence in wall-bounded flows: Where do we stand? *Mech. Eng. Rev.* **2016**, *3*, 15-00684. [CrossRef]

32. Manneville, P. Laminar-turbulent patterning in transitional flows. *Entropy* **2017**, *19*, 316. [CrossRef]

33. Tuckerman, L.S.; Chantry, M.; Barkley, D. Patterns in wall-bounded shear flows. *Annu. Rev. Fluid Mech.* **2020**, *52*, 343–367. [CrossRef]

34. Ishida, T.; Duguet, Y.; Tsukahara, T. Transitional structures in annular Poiseuille flow depending on radius ratio. *J. Fluid Mech.* **2016**, *794*, R2. [CrossRef]

35. Ishida, T.; Duguet, Y.; Tsukahara, T. Turbulent bifurcations in intermittent shear flows: From puffs to oblique stripes. *Phys. Rev. Fluids* **2017**, *2*, 073902. [CrossRef]

36. Ishida, T.; Tsukahara, T. Friction factor of annular Poiseuille flow in a transitional regime. *Adv. Mech. Eng.* **2016**, *9*, 1–10. [CrossRef]

37. Kunii, K.; Ishida, T.; Duguet, Y.; Tsukahara, T. Laminar–turbulent coexistence in annular Couette flow. *J. Fluid Mech.* **2019**, *879*, 579–603. [CrossRef]

38. Takeda, K.; Duguet, Y.; Tsukahara, T. Intermittency and critical scaling in annular Couette flow. *Entropy* **2020**, *22*, 988. [CrossRef]

39. Lemoult, G.; Shi, L.; Avila, K.; Jalikop, S.V.; Avila, M.; Hof, B. Directed percolation phase transition to sustained turbulence in Couette flow. *Nat. Phys.* **2016**, *12*, 254–258. [CrossRef]

40. Sano, M.; Tamai, K. A universal transition to turbulence in channel flow. *Nat. Phys.* **2016**, *12*, 249–253. [CrossRef]

41. Chantry, M.; Tuckerman, L.S.; Barkley, D. Universal continuous transition to turbulence in a planar shear flow. *J. Fluid Mech.* **2017**, *824*, R1. [CrossRef]

42. Hiruta, Y.; Toh, S. Subcritical laminar–turbulent transition as nonequilibrium phase transition in two-dimensional Kolmogorov flow. *J. Phys. Soc. Jpn.* **2020**, *89*, 044402. [CrossRef]

43. Kuroda, A.; Kasagi, N.; Hirata, M. Direct numerical simulation of turbulent plane Couette-Poiseuille flows: Effect of mean shear rate on the near-wall turbulence structures. In *Turbulent Shear Flows 9*; Springer: Berlin/Heidelberg, Germany, 1995; pp. 241–257.

44. Pirozzoli, S.; Bernardini, M.; Orlandi, P. Large-scale motions and inner/outer layer interactions in turbulent Couette–Poiseuille flows. *J. Fluid Mech.* **2011**, *680*, 534–563. [CrossRef]

45. Orlandi, P.; Bernardini, M.; Pirozzoli, S. Poiseuille and Couette flows in the transitional and fully turbulent regime. *J. Fluid Mech.* **2015**, *770*, 424. [CrossRef]

46. Kun, Y.; Lihao, Z.; Andersson, H.I. Turbulent Couette-Poiseuille flow with zero wall shear. *Int. J. Heat Fluid Flow* **2017**, *63*, 14–27.

47. Nakabayashi, K.; Kitoh, O.; Katoh, Y. Similarity laws of velocity profiles and turbulence characteristics of Couette–Poiseuille turbulent flows. *J. Fluid Mech.* **2004**, *507*, 43–69. [CrossRef]

48. Klotz, L.; Wesfreid, J.E. Experiments on transient growth of turbulent spots. *J. Fluid Mech.* **2017**, *829*. [CrossRef]

49. Klotz, L.; Lemoult, G.; Frontczak, I.; Tuckerman, L.S.; Wesfreid, J.E. Couette-Poiseuille flow experiment with zero mean advection velocity: Subcritical transition to turbulence. *Phys. Rev. Fluids* **2017**, *2*, 043904. [CrossRef]

50. Liu, T.; Semin, B.; Klotz, L.; Godoy-Diana, R.; Wesfreid, J.E.; Mullin, T. Anisotropic decay of turbulence in plane Couette–Poiseuille flow. *arXiv* **2020**, arXiv:2008.08851.

51. Wong, A.W.; Walton, A.G. Axisymmetric travelling waves in annular Couette–Poiseuille flow. *Q. J. Mech. Appl. Math.* **2012**, *65*, 293–311. [CrossRef]

52. Kumar, R.; Walton, A. Self-sustaining critical layer/shear layer interaction in annular Poiseuille–Couette flow at high Reynolds number. *Proc. R. Soc.* **2020**, *476*, 20190600. [CrossRef]

53. Abe, H.; Kawamura, H.; Matsuo, Y. Direct numerical simulation of a fully developed turbulent channel flow with respect to the Reynolds number dependence. *J. Fluids Eng.* **2001**, *123*, 382–393. [CrossRef]

54. *Turbulence and Heat Transfer Laboratory (THTLAB; Nobuhide Kasagi Lab.)*; The University of Tokyo: Tokyo, Japan. Available online: http://thtlab.jp/ (accessed on 30 November 2020).

55. Fukuda, T.; Tsukahara, T. Heat transfer of transitional regime with helical turbulence in annular flow. *Int. J. Heat Fluid Flow* **2020**, *82*, 108555. [CrossRef]

56. Avila, M.; Willis, A.P.; Hof, B. On the transient nature of localized pipe flow turbulence. *J. Fluid Mech.* **2010**, *646*, 127–136. [CrossRef]

57. Shi, L.; Avila, M.; Hof, B. Scale invariance at the onset of turbulence in Couette flow. *Phys. Rev. Lett.* **2013**, *110*, 204502. [CrossRef] [PubMed]

Publisher's Note: MDPI stays neutral with regard to jurisdictional claims in published maps and institutional affiliations.

Article

Transitional Channel Flow: A Minimal Stochastic Model

Paul Manneville [1,*] and Masaki Shimizu [2]

[1]   LadHyX, École Polytechnique, CNRS, Institut Polytechnique de Paris, 91128 Palaiseau, France
[2]   Graduate School of Engineering Science, Osaka University, Toyonaka 560-0043, Japan; shimizu@me.es.osaka-u.ac.jp
*   Correspondence: paul.manneville@ladhyx.polytechnique.fr; Tel.: +33-689-069-021

Received: 2 November 2020; Accepted: 24 November 2020; Published: 29 November 2020

Abstract: In line with Pomeau's conjecture about the relevance of directed percolation (DP) to turbulence onset/decay in wall-bounded flows, we propose a minimal stochastic model dedicated to the interpretation of the spatially intermittent regimes observed in channel flow before its return to laminar flow. Numerical simulations show that a regime with bands obliquely drifting in two stream-wise symmetrical directions bifurcates into an asymmetrical regime, before ultimately decaying to laminar flow. The model is expressed in terms of a probabilistic cellular automaton of evolving von Neumann neighborhoods with probabilities educed from a close examination of simulation results. It implements band propagation and the two main local processes: longitudinal splitting involving bands with the same orientation, and transversal splitting giving birth to a daughter band with an orientation opposite to that of its mother. The ultimate decay stage observed to display one-dimensional DP properties in a two-dimensional geometry is interpreted as resulting from the irrelevance of lateral spreading in the single-orientation regime. The model also reproduces the bifurcation restoring the symmetry upon variation of the probability attached to transversal splitting, which opens the way to a study of the critical properties of that bifurcation, in analogy with thermodynamic phase transitions.

Keywords: transition to/from turbulence; wall-bounded shear flow; plane Poiseuille flow; spatiotemporal intermittency; directed percolation; critical phenomena

---

1. Context

How laminar flow becomes turbulent, or the reverse, when the shearing rate changes, is a problem of great conceptual interest and practical importance. This special issue is focused on the case when the transition is characterized by the fluctuating coexistence of domains either laminar or turbulent in physical space at a given Reynolds number Re (control parameter), a regime called spatiotemporal intermittency, relevant to wall-bounded flows in particular. Several years ago, Y. Pomeau [1] placed that problem in the realm of statistical physics by proposing its approach in terms of a non-equilibrium phase transition called directed percolation (DP). This process displays specific statistical properties defining a universality class liable to characterize systems with two competing local states, one active, the other absorbing, with remarkably simple dynamical rules: any active site may contaminate a neighbor and/or decay into the absorbing state, and an absorbing state cannot give rise to any activity [2]. The coexistence is regulated by the contamination probability, and a critical point can be defined above which the mixture of active and absorbing states is sustained and below which the active state recedes, leaving room for a globally absorbing state. The fraction of active sites is a measure of the global status of the system. The subcritical context typical of wall-bounded flows, initially pointed out by Pomeau, seems an interesting testbed for universality [3,4]. Here, turbulence plays the role of the active state and laminar flow, being linearly stable, represents the absorbing state. DP has indeed been shown relevant to simple shear between parallel plates (Couette flow) [5] and its stress-free version (Waleffe flow) [6]. The most recent contributions to the field can be found in [7]. In this paper we will be interested in plane channel flow (also called plane Poiseuille flow), the flow driven by a pressure gradient between two parallel plane plates, which is not fully understood despite recent advances.

In this context, universal properties are notably difficult to extract from experiments, since they relate to the thermodynamic limits of asymptotically large systems in the long time limit, whereas what plays the role of microscopic scales involves already macroscopic agents, e.g., roll structures in convection or turbulent streaks in open flows, and the turnover time associated with such structures. However, universality focuses on quantitative aspects of systems sharing the same qualitative characteristics, in particular symmetries and the effective space dimension D in which these systems evolve. Delicate questions can thus be attacked by modeling attempts that implement these traits appropriately. This approach involves simplifications from the primitive equations governing the problem, here the Navier–Stokes equations, to low-order differential models implementing the building blocks of the dynamics [8], to coupled map lattices (CML) in which the evolution is rendered by maps and space is discretized [9,10], to cellular automata for which local state variables are also discretized, and ultimately to probabilistic cellular automata (PCA), where the evolution rule itself becomes stochastic [11]. The absence of a rigorous theoretical method supporting the passage from one modeling level to the next, such as multi-scale expansions or Galerkin approximations, makes the simplification rely on careful empirical observations of the case under study, which somehow comes and limits the breadth of the conclusions drawn.

1.1. Physical Context: Plane Channel Flow

Of interest here, the transitional range of plane channel flow displays a remarkable series of steps at decreasing Re from large values where a regime of featureless turbulence prevails. It has been the subject of numerous studies and references to them can be found in the article by Kashyap, Duguet, and Dauchot in this special issue [12]; see also [13]. Our own observations based on numerical simulations are described in [14,15] and summarized in Figure 1.

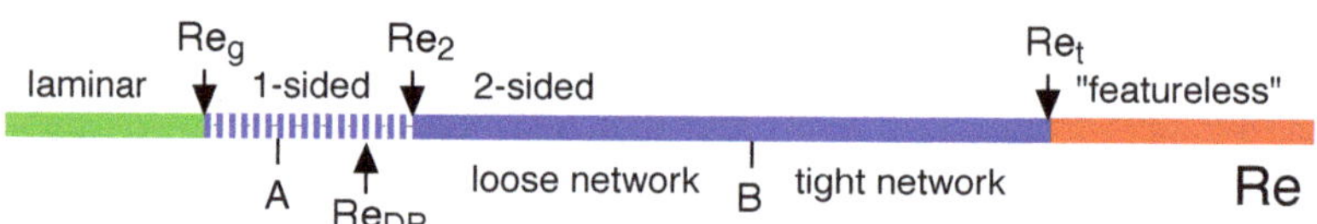

Figure 1. Bifurcation diagram of plane channel flow after [14]. $\mathrm{Re_g} \approx 700$. Transversal splitting sets in at $\mathrm{Re} \sim 800$ (event A). The extrapolated 2D-DP threshold is $\mathrm{Re_{DP}} \simeq 984$. The "one-sided $\rightarrow$ two-sided" transition takes place at $\mathrm{Re_2} \simeq 1011$. localized turbulent bands (LTBs) exist up to $\mathrm{Re} \approx 1200$ (event B), beyond which a continuous laminar–turbulent oblique pattern prevails up to the threshold for featureless turbulence $\mathrm{Re_t} \approx 3900$.

The Reynolds number used to characterize the flow regime is defined as $\mathrm{Re} = U_c h / \nu$, where $2h$ is the gap between the plates, U_c is the mid-gap stream-wise speed of a supposedly laminar flow under the considered pressure gradient, and ν the kinematic viscosity. This definition using U_c is appropriate for our numerical simulations under constant pressure-gradient driving. Other definitions involve the friction velocity U_τ, or the stream-wise speed averaged over the gap U_b. They are related either empirically, vis., U_b vs. U_c, or theoretically, vis., $\mathrm{Re_\tau} = \sqrt{2\mathrm{Re}}$ to be used in particular for connecting to the work presented in [12], and some other articles. See [14] for details. Below a first threshold $\mathrm{Re_t}$, featureless turbulence leaves room for a laminar–turbulent, oblique, patterned regime (upper transitional range) that next turns into a sparse arrangement of localized turbulent bands (LTBs) propagating obliquely along two directions symmetrical with respect to the general stream-wise flow direction, experiencing collisions and splittings ("two-sided" lower transitional regime). Event B in

Figure 1 corresponds to the opening of laminar gaps along the intertwined band arrangement observed in the tight laminar–turbulent network regime, and the simultaneous prevalence of downstream active heads (DAHs) driving the LTBs. Upon decreasing Re further, a symmetry-breaking bifurcation takes place at a second threshold Re_2, below which a single LTB orientation prevails. Figure 2 displays snapshots of the flow illustrating these last two stages.

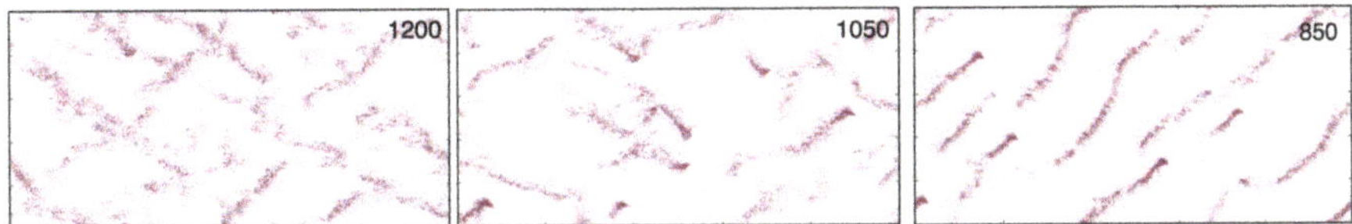

Figure 2. Illustration of the different regimes featuring the wall-normal velocity component at the mid-gap; turbulent/laminar flow is pink/white, after data in Figure 1 of [14]. The domain size is 250×500 (span-wise $\times$ stream-wise). The flow is from left to right. **Left**: Strongly intermittent loose continuous LTB network at Re = 1200 ($\sim$event B). **Centre**: Two-sided regime at Re = 1050 (Re $\gtrsim$ Re_2). **Right**: One-sided regime at Re = 850. Downstream active heads (DAHs) are easily identified in the two right-most panels; a single one is visible in the upper left corner of the left image, marking the transition between sustained regular patterns and loose intermittent ones. Images here and in Figures 2 and 3 are adapted from snapshots taken out of the supplementary material of reference [14].

A significant result in [14] was that the decrease of turbulence intensity with Re below event B followed expectations for directed percolation in two dimensions but that, controlled by the decreasing probability of transversal splitting, the bifurcation at Re_2 prevented the flow to reach the corresponding threshold. The latter could nevertheless be extrapolated to a value $Re_{DP} < Re_2$. The ultimate decay stage takes place at Reynolds numbers below the point whereat transversal splitting ceases to operate. Figure 3 illustrates an extremely rare occurrence of transversal splitting at a Reynolds number roughly corresponding to event A in Figure 1.

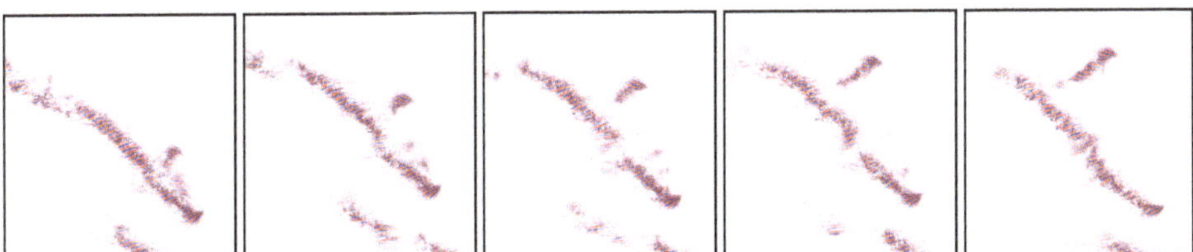

Figure 3. First observed occurrence of transversal splitting during a simulation at Re = 800 for $t \in (17100{:}100{:}17500)$. The stream-wise direction is horizontal and the flow is from left to right.

At lower Re, deprived of the possibility to nucleate daughters' LTBs of opposite propagation orientation, LTBs are forcibly maintained in the "one-sided" regime that eventually decays below a third threshold Re_g, marking the global stability of the laminar flow. Corresponding flow patterns are illustrated in Figure 4, the right panel of which displays the surprising result that the turbulent fraction decreases as a power law with an exponent β of the order of that for directed percolation in one dimension, despite the fact that the flow develops in two dimensions [16].

The objective of the present work is the design of a minimal PCA model for these two last stages that is applicable to flow states for Re below event B, incorporates the anisotropy features visible in Figures 2–4, and accounts for the specific role transversal splitting above event A, in view of providing clues to their statistical properties in relation to dimensionality and universality issues.

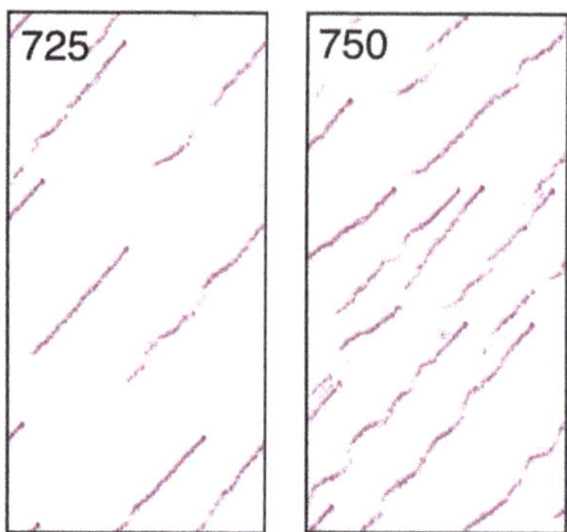

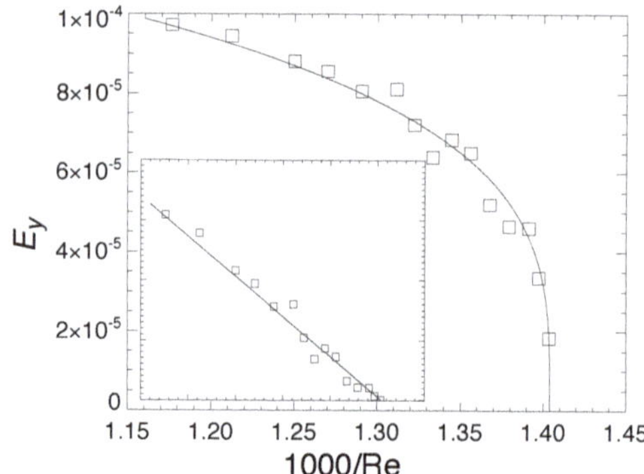

Figure 4. (**Left**): One-sided flow at Re = 725, 750; same representation as in Figure 2. The domain size is now 500×1000, the stream-wise direction is vertical, and the flow upwards. (**Right**): Used as a proxy for the turbulent fraction, $E_y = V^{-1} \int u_y^2 \, dV$ is displayed as a function of $1/Re$; inset: same data raised at power $1/\beta$ with $\beta = 0.28$ suggesting decay according to the DP scenario in 1D, adapted from [16].

1.2. Modeling Context: Directed Percolation, Probabilistic Cellular Automata, and Criticality Issues

Various modeling approaches to transitional wall-bounded flows have received considerable attention recently, from low-order Galerkin expansions of the primitive equations [17,18], to phenomenological theories based on a deep physical analysis of the processes involved in a reaction-diffusion context [19], to analogical systems expressed in terms of deterministic coupled map lattices [6,10], and to more conceptual models implementing the dynamics of cellular automata with probabilistic evolution rules (PCA) [20–22]. The model developed below belongs to this last category, implementing rules that focus on the main qualitative features seen in experiments. Such models are based on the conventional modeling of DP [2] which is most appropriate to account for the absorbing versus active character of local states.

Let us briefly recall the PCA/DP framework. In the most general case, the activity at site j at time $t + 1$, call it $S_j \in \{0, 1\}$, depends on the activity at sites in a full D-dimensional neighbor V_j of that site at time t and the status of the links, permitting or not the transfer of activity within the neighborhood. For convenience a $(D + 1)$-dimensional lattice is defined with one-way (directed) bonds in the direction corresponding to time so that D-dimensional directed percolation is often presented as a special $(D + 1)$-dimensional percolation problem. In the simplest case of one space dimension ($D = 1$), the neighborhood of a lattice site at j is the set of sites with $j' \in [j - r_1, j + r_2]$, comprising $r_2 + r_1 + 1$ sites, and it is supposed that contamination of the state at j at time $t + 1$ depends on the status of full configuration, the sites' activity, and the bonds' transfer properties ("bond–site" percolation [23]). In some systems, the propagation rule is totalistic in the sense that the output only depends on the number of active sites in the neighborhood and not on their positions, i.e., $\varsigma_j = \sum_{j' \in V_j} S_{j'}$; an interesting example is given in [24].

In view of future developments, let us discuss bond directed percolation in one dimension ($D = 1$) with two neighbors ($r_1 = 0$ or $r_2 = 0$), only depending on the probability p that bonds transfer activity. The evolution rule $S_j' = \mathcal{R}(S_j, S_{j+1})$, where S_j' denotes the state at node j and time $t + 1$, is totalistic. With $\varsigma_j = S_j + S_{j+1}$, we have (a) $\mathcal{R}(\varsigma = 0) = 0$ with probability 1 (a site connected to two absorbing parents never gets active whatever the links) and (b) $\mathcal{R}(\varsigma = 1) = 1$ with probability p (closed link transmitting activity), so that (a') $\mathcal{R}(\varsigma = 1) = 0$ with probability $1 - p$ (open link preventing transmission), (c) $\mathcal{R}(\varsigma = 2) = 0$ with probability $(1 - p)^2$ (absorbing since the two links are open), and (d) $\mathcal{R}(\varsigma = 2) = 1$ with probability $1 - (1 - p)^2 = p(2 - p)$, the complementary case.

The question is whether, depending on the value of p, once initiated, activity keeps continuing in the thermodynamic limit of infinite times in an infinitely wide system. An answer is readily obtained in the mean-field approximation where actual local states are replaced by their mean value, neglecting the effect of spatial correlations and stochastic fluctuations (we follow the presentation of [20]). The spatially-discrete Boolean variables S_j are, therefore, replaced by their spatial averages $S = \langle S_j(t) \rangle$ and this mean value is just the probability that any given site is active. It is then argued that the probability to get a future absorbing state, $1 - S'$, is given by activity not being transmitted $(1 - pS)^2$, which yields the mean-field equation:

$$1 - S' = (1 - pS)^2 = 1 - 2pS + p^2 S^2, \quad \text{i.e.,} \quad S' = 2pS - p^2 S^2. \tag{1}$$

Equilibrium states correspond to the fixed points of (1): $S' = S = S*$, which gives a nontrivial activity level $S_* = (2p - 1)/p^2$ when $p \geq p_c = 1/2$. Close to threshold, defining $\varepsilon = (p - p_c)/p_c = 2p - 1$ one gets $S_* \approx 4\varepsilon$. In the mean-field (MF) approximation S_* is the order parameter of the transition supposed to vary as ε^β, which defines the critical exponent β, here $\beta_{\mathrm{MF}} = 1$. Directed percolation is the prototype of non-equilibrium phase transitions and, as such, is associated with a set of critical exponents (see [2]). Both the critical probability p_c and the mean activity S_* are affected by the effects of fluctuations, with $p_c \approx 0.6445 > 1/2$ expressing that a probability larger than the mean-field estimate is necessary to preserve activity, and $\beta_{\mathrm{DP}} \approx 0.276$ when $D = 1$. The simple mean-field argument is not sensitive to the value of D in contrast with reality: $\beta_{\mathrm{DP}} \approx 0.584$ when $D = 2$, ≈ 0.81 when $D = 3$, and trends upwards to 1 reached at $D = 4 = D_c = 4$, called the upper critical dimension (see [2] for a review). Quite generally, mean-field arguments are valid for $D > D_c$. We are interested in another critical exponent, α. When starting from a fully active system exactly poised at p_c, the turbulent fraction is observed to decrease with time (the number of iteration steps) as $\langle S \rangle \propto t^{-\alpha}$ with $\alpha \approx 0.159$ when $D = 1$ and 0.451 when $D = 2$, whereas the mean-field prediction, easily derived from (1), is $\alpha_{\mathrm{MF}} = 1$. Scaling theory shows that $\alpha = \beta/\nu_\parallel$, where $\nu_\parallel$ is the exponent accounting for the decay of time correlations while $\nu_\perp$ describes the decay of space correlations [2].

Universality is a key concept in the field of critical phenomena characterizing continuous phase transitions. It leads to the definition of universality classes expressing the insensitivity of critical properties to specific characteristics of the systems and retaining only properties linked to the symmetries of the order parameter and the dimension of space. For directed percolation, universality is conjectured to be ruled by a few conditions put forward by Grassberger and Janssen: that the transition is continuous into a unique absorbing state and characterized by a positive one-component order parameter, and that the processes involved are short-range and without weird properties such as quenched randomness; see [2]. Universality issues are discussed at length elsewhere in this special issue, in particular by Takeda et al. [25].

In this first approach, we shall examine how universality expectations hold for the ultimate decay stage of transitional channel flow at Re_g, as described in Section 1.1, and limit the discussion to the consideration of exponents β and α. This will be done in Section 3, the next section being devoted to the derivation of the model and its mean-field study. Section 4 focuses on its ability to account for the symmetry-breaking bifurcation at Re_2, and our conclusions are presented in Section 5.

2. Description of the Model

2.1. Context

The approach to be developed is not new in the field of transitional flows. For example, studying plane channel flow, Sano and Tamai [21] introduced a plain 2D-DP model dedicated to support their experimental results, with a simple spatial shift implementing advection and a uniformly turbulent state upstream corresponding to their setup. Earlier, a similarly conceptual model was examined by Allhoff and Eckhardt [20], who introduced a PCA with two parameters accounting for persistence and lateral spreading appropriate for the symmetries of plane Couette flow, developed its mean-field treatment, and performed simulations to illustrate the spreading of spots and decay of turbulence in agreement with expectations. In a similar spirit but introducing more physical input,

Kreilos et al. [22] analyzed the development of turbulent spots in boundary layers as a function of the residual turbulence level upstreams, separating a deterministic transport step from a stochastic growth/decay step with probabilities extracted from a numerical experiment, gaining insight into the statistics of boundary layer receptivity.

Following the lines of research suggested by those works, we developed a 2D model designed to interpret the decay of channel in the LTB regimes from two-sided to one-sided at decreasing Re, just qualitatively proposing a plausible variation of probabilities introduced as functions of Re. In our approach, the elementary agents are the LTBs themselves either propagating to the left or to the right of the stream-wise direction. To them we attach variables analogous to spins in magnetic phase transitions problems. Even if in computations, numerical values $S = \pm 1$ will be used, for descriptive and graphical convenience we shall associate them with colors—specifically: blue (B) and red (R) for right- and left-propagating LTBs, respectively. Laminar sites will be denoted using the empty-set symbol $\varnothing$, will have value 0, and will be graphically left blank. These agents will be seated at the nodes of a square lattice with coordinates (i, j), i.e., $S_{(i,j)}$ with $S \mapsto \{R, B, \varnothing\}$ at the given site. As seen in Figure 5a, we place the stream-wise direction along the first diagonal of the lattice so that the LTBs will move along the horizontal and vertical axes; see Figure 5b.

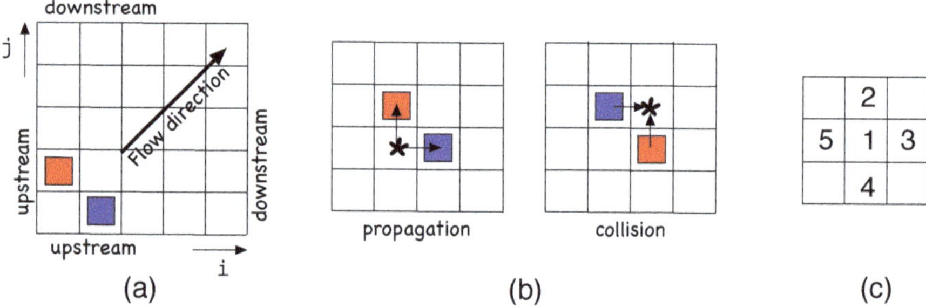

Figure 5. (a) Cellular automata lattice with the two types of active states, B and R; the state at an empty node is denoted $\varnothing$ and left blank. (b) Left: the two possible kinds of propagation from an initial position marked with the "*". Right: collision configuration to the point marked with the "*". (c) Labeling of the von Neumann neighborhood used to account for the dynamics.

A strong assumption is that an LTB as a whole corresponds to a single active state, while the discretization of space coordinates $(i, j) \in \mathbb{Z}^2$, and time $t \in \mathbb{N}$ tacitly refers to an appropriate rescaling of time and space. Furthermore, interactions are taken as local, with configurations limited to nearest neighbors in each space direction. Accordingly, the dynamics at a site (i, j) only depend on the configuration of its von Neumann neighborhood $\mathcal{V}_{(i,j)} := \{(i, j), (i \pm 1, j), (i, j \pm 1)\}$, Figure 5c, while evolution is driven by a random process. We now turn to the definition of rules that mimic the actual continuous space-time, subcritical and chaotic, Navier–Stokes dynamics governing the LTBs' propagation, decay, splitting, and collisions, via educated guesses from the scrutiny of simulation results, in particular those in the supplementary material attached to [14].

2.2. Design of the Model

Let us first give a brief description of the processes to be accounted for. Below Re ≈ 800 (event A) only decay and longitudinal splittings are possible. Not visible in the snapshots of Figure 4 (left) but observable in the movies is the fact that a daughter LTB resulting from longitudinal splitting runs behind its mother along a track that may be slightly shifted upstream. This shift is negligible when Re is small (in-line longitudinal splitting) but as Re increases it becomes more and more visible while the general propagation direction is unchanged (off-aligned longitudinal splitting). On the other hand, Figure 4 clearly illustrates the fact that, upon transversal splitting, the new-born LTB systematically develops on the downstream side of its parent. Importantly, the propagation of LTBs is a dynamical feature different from advection treated as a deterministic step in [22]. Accordingly, it will be understood as a statistical propensity to move in a given direction resulting from an imbalance of stochastic "forward" and "backward" processes along their directions of motion. Other complex processes also seen in the simulations, such as fluctuating propagation with acceleration, slowing down, or lateral wandering, will be included only in so far as they can be decomposed into such more elementary events. All the events to be included in the model can be translated into the language of reaction–diffusion processes, persistence or death, offspring production, and coalescence, common in the field of DP theory [2].

On general grounds the governing equation reads:

$$S'(i,j) = \sum_{\mathcal{C}'} R_{\mathcal{C}'} \delta_{\mathcal{C}'\mathcal{C}(i,j)}, \tag{2}$$

where $\mathcal{C}(i,j)$ is the neighborhood configuration of site (i,j) at time t, $\mathcal{C}'$ one of the possible configurations, and $R_{\mathcal{C}'}$ a stochastic variable taking value 1 with probability $p_{\mathcal{C}'}$ corresponding to configuration $\mathcal{C}'$ and value 0 with probability $1 - p_{\mathcal{C}'}$. The Kronecker symbol $\delta_{\mathcal{C}'\mathcal{C}}$ is here to select the configuration $\mathcal{C}'$ that matches $\mathcal{C}$. Depending on $\mathcal{C}$ and $\mathcal{C}'$, the output $S'(i,j)$ can be B or R.

Figure 6 illustrates the set of possible single-colored neighborhoods, either B (upper line) or R (lower line). Following the indexation in Figure 5c, the order of the columns is based on the physical condition and respects the upstream/downstream distinction illustrated in Figure 5a, making configurations with the same index physically equivalent.

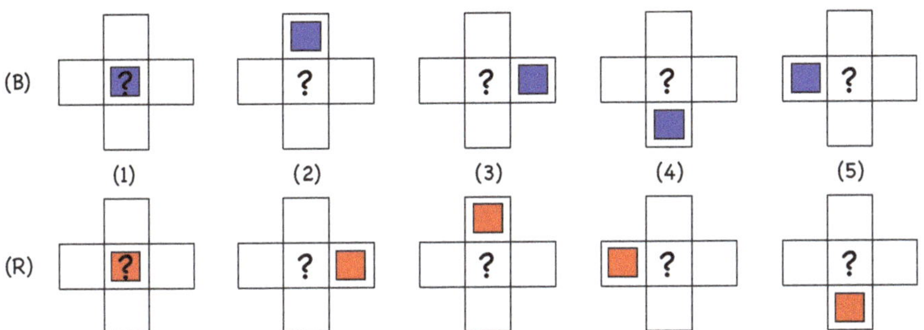

Figure 6. Single-color configurations: from the overall geometry depicted in Figure 5a, the downstream side of a state is to the top for B states and to the right for R states. Each colored square indicates the active state in the configuration at time t of site (i,j) at the center. The question mark features the probabilistic outcome (time $t+1$).

These single-color elementary configurations will be denoted as $\mathcal{C}_i$ with $i \in [1{:}5]$. They will be described as $[SSSSS]$ with $S = B$, R, or $\varnothing$. Hence $\mathcal{C}_3 \equiv [\varnothing\varnothing B\varnothing\varnothing]$ or $[\varnothing R\varnothing\varnothing\varnothing]$. Later, more complicated configurations will not be given a name but just a description following the same rule, e.g., $[\varnothing BBR\varnothing]$.

Importantly, we make the assumption that the future state at a given node, the question marks in Figure 6, is the result of the probabilistic combination of the independent contributions of elementary configurations involving a single active state in its neighborhood.

First of all, the void configuration $\mathcal{C}_0 \equiv [\varnothing\varnothing\varnothing\varnothing\varnothing]$ obviously generates an empty site with probability 1, hence an occupied site with probability $p_{\mathcal{C}_0} = 0$, in order to preserve the absorbing character of the dynamics. All the other configurations evolves according to probabilities that are free parameters just constrained by empirical observations. Let us now interpret probabilities associated with the five situations depicted in Figure 6, focusing on the case of B states:

1. $C_5 \equiv [\varnothing\,\varnothing\,\varnothing\,\varnothing\,B]$ corresponds to the natural propagation of the active state along its own motion direction. Accordingly, the active site B at $(i-1,j)$ is expected to be found at (i,j) and time $t+1$ with a high probability, $p_{C_5} = p_5 \lesssim 1$, which corresponds to the near-deterministic propagation of an active state as observed for $\mathrm{Re} \geq \mathrm{Re_g}$. With probability $1 - p_5 \ll 1$, site (i,j) will not turn active, which means that the LTB has decayed or experienced a speed fluctuation that delayed its propagation. The corresponding R configuration is $C_5 \equiv [\varnothing\,\varnothing\,\varnothing\,R\,\varnothing]$.

2. Configuration $C_1 \equiv [B\,\varnothing\,\varnothing\,\varnothing\,\varnothing]$ corresponds to an active site B at (i,j) that is not supposed to stay in place but move to $(i+1,j)$ with probability p_5 and leave site (i,j) empty at time $t+1$. The probability p_1 that (i,j) is still active at time $t+1$, therefore, generally corresponds to the creation of a novel active state by in-line longitudinal splitting at the rear of the active state that has effectively moved. Persisting activity at (i,j) and time $t+1$ can also be the result of state at (i,j) and time t experiencing a speed fluctuation leaving it stuck at the same place with probability $1 - p_5$ as argued above for configuration C_5. The presence of parameter p_1 undoubtedly makes the dynamics richer. The corresponding R-configuration is $C_1 \equiv [R\,\varnothing\,\varnothing\,\varnothing\,\varnothing]$.

3. Configuration $C_2 \equiv [\varnothing\,B\,\varnothing\,\varnothing\,\varnothing]$ corresponds to an active state B at site $(i,j+1)$ that contaminates backwards and laterally upstream the site at (i,j) in addition to its likely propagation to $(i+1,j+1)$ with probability p_5. This is precisely what is sometimes observed for longitudinal splitting, where the daughter follows a track parallel to that of the mother but slightly shifted upstream, i.e., off-aligned. Configurations C_1 and C_2 both account for longitudinal splitting but the latter hence introduces some lateral diffusion. Along this line of thought, numerical simulation results in [14], illustrated in Figure 4, suggest that probability p_2 is tiny close to $\mathrm{Re_g}$ but increases with Re. The corresponding R-configuration is $C_2 \equiv [\varnothing\,\varnothing\,R\,\varnothing\,\varnothing]$.

4. In configuration $C_3 \equiv [\varnothing\,\varnothing\,B\,\varnothing\,\varnothing]$, the active site B at $(i+1,j)$ is supposed to advance further at $(i+2,j)$ with probability p_5. Persisting activity at (i,j), therefore, means longitudinal splitting ahead but now with the opening of a wide laminar gap between the offspring left behind at (i,j) and the parent that has advanced, with probability p_5, at $(i+2,j)$. Else, activity at (i,j) and $t+1$ could result from activity at $(i+1,j)$ and time t propagating backwards to (i,j) at time $t+1$. These circumstances have not been observed and appears unlikely or impossible, which suggests to take $p_3 = 0$. The corresponding R-configuration is $C_3 \equiv [\varnothing\,R\,\varnothing\,\varnothing\,\varnothing]$.

5. In configuration $C_4 \equiv [\varnothing\,\varnothing\,\varnothing\,B\,\varnothing]$, the state B at $(i,j-1)$ and time t is expected to be at $(i+1,j-1)$ at time $t+1$. State at (i,j) being active at $t+1$ means contamination backwards and laterally downstream, which is never observed in the simulations; hence, $p_4 = 0$. The corresponding R-configuration is $C_4 \equiv [\varnothing\,\varnothing\,\varnothing\,\varnothing\,R]$.

6. Still about configuration C_4, the situations described in the previous items all imply single-colored evolution, which is guaranteed below the onset of transversal splitting, i.e., $R \lesssim 800$. When $\mathrm{Re} \gtrsim 800$, as illustrated in Figure 3, this splitting produces an R offspring at (i,j) out of a B parent at $(i,j-1)$ or B offspring from an R parent at $(i-1,j)$, as sketched in Figure 7 (left). A probability $p_4' \neq 0$ will be associated with it, where the prime is meant to recall that it involves states of different colors.

To summarize, as it stands the model involves four parameters: p_1 mainly governs longitudinal splitting and p_2 additional lateral diffusion, p_5 is for propagation, and p_4' for transversal splitting. The propagation of active states along their own direction involves probabilities associated with elementary configurations C_1 and C_5 while the overwhelming contribution of p_5 favors one direction. Configuration C_3 that could have contributed to the balance is empirically found negligible, saving one parameter as indicated above.

Neighborhoods with more than one active site are treated by assuming that the future state S' of the central node (i,j) is the combined output of its elementary ingredients, each contribution being considered as independent of the others, i.e., without memory of the anterior evolution, of which the considered configuration is the outcome. The computation of the probability attached to the output of a given single-colored neighborhood is then straightforward. The argument follows the lines given for directed percolation, bearing on the probability that the state at the node will be absorbing (empty) and leading to Equation (1) in the mean-field approximation [20,24]. Things are a little more complicated when the neighborhood is two-colored since in all mixed-colored cases some configurations correspond to collisions and others allow for the nucleation of a differently colored offspring when $p'_4 \neq 0$.

For an elementary configuration, non-contamination of site (i,j) from an active neighboring state in position $k \in [1{:}5]$ takes place with probability $(1 - p_k)$ and of course with probability 1 if the corresponding site is empty. This gives the general formula $(1 - p_k S_k)$, where $S_k = 1$, when the site is active, either B or R, and $S_k = 0$ when it is absorbing ($\varnothing$). For a configuration $C_x = [S_1, S_2, S_3, S_4, S_5]$, where $S = B$, R, or $\varnothing$, the probability to get an absorbing state is $(1 - p_{C_x}) = \prod_k (1 - p_k S_k)$ hence for the node to be activated $p_{C_x} = 1 - \prod_k (1 - p_k S_k)$. To deal with two-colored neighborhoods properly, we must be a little more specific and write the probability of the state S' of a given color S as

$$p_{[S_1, S_2, \bar{S}_4, S_5]} = 1 - (1 - p_1 S_1)(1 - p_2 S_2)(1 - p'_4 \bar{S}_4)(1 - p_5 S_5) \tag{3}$$

where it is understood that if $S = B$, then $\bar{S} = R$ or the reverse, and $S_j = 0$ for $j = 1, 2, 5$, or $\bar{S}_4 = 0$ if the corresponding states are $\varnothing$. Figure 7 (right) illustrates the most interesting two-state configurations with different colors corresponding to collisions (C1) and offspring generation (C2). Such a situation is dealt with by adding a supplementary rule:

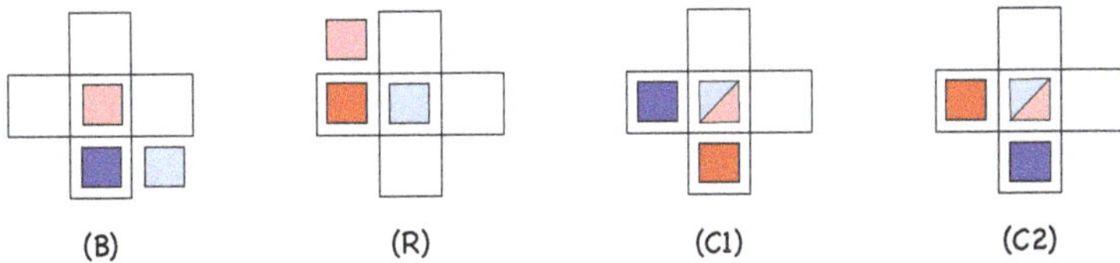

Figure 7. Modeling of transversal splitting for states of type (B) propagating horizontally and (R) propagating vertically, the base flow being along the diagonal ($\nearrow$). Heavy colors indicate states present at time t and, playing the role attributed to question marks in Figure 6; light colors stand for states possibly present at time $t + 1$ according to probabilities p_5 (propagation) and p'_4 (transversal splitting). Conflicting configurations are (C1) ($[SSSRB]$ corresponding to propagation leading to a collision and (C2) $[SSSBR]$ corresponding to simultaneous transversal splittings, respectively (here $S = \varnothing$ for clarity).

7. When the general expression (3) gives non-zero probabilities to S' and $\bar{S}'$ the resulting superposition of states is not allowed and a choice has to be made. It might seem natural to keep the state with the maximum probability but, depending on circumstances hard to decipher, collisions sometimes appear to cause the decay of both protagonists or else reinforce the dominance of one color in a given region of space. A similar bias can affect transversal splitting. These peculiarities are not taken into account here: for simplicity, in all conflicting cases, we make the assumption that the result is non-empty and random with probability $1/2$.

The model is now complete with parameters clearly related to empirical observations, plausible relative orders of magnitude and sense of variation: Probability p_5 is the main ingredient for the built-in propagation of the two families of LTBs (active states). In turn p_1 is obviously related to the behavior of the system close to decay at and slightly above Re_g. The value given to probability p_2 will appear crucial to the 1D reduction of DP in a 2D medium as observed experimentally (Figure 4, right). Finally, we can anticipate that probability p'_4 will control the one-sided/two-sided symmetry-restoring bifurcation, as it continuously grows from 0 beyond Event A at $R \approx 800$.

2.3. Mean-Field Approach

The explanatory potential of the model is first examined by means of a mean-field approximation which mainly relies on the replacement of fluctuating quantities by space-averaged values and the neglect of correlations. The observables involved in the mean-field expressions are the ensemble averages of the microscopic states $\langle S(i,j) \rangle$. Their values at $t+1$ are obtained by taking averages of the governing Equation (2) using the expression of the configurational probabilities given in (3). By assumption/definition $\langle S' \rangle$ is the mean outcome of p_{C_x} averaged over all the possible configurations, where space dependence (i,j) is temporarily kept: $\langle S'(i,j) \rangle = \langle p_{[S_1,S_2,S_4,S_5]} \rangle$. This gives a set of two equations:

$$\langle B'(i,j) \rangle = 1 - \langle (1 - p_1 B(i,j))(1 - p_2 B(i,j+1))(1 - p'_4 R(i-1,j))(1 - p_5 B(i-1,j)) \rangle, \quad (4)$$

$$\langle R'(i,j) \rangle = 1 - \langle (1 - p_1 R(i,j))(1 - p_2 R(i+1,j))(1 - p'_4 B(i,j-1))(1 - p_5 R(i,j-1)) \rangle. \quad (5)$$

The approximation now enters the evaluation of the products on the right hand side of the equation. Each variable is replaced by its average and the spatial dependence is dropped: $\langle B(i,j) \rangle \mapsto \langle B \rangle$ and $\langle R(i,j) \rangle \mapsto \langle R \rangle$. Further, correlations are neglected so that the average of a product is just the product of averages. The expansions of (4) and (5) in powers of $\langle B \rangle$ and $\langle R \rangle$ are then readily obtained. Forgetting for a moment the intricacy linked to transversal splitting/collisions, the general expression for the dummy variables $\langle S \rangle$ and $\langle S' \rangle$ reads:

$$\langle S' \rangle = \sum_k p_k \langle S_k \rangle - \sum_{k_1,k_2} p_{k_1} p_{k_2} \langle S_{k_1} \rangle \langle S_{k_2} \rangle + \text{h.o.t.} \quad (6)$$

with $p_k \in \{p_1, p_2, p'_4, p_5\}$ and where h.o.t. stands for the higher order terms, formally cubic, quartic, etc. The first sum in (6) corresponds to the contribution of the elementary configurations introduced in Figure 6, and the second sum to binary configurations, in particular the nontrivial ones corresponding to transversal splittings and collisions examined in Figure 8 (right). Orders of magnitude among the p_k, further support neglecting the contribution of configurations populated with three or more active sites, involving products of three or more probabilities p_k, and among contributions of a given degree, those not containing p_5 when compared to those that do, recalling the assumption $p_5 \lesssim 1$ and $\{p_1, p_2\} \ll 1$ implied by the nearly deterministic propagation of states in position 5 of Figure 6. A number of terms can, therefore, be neglected in the expanded forms of (4) and (5), which after simplification read:

$$\langle B' \rangle = (p_1 + p_2 + p_5)\langle B \rangle + p'_4 \langle R \rangle - p_5(p_1 + p_2)\langle B \rangle^2 - p_5{}^2 \langle B \rangle \langle R \rangle, \quad (7)$$

$$\langle R' \rangle = (p_1 + p_2 + p_5)\langle R \rangle + p'_4 \langle B \rangle - p_5(p_1 + p_2)\langle R \rangle^2 - p_5{}^2 \langle R \rangle \langle B \rangle. \quad (8)$$

This system presents itself as the discrete time counterpart of the differential system introduced in [14] to interpret the symmetry-breaking bifurcation observed at decreasing Re in the simulations. As a matter of fact, subtracting $\langle B \rangle$ and $\langle R \rangle$ on both sides of (7) and (8) respectively, one gets:

$$\langle B' \rangle - \langle B \rangle \approx \frac{d\langle B \rangle}{(dt \equiv 1)} = (p_1 + p_2 + p_5 - 1)\langle B \rangle + \ldots \quad (9)$$

$$\langle R' \rangle - \langle R \rangle \approx \frac{d\langle R \rangle}{(dt \equiv 1)} = (p_1 + p_2 + p_5 - 1)\langle R \rangle + \ldots \quad (10)$$

to be compared with system (1,2) in [14], reproduced here for convenience:

$$\frac{dX_+}{dt} = aX_+ + cX_- - bX_+^2 - dX_+X_-, \quad (11)$$

$$\frac{dX_-}{dt} = aX_- + cX_+ - bX_-^2 - dX_-X_+, \quad (12)$$

where $X_\pm$ represents what are now the densities $\langle B \rangle$ and $\langle R \rangle$. The coefficients in (11) and (12) are then related to the probabilities introduced in the model as $a \propto p_1 + p_2 + p_5 - 1$, $b \propto p_5(p_1 + p_2)$, $c \propto p'_4$, and $d \propto p_5^2$. By omitting the common proportionality constant that accounts for the time-stepping inherent in the discrete time reduction (featured by the denominator of left-hand sides in (9) and (10) as "($\mathrm{d}t \equiv 1$))," constants a, b, c, and d will serve as short-hand notation for the corresponding full expressions in terms of the probabilities p_k.

Since fixed points given by the condition $\langle S' \rangle = \langle S \rangle$ is strictly equivalent to $\mathrm{d}X_\pm / \mathrm{d}t = 0$, we can next take advantage of the analysis performed in [14] and predict a supercritical symmetry-breaking bifurcation for an order parameter $|\langle B \rangle - \langle R \rangle|$ (denoted "A" in [14]) at a threshold given by $c^{\mathrm{cr}} = a(d - b)/(d + 3b)$. This symmetry-breaking bifurcation takes place for $p'_4 = c > 0$, but the model can deal with the regime below event A at $\mathrm{Re} \approx 800$ for which $p'_4 \equiv 0$. In that case the bifurcation corresponding to global decay at $\mathrm{Re_g}$ takes the form of two coupled equations generalizing (1) for DP. Using the abridged notation, these equations read:

$$\langle B' \rangle = (a + 1)\langle B \rangle - b\langle B \rangle^2 - d\langle B \rangle\langle R \rangle, \qquad \langle R' \rangle = (a + 1)\langle R \rangle - b\langle R \rangle^2 - d\langle R \rangle\langle B \rangle. \tag{13}$$

In addition to the trivial solution $\langle R \rangle_0 = \langle B \rangle_0 = 0$ corresponding to laminar flow, we have two kinds of non-trivial solutions, either single-sided ($*$) with $\langle R \rangle \neq 0$ and $\langle B \rangle = 0$ or $\langle B \rangle \neq 0$ and $\langle R \rangle = 0$, the non-vanishing solution being $\langle S \rangle_* = a/b$, with $S = R$ or B, or double-sided ($**$) with $\langle B \rangle_{**} = \langle R \rangle_{**} = a/(b + d)$. A straightforward stability analysis of the fixed points of iterations (13) shows that the one-sided solution is stable when $b < d$ and unstable otherwise whereas the reversed situation holds for the two-sided solution. Returning to probabilities, the global stability threshold is thus given for $a = 0$; hence, $(p_1 + p_2 + p_5)^{\mathrm{cr}} = 1$ and the one-sided solution is expected when $b < d$; i.e., $p_1 + p_2 < p_5$. Results of the mean-field approach adapted from [14] to the present formulation will be illustrated in Figure 14 below.

2.4. Numerical Simulations

While serving as a guide to the exploration of a vast range of parameters, the simplified mean-field theory developed above is not expected to give realistic results relative to the critical properties expected near the transition point, whether decay at $\mathrm{Re_g}$ or symmetry restoration above $\mathrm{Re_2}$. For example, observations suggest that LTB propagation is a dominant feature; hence, $p_5 \lesssim 1$ and $\{p_1, p_2\}$ is small, leading us to expect stable one-sided solutions systematically. This conclusion, however, strongly relies on neglecting all terms beyond second degree in (4) and (5) in the evaluation of the contribution of densely populated configurations, leading to (7) and (8). This is legitimate only when $\langle S \rangle^n \ll \langle S \rangle^2$, i.e., $\langle S \rangle \ll 1$, that is, close to decay in the case of a continuous (second-order) transition but not necessarily elsewhere in the parameter space, in particular at the one-sided/two-sided bifurcation where both $\langle R \rangle$ and $\langle B \rangle$ are of the same order of magnitude but may be large. Even when keeping the assumption of independence of contributions to the future state at a given lattice node, this problem is not easily addressed and, at any rate, has to be properly accounted for in the presence of stochastic fluctuations, which will be done numerically.

The translation of the probabilistic rules introduced in Section 2.2 using Matlab® is straightforward once the "$B/R/\varnothing$" convention is appropriately translated into "$+1/-1/0$". No assumption is made other than the independence of the contributions of the different configurations to the outcome at a given lattice node, by strict application of the rules expressed through (2) and (3). In particular, computations involve the contribution of all configurations and not only the unary or binary ones, as presumed to derive the mean-field equations. Periodic boundary conditions have been applied to 2D lattices of various dimensions ($N_B \times N_R$), where N_B (N_R) is the number of sites in the propagation direction of B (R) active states, with ordinarily $N_B = N_R$. At each simulation step, we shall measure the mean activity of B and R states denoted $\langle B \rangle$ and $\langle R \rangle$ above and from now on called turbulent fractions, as $F_t(B) = (N_B N_R)^{-1}\#(B)$ and $F_t(R) = (N_B N_R)^{-1}\#(R)$ where $\#(B)$ and $\#(R)$ are the numbers of sites in the corresponding active state.

A preliminary study of the model in a small domain has shown that the different transitional regimes and the symmetry-breaking bifurcation were indeed present as expected from the simplified mean-field approach. (We remind that the model contains nothing appropriate for organized laminar–turbulent regimes for Re > 1200 and is relevant only for the strongly intermittent sparse LTB networks pictured in Figures 2 and 3). In [14], we argued that the onset of transversal splitting was the source of genuinely 2D behavior. Accordingly we shall consider the stochastic model in two steps, below and above the onset of transversal splitting, here associated with $p'_4 \equiv 0$ and $p'_4 > 0$ respectively. Furthermore, in the simulations the LTBs were seen to propagate obliquely with respect to the background downstream current. This propagation is nearly all contained in the probability attached to configuration $\mathcal{C}_5$ (p_5 for propagation and $1 - p_5$ for decay or slowing-down), and to a lesser extent influenced by the contribution of configuration $\mathcal{C}_1$, mostly associated with in-line longitudinal splitting. We shall account for the limited sensitivity of the propagation speed to the value of Re to fix p_5 constant and close to 1, more specifically $p_5 = 0.9$, and let other parameters vary. The role of p_2 and p'_4, both related to 2D features, will be studied separately in the two next sections.

3. Before Onset of Transversal Splitting, $P'_4 = 0$

3.1. Coarsening from Two-Sided Initial Conditions

In the absence of transversal splitting, changes in the population of each state only comes from transversal collisions. As documented in [14], when starting from an initial condition with two similarly represented orientations, collisions lead to the formation of domains uniformly populated by one of each species, following from a majority rule, with interactions limited to the domain boundaries. A coarsening takes place with one species progressively disappearing to the benefit of the other, leaving a single-sided state at large times. The process is illustrated here using simulations of the model with $p_5 = 0.9$, $p_1 = 0.1$, $p_2 = 0.07$, values known from the preliminary study to produce a sustained nontrivial final state.

The decay from a fully active state populated with a random distribution of B and R states in equal proportions is scrutinized in a 256×256 domain with periodic boundary conditions. Figure 8 illustrates a particularly long transient displaying the different stages observed during a typical experiment.

The upper panel displays the time series of the turbulent fractions for each species, B and R, for a two-sided high-density initial condition, $F_t(B) + F_t(R) = 1$, $F_t(B) \simeq F_t(R) \simeq 0.5$. Contrasting with the monotonic variation observed when starting from one-sided initial conditions, either increasing from a low density of active states ($F_t = 0.05$) or decreasing from a fully active configuration ($F_t = 1$), the turbulent fractions change in a more complicated way that is easily understood when looking at the bottom line of snapshots. The total turbulent fraction first decreases due to the dominant effect of collisions. These collisions tend to favor a spatial modulation of the activity amplifying inhomogeneities in the initial conditions. This distribution results from the majority effect expressing the local stability of one-sided states predicted by the mean-field analysis. A periodic pattern already appears at $t = 100$, with bands oriented parallel to the second diagonal of the square domain. B states move right along the horizontal axis, and R states up along the vertical axis, at the same average speed so that the pattern drifts along the first diagonal of the domain. Regions where B or R dominate are locally stable against destructive collisions and activity is limited to B/R interfaces. After a while, splittings begin to counteract collisions and an overall activity recovers, here for $t \approx 250$. The local density of B and R states increases inside bands that become better defined, reaching a sustained regime with two R–B alternations, wide and narrow, at $t \simeq 1500$. This configuration is nearly stable and slowly evolves only due to the erosion of narrowest bands at the R/B interfaces. At $t \approx 5500$ these bands disappear by merging, leaving two bands, B wide and R narrow. The same slow erosion process leads to the final homogeneous B regime by decay of the R band at $t \sim 96{,}000$. The two successive band decays take place at roughly constant total turbulent fraction with fast adjustment at the band decay, up to the final single-sided turbulent fraction. The asymptotic state is independent of the way it has been obtained, from one-sided or two-sided initial conditions.

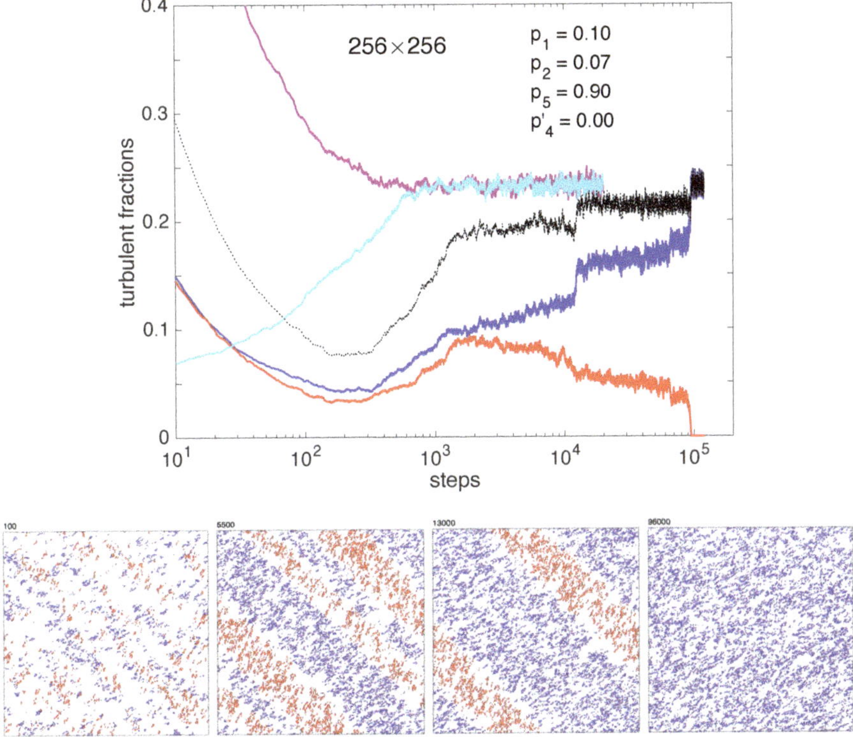

Figure 8. (**Top**): Time series of the turbulent fractions for a simulation from a fully active initial configuration with B and R states in equal proportions—blue and red in graphs, respectively; the dotted black trace is for the total turbulent fraction. Two simulations starting from low ($F_t = 0.05$, cyan) and high ($F_t = 1$, magenta) density one-sided states are displayed for comparison. (**Bottom**): Snapshot of state during the simulations from the two-sided initial condition, at $t = 100$ during initial decay, at $t = 5500$ with two pairs of active bands of each color, at $t = 13{,}000$ when the narrowest bands merge and disappear, at $t = 96{,}000$ when the R active band disappears, leaving a uniform B state.

The long duration of the transient taken as an example is due to the near stability of the rather regular pattern building up after the initial fast decay. This property is in fact the result of a geometrical peculiarity of the square domain: B and R states travel statistically at the same speed through the domain, horizontally and vertically, respectively, so that the band integrity is maintained despite propagation and the evolution controlled by collisions at the B–R and R–B interfaces only. The observed slow erosion process only results from large deviations among collisions. In rectangular domains, the propagation times become different and the symmetry of the two interfaces is lost. A bias results, which induces a systematic erosion of bands and a shorter transient duration. Whatever the aspect ratio, one of the states is always ultimately eliminated and the last stage of the transient corresponds to a trend toward a statistically uniform saturated one-sided regime with a turbulent fraction strictly independent of the shape. Accordingly, to save the time corresponding to the transient, in the next section we will study the decay of the one-sided regime by starting from random one-sided initial conditions.

All these features nicely fit the empirical observations discussed at length in [14] where similar transients were obtained below the onset of transversal splitting—in much smaller effective domains and with far fewer interacting LTBs, however (Figure 2, right panel, and Figure 4, left panels).

3.2. Decay: 1D vs. 2D

The model is designed to exemplify a decay according to the DP scenario in a two-dimensional setting, with specificities linked to the anisotropic propagation properties of the LTBs in transitional channel flow, and, in particular, propose an interpretation for the observation of 1D-DP exponents in the absence of transversal splitting ($p'_4 = 0$). Accordingly, we examine the role of transverse diffusion (parameter p_2) modeling the slight upstream shift that may affect LTBs at longitudinal splitting. We focus on a set of experiments with $p_5 = 0.9$, p_2 fixed, and control parameter p_1. When p_2 cancels exactly, it is easily understood that transversal expansion is forbidden: An active B state at $(i, j + 1)$ or R state at $(i + 1, j)$ at time t cannot give birth to an active state of the same kind at (i, j) at $t + 1$. The evolution stems from processes associated with configuration $\mathcal{C}_5$ with probability p_5 or $\mathcal{C}_1$ with probability p_1. These processes change occupancy only along direction i for B states, and j for R states, precisely in the direction corresponding to the single-sided regime considered (after termination of the transient). The dynamics are, therefore, strictly one-dimensional and decay is expected to follow the 1D-DP scenario. In contrast, introducing some transverse diffusion ($p_2 \neq 0$) immediately gives some 2D character to the dynamics. This is illustrated in Figures 9–12.

We consider first p_2 non-zero and relatively large $p_2 = 0.1$. Figure 9 displays the behavior of the turbulent fraction as a function of p_1.

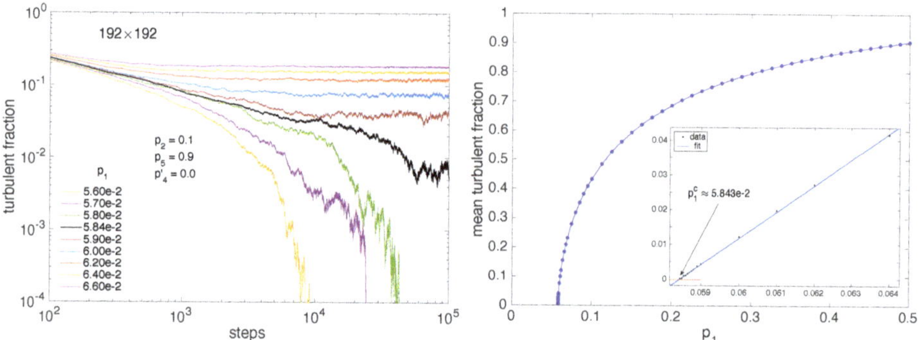

Figure 9. (**Left**): Time series of the turbulent fraction at different values of p_1; average over 5 (10) independent simulations ($p_1 = 0.0584 \approx p_1^c$, black trace). (**Right**): Mean value of the turbulent fraciton at stationary state as a function of p_1 (original data). Inset: once raised to power $1/\beta$, with $\beta = 0.584 \approx \beta_{DP}$ for $D = 2$, the mean turbulent fraction tends to 0 linearly with an extrapolated threshold $p_1^c = 0.05843$.

The left panel illustrates the decrease of the turbulent fraction with the number of steps from a uniformly fully turbulent single-sided state ($F_t = 1$ at $t = 0$) in a domain $\mathcal{D} = (192 \times 192)$, showing the saturation to a finite value $\langle F_t \rangle$ above threshold, a near power-law decay close to threshold, and an exponential decay below. The right panel presents the mean of F_t after elimination of an appropriate transient as a function of p_1, for simulations in domains up to 512×512 for the lowest values of F_t. Once fitted in the range $p_1 \in [0.058, 0.064]$ against the expected power law behavior $\langle F_t \rangle = a(p_1 - p_1^c)^\beta$ one gets $a = 3.213$ $(2.936, 3.489)$, $p_1^c = 0.05844$ $(0.05842, 0.05845)$, $\beta = 0.5811$ $(0.566, 0.5962)$, in very good agreement with the value $\beta_{DP} \approx 0.584$ when $D = 2$ [2]. This is confirmed in the inset of Figure 9 (right) showing $\langle F_t \rangle^{1/0.584}$ as a function of p_2 for F_t small, the linear variation of which extrapolates to zero for $p_1 \approx 0.05843$.

Having a good estimate of the threshold one can next consider the decay of the turbulent fraction, which is supposed to decrease as a power law at criticality, $p_1 = p_1^c$: $F_t \sim t^{-\alpha_{DP}}$ with $\alpha_{DP} = \beta_{DP}/\nu_{\parallel DP}$ where $\nu_{\parallel DP} \approx 1.295$; hence, $\alpha_{DP} \approx 0.451$ [2]. Figure 10 (left) shows that this is indeed the case for the compensated turbulent fraction $F_t \times t^{\alpha_{DP}}$, up to the moment when fluctuations become too important due to size effects and lack of statistics. When p_1 is different from p_1^c but stays sufficiently close to it, the variation of the turbulent fraction keeps trace of the critical situation, except that the number of steps needs to be rescaled by the distance to threshold due to critical slowing down: the time scale τ diverging as $(p_1 - p_1^c)^{-\nu_\parallel}$, number of steps is rescaled upon multiplying it by $(p_1 - p_1^c)^{\nu_\parallel}$. Figure 10 (right) indeed shows a good collapse of the compensated curves as a function of the rescaled number of steps when using the exponents corresponding to 2D-DP, $\alpha_{DP} \approx 0.451$ and $\nu_{\parallel DP} \approx 1.295$ [2].

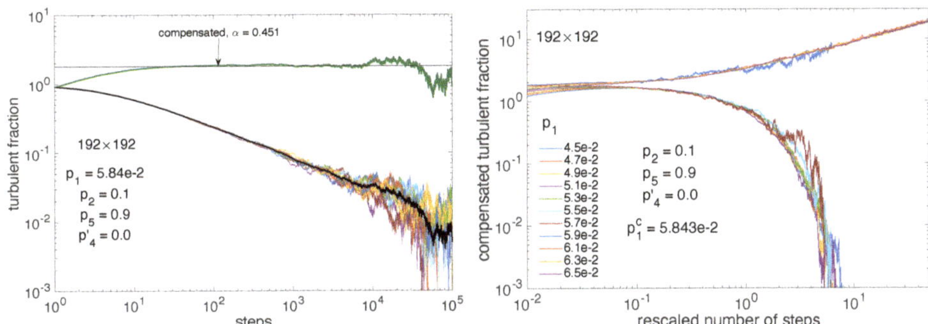

Figure 10. (**Left**): Power law decay of the turbulent fraction at $p_1 = 0.0584 \approx p_1^c = 0.05843$: compensation with α_{DP} confirms the 2D nature of the process. (**Right**): Critical behavior near threshold: compensated turbulent fraction $\langle F_t \rangle \times t^{\alpha_{DP}}$ as a function of the number of steps rescaled by $(p_1 - p_1^c)^{\nu_{\parallel DP}}$ for $p_1 \in [4.5{:}0.2{:}6.5] \times 10^{-2}$ surrounding the presumed critical value p_1^c, with the exponents corresponding to DP for $D = 2$.

We now consider $p_2 = 0$ which, as argued earlier, should fit the critical behavior of directed percolation when $D = 1$. In that case, when using square or nearly-square rectangular domains, size effects turn out to be particularly embarrassing as will be illustrated quantitatively soon. However, we can take advantage of the fact that, assuming propagation in the one-sided regime, e.g., along the direction for B active states, N_B being the corresponding number of sites involved, the computed turbulent fraction is, in fact, the average of the activity over N_R independent lines in the complementary direction, while still being sensitive to size effects controlled by N_B. Accordingly, at given computational load (proportional to $N_B \times N_R$), one can freely increase the size artificially in considering a strongly elongated domain $\mathcal{D}' = [(N_B \times k) \times (N_R/k)]$, with k sufficiently large that the average over N_R/k independent lines still make sense from a statistical point of view, while postponing size effects. With reference to a (192×192) domain, we have obtained good results with $k = 16$, i.e., 3072×12 up to $k = 64$, i.e., 12288×3.

Though this choice is a bit extreme, we present here results about 1D-DP criticality with the 12288×3 domain in Figure 11. The left panel displays the variation of the mean turbulent fraction with p_1, which has been fitted against the expected power law, $\langle F_t \rangle = a(p_1 - p_1^c)^\beta$. One gets $a = 1.473$ $(1.446, 1.5)$, $p_1^c = 0.2682$ $(0.2682, 0.2683)$, $\beta = 0.2701$ $(0.2664, 0.2738)$. This value of β is quite compatible with the value $\beta_{DP} \approx 0.276$ when $D = 1$ [2]. Furthermore, accepting this value, a linear fit of $\langle F_t \rangle^{1/\beta}$ with p_1 then provides an extrapolated threshold $p_1^c = 0.26817$. As seen in the right panel of Figure 11, in the neighborhood of p_1^c a good collapse is obtained for the compensated turbulent fraction as a function of the rescaled number of steps when using the exponents $\alpha = 0.159$ and $\nu_\parallel = 1.734$ corresponding to 1D-DP [2].

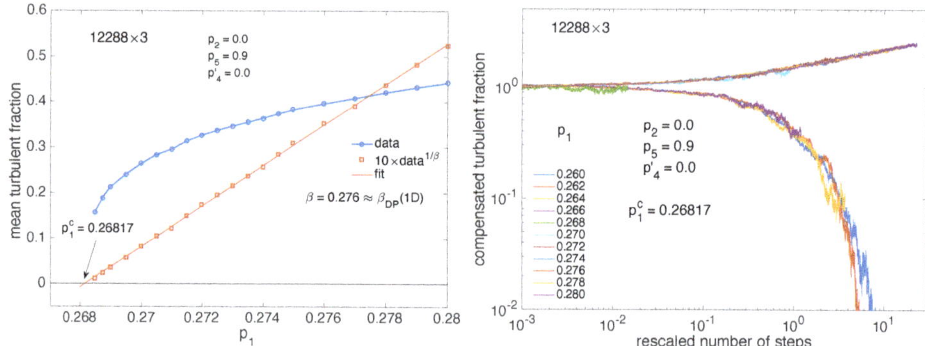

Figure 11. (**Left**): Mean turbulent fraction at stationary state as a function of p_1. Once raised at power $1/\beta$ with $\beta = 0.276 \approx \beta_{DP}$ for $D = 1$, the mean turbulent fraction tends to 0 linearly with an extrapolated threshold $p_1^c = 0.26817$. (**Right**): Critical behavior near threshold: compensated turbulent fraction $\langle F_t \rangle \times t^{\alpha_{DP}}$ as a function of the number of steps rescaled by $(p_1 - p_1^c)^{\nu_{\parallel DP}}$ for $p_1 \in [0.260{:}0.002{:}0.280]$ surrounding the presumed critical value p_1^c, with the exponents corresponding to DP for $D = 1$.

Size effects already alluded to above are illustrated in Figure 12.

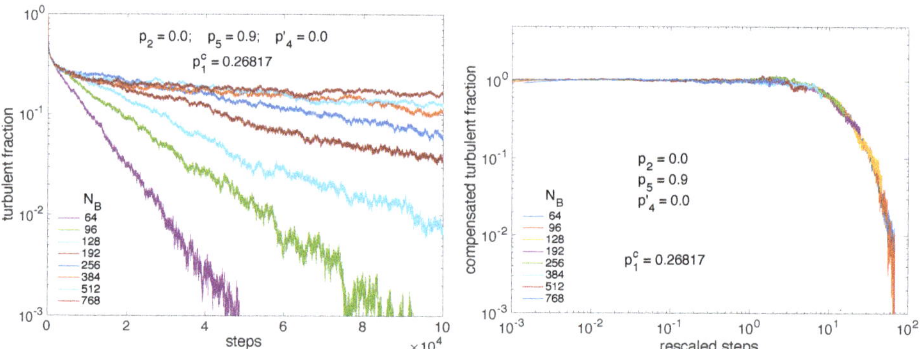

Figure 12. Size effects when $p_2 = 0$. (**Left**): Raw data showing late exponential decay and progressive prevalence of power-law decay as N_B grows. (**Right**): Rescaled data. According to scaling theory, the appropriate scale for the number of steps (time t) is $N_B^{z/D}$; hence, $t \mapsto t/N_B^{z/D}$, with $z \approx 1.58$ when $D = 1$, while the turbulent fraction has to be compensated for decay as $F_t \times t^\alpha$ with $\alpha = 0.159$. The collapse of traces illustrates universality with respect to 1D-DP.

Displaying the turbulent fraction as a function of the number of steps for linear size N_B from small systems to relatively large ones ($N_B = 64$ up to 768) in lin-log scale, the left panel illustrates the late stage of decay right at criticality as obtained from the previous study summarized in Figure 11. It is seen that, in the time-window considered $(0, 10^5)$ the exponential dependence observed at small sizes is progressively replaced by the power-law behavior expected at criticality at infinite size. Size effects are also ruled by scaling theory; see, e.g., [2] for DP. They relate to correlations in physical space that are associated with exponent $\nu_\perp$. The ratio $z = \nu_\parallel / \nu_\perp$ is called the dynamical exponent and theory predicts that, for finite size systems, scaling functions depend on time with the number of sites as $t^{D/z}/N$ where N is the total number of sites. In the (quasi-)one-dimensional regime we are interested in, $D = 1$, N is just N_B and $z = 1.58$ [2]. The right panel of Figure 12 indeed shows extremely good collapse of the traces corresponding to those in the left panel, once the number of steps is rescaled as

$t/N_B^{1.58}$ and the turbulent fraction is compensated for decay as $F_t(t) \times t^{0.159}$, both exponents taking on the 1D-DP values already mentioned.

Of interest in the context of channel flow decay, the crossover from 2D behavior for p_2 sizable (e.g., $p_2 = 0.1$, Figures 9 and 10) to 1D behavior for $p_2 = 0$ is of interest since p_2 is associated with the progressive importance of off-aligned longitudinal splitting as Re increases. A series of values of p_2, decreasing to zero roughly exponentially, has been considered and the corresponding DP threshold has been determined as given in Table 1 and shown in Figure 13 (left).

Table 1. Values of p_1 at criticality at given p_2 ($p_5 = 0.9$ and $p'_4 = 0$).

p_2	0.0	0.0001	0.0002	0.0005	0.001	0.002	0.005	0.01	0.02	0.05	0.1
p_1^c	0.2682	0.2585	0.2548	0.2476	0.2404	0.2302	0.2111	0.1907	0.1629	0.1109	0.0584

Except for $p_2 = 0$ determined as explained above (Figure 11), these values have been obtained in domains 192×192 with averaging over 10 independent experiments. Figure 13 (right) displays the averaged time-series of the turbulent fraction at criticality for each of these values of p_2, once compensated for decay according to 2D-DP ($\alpha = 0.451$).

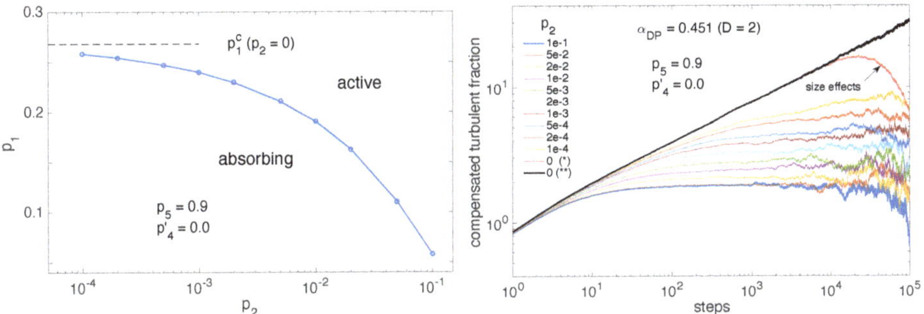

Figure 13. Crossover $p_2 \to 0$. (**Left**): Criticality condition separating the sustained active regime from the absorbing regime. (**Right**): A DP-like process governs the decay when the line is crossed, the characteristics of which can be understood from the asymptotic power-law decrease of the turbulent fraction as a function of time, here compensated by t^α, with $\alpha = \alpha_{DP}(D = 2)$.

The results for $p_2 = 0$, evolving as $t^{\alpha_{2D} - \alpha_{1D}}$ are marked with (*) and (**) are obtained in the 192×192 domain and in the 3072×12 quasi-1D domain, respectively. In the time span considered here, the latter is free from finite-size effects which is not the case of the former with the corresponding compensated data decaying exponentially at the largest times. It is easily seen that, except for $p_2 = 0$, the compensated time series display a wide plateau indicating that 2D behavior holds for a certain amount of time. Whereas traces for $p_2 = 0.1$ and $p_2 = 0.05$ cannot be distinguished, for smaller values of p_2 the plateau regime starts at larger and larger times and develops after having followed the 1D trace for longer and longer durations, clearly indicating the influence of the anisotropy controlling the effective dimensional reduction. A similar consequence of the crossover affects the decrease of the mean turbulent fraction with the distance to threshold but, apart from this qualitative observation, no reliable information can be obtained on exponent β owing to the difficulty to reach the relevant critical regime.

We shall not document the case when $p_1 = 0$ and p_2 varies. This situation is not observed in the simulations since off-aligned longitudinal splitting is conspicuous only sufficiently above Re_g, in the vicinity of which decay is fully accounted for by in-line longitudinal splitting modeled by a variable $p_1 \neq 0$, but the possibility remains, at least conceptually. The decay when $p_1 = 0$

happens to follow the same 1D-DP scenario though the argument is slightly less immediate. It relies on the observation that no growth is possible in the propagation direction of a given LTB species, whereas off-aligned longitudinal splitting ($p_2 \neq 0$) permits growth and diffusion in the transverse direction. Under the combined effects of transversal diffusion (p_2 small) and propagation (p_5 large), near-threshold, the sustained turbulent regime is made of quasi-1D clusters that are aligned with and drift along the diagonal of the lattice, i.e., the stream-wise direction, and get thinner and thinner when decaying, supporting the reduction to a "$D = 1$" scenario. Here, the trick used for $p_2 = 0$ does not work, and simulations in square domains are necessary with no escape for size effects which hinders the observation of the critical regime. Nevertheless, p_2^c when $p_1 = 0$ seems close to p_1^c when $p_2 = 0$, suggesting some symmetry between p_1 and p_2.

The relevance of the results with $p_4' \equiv 0$ to transitional channel flow will be discussed in the concluding section. We now turn to the general two-sided case with transversal collisions and splittings.

4. Beyond Onset of Transversal Splitting, $P_4' > 0$

In statistical thermodynamics systems, critical properties at a second order phase transition leads to define a full set of exponents governing the variation of macroscopic observables close to criticality [26]. The concept of universality was introduced to support the observation that these systems can be classified according to the value of their exponents depending on a few qualitative characteristics, the most prominent ones being the symmetries of the order parameter and the dimension of physical space. This viewpoint can be extended to far-from-equilibrium systems such as coupled map lattices (CMLs) displaying nontrivial collective behavior. The associated ordering properties present many characteristics of thermodynamical critical phenomena at equilibrium. Universality classes beyond those known from equilibrium thermodynamics have been shown to exist with different sets of exponents. An additional criterion, the synchronous or asynchronous nature of the dynamics, has been found relevant to distinguish among them [27]. In the context of the present model, as soon as probability p_4' grows from zero, fully one-sided configurations previously reached after the termination of a possibly long transient are now unstable against the presence of states with the complementary color. The stationary regime that develops in the long term can be, either ordered, i.e., one-sided with one dominant active state (B or R), or disordered, i.e., two-sided with statistically equal fractions of each active state (B and R). Furthermore, a transition at some critical value $p_4'^c$ is expected to take place on general grounds. This gives us the motivation to study the response of the model to the variation of p_4' as a critical phenomenon beyond the mean-field expectations of Section 2.3.

The results of the mean-field approach, system (11) and (12), rephrased from [14], are depicted in Figure 14 (left). Upon variation of parameter c representing p_4' up to an unknown rescaling factor, all along the one-sided regime ($c < c^{cr}$), the total turbulent fraction is seen to decrease while the order parameter measuring the lack of symmetry similarly decreases to zero according to the usual Landau square-root law. Obviously symmetrical, the two-sided regime ($c > c^{cr}$) is then characterized by a total turbulent fraction that regularly grows due to the contribution of splitting, whatever the type of active state.

From now on, we shall simply refer to the turbulent fractions and other statistical quantities as their time average over a sufficiently long duration, up to 2×10^6 simulation steps, after elimination of an appropriate transient, up to 10^5 steps, the largest values being necessary close to the transition point owing to the well-known critical slowing down. On the one hand, the total turbulent fraction is obviously defined as $\overline{F_t(B)} + \overline{F_t(R)}$, where the over-bar denotes the time averaging operation. (Later on, we shall omit this over-bar when no ambiguity arises between the instantaneous value of a quantity and its time average, especially for the axis labelling in figures.) On the other hand, the lack of symmetry can be measured by the signed difference averaged over time $\overline{F_t(B) - F_t(R)}$, able to distinguish global B orientation from its R counterpart, or rather its absolute value $\left| \overline{F_t(B) - F_t(R)} \right|$ since we are only interested in the amplitude of the asymmetry (called 'A' in [14]) and not in which orientation is dominant, the two being equivalent a priori for symmetry reasons. However, due to the finite size of the system, in the symmetry-broken regime close to threshold, orientation reversals can be observed as illustrated later (Figure 15), so that blind statistics in the very long durations are no longer representative of the actual ordering. Like in thermal systems [28] or their non-equilibrium counterparts [27], it is thus preferable to define the order parameter through the mean of the unsigned difference: $\overline{|F_t(B) - F_t(R)|}$. Corresponding simulation results are displayed in Figure 14 (right) for a system of size (256×256). The general agreement between the two diagrams is remarkable, up to an unknown multiplicative factor translating c into p'_4, as discussed earlier. One can notice that the order parameter is minimal but not zero in the two-sided regime, which is due to fluctuations and the fact that the two operations of averaging over time and taking the absolute value do not commute. Finite-size effects are also apparent as a rounding of the graph at the location of the would-be critical point in the thermodynamic limit.

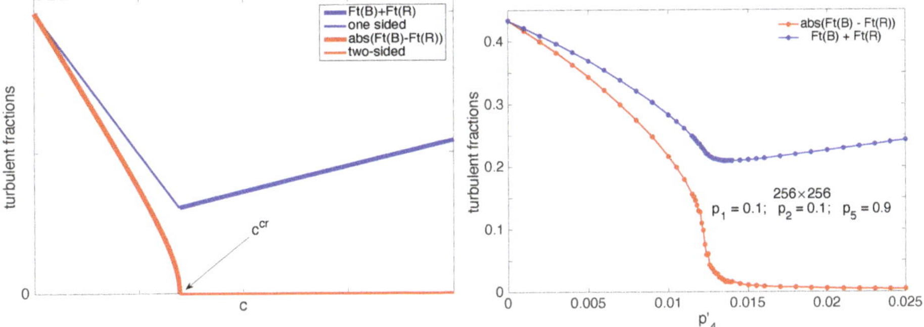

Figure 14. (**Left**): Bifurcation diagram of system (11) and (12) after [14]. The total turbulent fraction is $F_t(B) + F_t(R)$ and the order parameter characterizing the transition is $\mathrm{abs}(F_t(B) - F_t(R))$. A standard supercritical bifurcation is expected for this quantity with $\mathrm{abs}(F_t(B) - F_t(R)) \propto (c^c - c)^{1/2}$ in the one-sided regime, whereas $F_t(B) = F_t(R)$ in the two-sided regime. (**Right**): Time average of turbulent fractions as functions of the control parameter p'_4 after elimination of an appropriate transient as obtained from simulations of the stochastic model.

The current justification for taking the absolute value is that the time between orientation reversals diverges with the system size and the phase transition only takes place once we have taken the thermodynamic limit of infinitely large systems studied over asymptotically long durations [28]. Accordingly, very long well-oriented intermissions can be considered as representative of the symmetry-broken regime. The problem is illustrated in Figure 15 displaying the time series of $F_t(B) - F_t(R)$ and histograms of $|F_t(B) - F_t(R)|$ for $p'_4 = 0.0121$, still in the one-sided but already alternating regime, next for $p'_4 = 0.0125$ and 0.0126, where one can notice a change in the shape of the histogram, and finally for $p'_4 = 0.0140$, sufficiently deep inside the two-sided regime where the histogram displays a sharp maximum at the origin. On this basis one could use the histograms of the "order parameter" and determine the threshold from the position of its most probable value, whether non-zero in the symmetry-broken state or at the origin when symmetry is restored. This procedure would give $p'^c_4 \approx 0.01255$.

The symmetry-breaking bifurcation can now be studied beyond the mean-field description as other collective phenomena studied in equilibrium and far-from-equilibrium statistical physics: In addition to the order parameter, the variation of which leads to the definition exponent β in the ordered regime, another observable of interest is the susceptibility measuring the response to an applied field conjugate to the order parameter, vis. $M = \chi H$ with the magnetization M coupled to magnetic field H in the case of magnets. The susceptibility diverges near the critical point, with leads to the definition of two exponents γ and γ' in the disordered and ordered regime, respectively.

Universality implies $\gamma = \gamma'$, as can already easily be derived in the mean-field framework. When a conjugate field cannot be defined, one uses the property that fluctuations take the instantaneous value of the order parameter away from its average value, which can be understood as resulting from the response to a conjugate field. This helps one to relate the susceptibility to the variance of fluctuations of the order parameter. The identification is up to a multiplication by the "volume" of the system that has to be introduced in order to compare the results from systems with different sizes. This is what will be done here; hence, $\chi = N_B N_R \times \mathrm{var}\left(|F_t(B) - F_t(R)|\right)$. As shown in Figure 16 (top), this quantity displays a sharp maximum, indicative of the singularity expected at the thermodynamic limit. In a finite-size but large system, the critical point is then estimated from the position of the maximum of the susceptibility. Here, this gives $p_4'^{\,c} \approx 0.0123$ slightly smaller but compatible with the value obtained above from the examination of the histograms. Unfortunately, this discrepancy due to size-effects forbids us to determine exponents β and γ with some confidence.

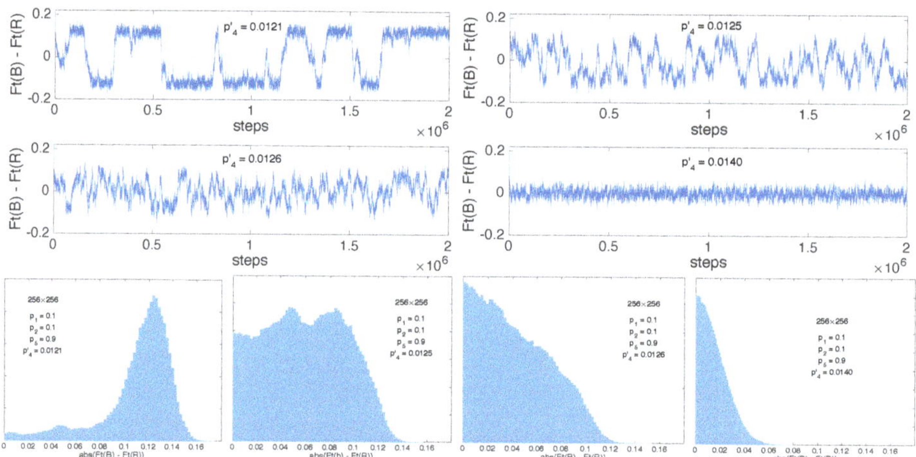

Figure 15. Time series of the instantaneous mean orientation as measured by $F_t(B) - F_t(R)$ in a 256×256 domain for $p_1 = p_2 = 0.1$ and $p_5 = 0.9$, at $p_4' = 0.0121$ below the onset of the one-sided regime, at $p_4' = 0.0125$ and 0.0126 near the bifurcation point, and at $p_4' = 0.0140$ in the two-sided regime. Bottom line: Corresponding histograms of $|Ft(B) - Ft(R)|$. The histograms were all built using 75 bins and contain the same number of points for $10^5 < t < 2 \times 10^6$, but the vertical scales are not identical.

Having in mind results of the mean-field approach, namely, $\beta = 1/2$ and $\gamma = 1$, we can, however, estimate the range where stochastic fluctuations have nontrivial effects. The bottom-left panel of Figure 16 displays the variation of the order parameter already shown in Figure 14 (right), but now squared in order to show that, far from the critical point, the system fulfils the mean-field square-root prediction to an excellent approximation, with an extrapolated threshold $p_4'^{\,\mathrm{MF}} \simeq 0.0133$, shifted upwards with respect to the estimates obtained from the simulations $p_4'^{\,c.} \simeq 0.0123$–$0.0125$.

In the same way, the divergence of the susceptibility with exponents $\gamma = \gamma' = 1$ expected from the mean-field argument shows up upon retreating the data already given in Figure 16 (top) and plotting $1/\chi$ as a function of p_4'. This is done in Figure 16 (bottom-right) showing the same linear variation of $1/\chi$ below and above the transition point, in agreement with the theory. The extrapolation of the linear fits on both sides of the transition yield $p_4'^{\,c} \approx 0.0127$ in reasonable agreement with the value obtained from the order parameter variation in the same conditions and definitely larger than the empirical values. Clearly, deviations seen in the boxed parts of these two figures warrant further scrutiny, motivating our current approach via finite-size scaling theory [28] in search for universality. On going work attempts at a full characterization of the critical regime through exponents determination. Though local agents do not behave as Ising spins, symmetries are basically identical, so that the equilibrium 2D Ising universality class or its non-equilibrium extension [27] might be relevant. We shall discuss this further below.

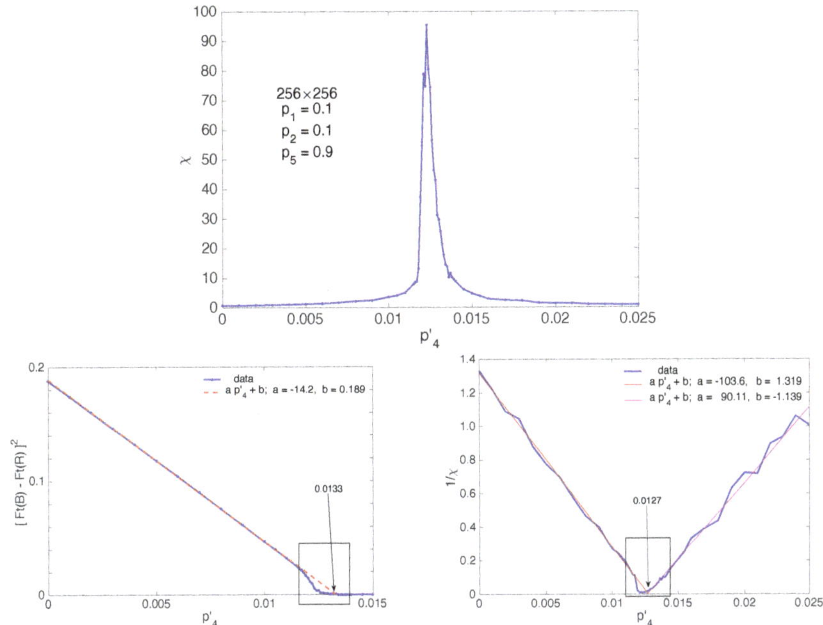

Figure 16. (**Top**): Variation of the susceptibility χ as a function of p'_4. (**Bottom**): Evidence of mean-field behavior away from the critical regime: The order parameter squared (**left**) and the inverse of the susceptibility (**right**) both vary linearly with p'_4 sufficiently far from the mean-field extrapolated threshold.

5. Discussion and Concluding Remarks

Coming long after a conjecture by Pomeau [1], empirical evidence is growing that the ultimate stage of decay of wall-bounded turbulent flows towards the laminar regime follows a directed-percolation scenario. The evidence comes from laboratory experiments and direct numerical simulation of the Navier–Stokes equations but this support is still far from a theoretical justification. The recognition of the globally subcritical character of nontrivial states away from laminar flow and the elucidation of the structure of coherent structures involved in these nontrivial states [29] were first steps in this direction. The next ones would be the elucidation of special phase space trajectories from sustained localized turbulence accounting for the decay to laminar regime, on one side, and to proliferation via splitting, on the other side, using specific algorithms for the detection of rare events and the determination of transition rates that can be attached to them (see [30] for an illustrative example and references). These are heavy, and possibly not much rewarding, tasks but it would be nice to be able to attach numbers to specific events such as the splittings illustrated in Figure 3 or Figure 4. We have chosen to short-circuit such studies through analogical modeling, by which seemed more appropriate to make further progress regarding the thermodynamic limit and associated universality issues. One should though consider this practice as providing hints and not a demonstration that the results will apply to the case under study.

In the present paper, the problem has been considered from this last viewpoint, assuming that the ultimate decay stages were amenable to the most abstract level of implementation in terms of probabilistic cellular automata [2], following [20,22]. We focussed on the specific case of channel flow that offers a particularly rich transitional range. Its upper part displays regular non-intermittent laminar–turbulent patterns that can better be described using the tools of pattern-forming theory [15,25,31]. The lower transitional range is characterized by their spatiotemporally intermittent disaggregation, to which the considered type of modeling is particularly relevant. The analogy alluded to above has, however, been severely constrained to fit the empirical observations. The main assumptions were the introduction of two types of active agents attached to each kind of localized turbulent bands propagating in one of the two possible orientations with respect to the stream-wise direction. Interactions were assumed local so that the probabilistic cellular automata evolved simple nearest neighbors von Neumann neighborhoods (Figures 5–7). Scrutiny of simulation results lead to the introduction of a certain number of probabilities governing the fate of single-occupancy neighborhoods. Multiple-occupancy was treated as a combination of single-occupancy configurations supposedly independent, reducing the number of parameters to be introduced and drastically simplifying the interactions (at any rate impractical to estimate in detail). A clear-cut physical interpretation was, however, given to each parameter in the set reduced to four, accounting for every possible stochastic event affecting the agents, namely, propagation, decay, and splitting, either longitudinal or transversal. A mean-field study of the model, neglecting the nontrivial effects of stochastic fluctuations, reproduced the empirical bifurcation diagram of channel flow at a qualitative level (Figure 14). Transitions have been studied quantitatively by numerical simulation of the stochastic model considering variations of these parameters as putative functions of the Reynolds number Re, highlighting three situations:

In the two first cases, the parameter p_4' associated with transversal splitting, i.e., the nucleation of a daughter with orientation opposite of its mother, was switched off, as inferred from observations for Re $\lesssim$ 800, where the single-sided regime is well established. The coarsening observed when starting from two-sided initial conditions was faithfully reproduced (Figure 8) and decay seen to follow the directed-percolation expectations. The specific conclusion was that, when parameter p_2 is no-zero, with p_2 attached to longitudinal but upstream-shifted splitting, the scenario is typical of a 2D system with a high level of confidence, whereas when it is strictly zero, i.e., the daughter strictly aligned with the mother, the decay is 1D. A cross-over is observed when p_2 is reduced, that manifests itself as a transient reminiscent of 1D behavior, the longest the closest p_2 is to zero. Simulations of channel flow have shown that exponent β controlling the ultimate decay of the turbulent fraction was that of 1D directed percolation [16]. Since parameter p_2 is attached to the slight upstream trajectory shift experienced by a daughter upon splitting from its mother, this observation strongly suggests that the trajectory shift is mostly irrelevant and that localized turbulent bands propagate along independent tracks so that the end result is just a mean over the direction complementary to their propagation direction.

The last situation we have considered corresponds to $p_4' \neq 0$, with transversal splitting on. This parameter measures the frequency of transversal splitting and is expected to increase with Re. Accordingly, the system can change from one-sided when p_4' is zero or small, to two-sided when it is large. The transition has indeed been observed and mean-field predictions were well observed far from the transition point. Unfortunately, while the effect of fluctuations close to that point was obvious, strong size effects have forbidden us to approach it and evaluate critical corrections. This is the subject of on-going work within the framework of finite-size scaling theory [2,27,28]. This follow-up should allow us to establish the universality class to which this transition belongs. Here, the left-right symmetry of localized turbulent bands with respect to the stream-wise direction is reminiscent of the up-down symmetry of magnetic systems at thermodynamic equilibrium, which may lead to conjecture the relevance of the 2D Ising class [26]. This class appears also applicable to coupled map lattices with the same up-down symmetry when updated asynchronously, one site after the other, close to randomization by thermal fluctuations. In contrast, another universality class is obtained with synchronous update [27]. Here, the situation is unclear: on the one hand, configurations are treated as a whole in a simulation step, which tips the scales in favor of a synchronous update model (in line with what is expected for a problem primitively formulated in terms of partial differential equations); on the other hand, spatial correlations generated by the deterministic dynamics governing the coupled map lattices are weakened by the independence of random drawings at the local scale, which can be viewed as a source of asynchrony in the probabilistic cellular automata. In its application to the symmetry-breaking bifurcation in channel flow, this uncertainty is, however, only of conceptual importance in view of size effects: owing to the large and unknown time-scale rescaling that allowed us to pass from flow structures to local agents in the model and to the narrowness of the region where critical corrections are expected, the mean-field interpretation developed in [14] appears amply sufficient.

In the three cases that were considered in detail (specific cuts in the parameter space), the transitions remained continuous. However, this may not always be the case since there are known example of similar systems displaying transitions akin to first-order ones [24]. Even while keeping the same general frame, a plethora of circumstances of physical interest can be mimicked: propagation can be made more stochastic by decreasing p_5, splitting rules not observed in channel flow can be considered, e.g., with p_3 or p_4 different from zero, etc., though it seems hard to anticipate situations where the universal features pointed out here would not hold. In contrast, when dealing with highly populated configurations, even in the simple nearest-neighbor von Neumann setting, rules can be made more complicated by introducing the neighborhood's degree of occupation. This introduction might help us to account also for the upper part of the transitional range of wall-bounded flows characterized by the emergence of regular patterns in the same stochastic framework [11]. The construction of the present model is, of course, fully adapted to the study of universality in the framework of the theory of critical phenomena in statistical physics, especially directed percolation. Still, we are confident that the kind of approach illustrated here brings a valuable contribution to the understanding of the transition to turbulence, by rationalizing its key ingredients in an easily accessible way.

Author Contributions: Conceptualization, P.M., M.S.; methodology, P.M.; software, P.M., M.S.; validation, M.S.; formal analysis, P.M.; investigation, P.M.; resources, M.S.; data curation, M.S., P.M.; writing—original draft preparation, P.M.; writing—review and editing, P.M., M.S. All authors have read and agreed to the published version of the manuscript.

Funding: This research received no external funding.

Acknowledgments: P.M. wants to thank H. Chaté (CEA-Saclay, Gif-sur-Yvette, France) for pointing out a model that could help one uncover the universal contents of the symmetry-breaking bifurcation on the same footing as the decay near $\mathrm{Re_g}$.

Conflicts of Interest: The authors declare no conflict of interest.

Abbreviations

1D/2D/3D	One/two-dimensional (depending on 1/2 *space* coordinates)
DP	Directed percolation (stochastic competition between decay and contamination)
LTB	Localized turbulent bands
DAH	downstream active head (part of an LTB controlling its propagation)
PCA	Probabilistic cellular automata
CML	Coupled map lattice (spatially-discrete iterative model)

References

1. Pomeau, Y. Front motion, metastability and subcritical bifurcations in hydrodynamics. *Phys. D* **1986**, *23*, 3–11. [CrossRef]
2. Hinrichsen, H. Non-equilibrium critical phenomena and phase transitions into absorbing states. *Adv. Phys.* **2000**, *49*, 815–958. [CrossRef]
3. Manneville, P. Transition to turbulence in wall-bounded flows: Where do we stand? *Mech. Eng. Rev. Bull. JSME* **2016**, *3*. [CrossRef]

4. Tuckerman, L.S.; Chantry, M.; Barkley, D. Patterns in Wall-Bounded Shear Flows. *Annu. Rev. Fluid Mech.* **2020**, *52*, 343–367. [CrossRef]
5. Lemoult, G.; Shi, L.; Avila, K.; Jalikop, S.V.; Avila, M.; Hof, B. Directed percolation phase transition to sustained turbulence in Couette flow. *Nat. Phys.* **2016**, *12*, 254–258. [CrossRef]
6. Chantry, M.; Tuckerman, L.S.; Barkley, D. Universal continuous transition to turbulence in a planar shear flow. *J. Fluid Mech.* **2017**, *824*, R1. [CrossRef]
7. 73rd Annual Meeting of the APS Division of Fluid Dynamics. Sessions Y03 and Z10. 2020. Available online: https://dfd2020chicago.org/ (accessed on 27 November 2020).
8. Waleffe, F. On a self-sustaining process in shear flows. *Phys. Fluids* **1997**, *9*, 883–900. [CrossRef]
9. Kaneko, K. Overview of coupled maps. *Chaos* **1992**, *2*, 279–283. [CrossRef]
10. Bottin, S.; Chaté, H. Statistical analysis of the transition to turbulence in plane Couette flow. *Eur. Phys. J. B* **1998**, *6*, 143–155. [CrossRef]
11. Vanag, V.K. Study of spatially extended dynamical systems using probabilistic cellular automata. *Phys. Uspekhi* **1999**, *42*, 413–434. [CrossRef]
12. Kashyap, P.V.; Duguet, Y.; Dauchot, O. Flow Statistics in the Transitional Regime of Plane Channel Flow. *Entropy* **2020**, *22*, 1001. [CrossRef]
13. Gomé, S.; Tuckerman, L.S.; Barkley, D. Statistical transition to turbulence in plane channel flow. *Phys. Rev. Fluids* **2020**, *5*, 083905. [CrossRef]
14. Shimizu, M.; Manneville, P. Bifurcations to turbulence in transitional channel flow. *Phys. Rev. Fluids* **2019**, *4*, 113903. [CrossRef]
15. Manneville, P.; Shimizu, M. Subcritical Transition to Turbulence in Wall-Bounded Flows: The Case of Plane Poiseuille Flow. Rencontre du Non-Linéaire, Paris. 2019. Available online: http://nonlineaire.univ-lille1.fr/SNL/media/2019/CR/ComptesRendusRNL2019.pdf (accessed on 27 November 2020).
16. Shimizu, M.; Manneville, P. Onset of sustained turbulence and (1+1)D directed percolation in channel flow. In Proceedings of the Annual Meeting of Fluid Mechanics Society of Japan, Tokyo, Japan, 13–15 September 2019.
17. Seshasayanan, K.; Manneville, P. Laminar-turbulent patterning in wall-bounded shear flow: A Galerkin model. *Fluid Dyn. Res.* **2015**, *47*, 035512. [CrossRef]
18. Chantry, M.; Tuckerman, L.S.; Barkley, D. Turbulent–laminar patterns in shear flows without walls. *J. Fluid Mech.* **2016**, *791*, R8. [CrossRef]
19. Barkley, D. Theoretical perspective on the route to turbulence in a pipe. *J. Fluid Mech.* **2016**, *803*, P1. [CrossRef]
20. Allhoff, K.T.; Eckhardt, B. Directed percolation model for turbulence transition in shear flows. *Fluid Dyn. Res.* **2012**, *44*, 031201. [CrossRef]
21. Sano, M.; Tamai, K. A universal transition to turbulence in channel flow. *Nat. Phys.* **2016**, *12*, 249–253. [CrossRef]
22. Kreilos, T.; Khapko, T.; Schlatter, P.; Duguet, Y.; Henningson, D.S.; Eckhardt, B. Bypass transition and spot nucleation in boundary layers. *Phys. Rev. Fluids* **2016**, *1*, 043602. [CrossRef]
23. Domany, E.; Kinzel, W. Equivalence of cellular automata to Ising models and directed percolation. *Phys. Rev. Lett.* **1984**, *53*, 311–314. [CrossRef]
24. Bagnoli, F.; Boccara, N.; Rechtman, R. Nature of phase transitions in a probabilistic cellular automaton with two absorbing states. *Phys. Rev. E* **2001**, *63*, 046116. [CrossRef] [PubMed]
25. Takeda, K.; Duguet, Y.; Tsukahara, T. Intermittency and Critical Scaling in Annular Couette Flow. *Entropy* **2020**, *22*, 988. [CrossRef]
26. Stanley, H.E. *Introduction to Phase Transitions and Critical Phenomena*; Oxford University Press: Oxford, UK, 1988.
27. Marcq, P.; Chaté, H.; Manneville, P. Universality in Ising-like phase transitions of lattices of coupled maps. *Phys. Rev. E* **1997**, *55*, 2606–2627. [CrossRef]
28. Privman, V. (Ed.) *Finite-Size Scaling and Numerical Simulation of Statistical Systems*; World Scientific: Singapore, 1990.
29. Kawahara, G.; Uhlmann, M.; van Veen, L. The significance of simple invariant solutions in turbulent flows. *Annu. Rev. Fluid Mech.* **2012**, *44*, 203–225. [CrossRef]

30. Rolland, J. Extremely rare collapse and build-up of turbulence in stochastic models of transitional wall flows. *Phys. Rev. E* **2018**, *97*, 023109. [CrossRef]

31. Manneville, P. Laminar-Turbulent Patterning in Transitional Flows. *Entropy* **2017**, *19*, 316. [CrossRef]

Publisher's Note: MDPI stays neutral with regard to jurisdictional claims in published maps and institutional affiliations.

Article

Spatiotemporal Intermittency in Pulsatile Pipe Flow

Daniel Feldmann *, Daniel Morón and Marc Avila

Center of Applied Space Technology and Microgravity (ZARM), University of Bremen, Am Fallturm 2,
28359 Bremen, Germany; daniel.moron@zarm.uni-bremen.de (D.M.); marc.avila@zarm.uni-bremen.de (M.A.)
* Correspondence: daniel.feldmann@zarm.uni-bremen.de

Received: 4 December 2020; Accepted: 24 December 2020; Published: 30 December 2020 

Abstract: Despite its importance in cardiovascular diseases and engineering applications, turbulence in pulsatile pipe flow remains little comprehended. Important advances have been made in the recent years in understanding the transition to turbulence in such flows, but the question remains of how turbulence behaves once triggered. In this paper, we explore the spatiotemporal intermittency of turbulence in pulsatile pipe flows at fixed Reynolds and Womersley numbers ($Re = 2400$, $Wo = 8$) and different pulsation amplitudes. Direct numerical simulations (DNS) were performed according to two strategies. First, we performed DNS starting from a statistically steady pipe flow. Second, we performed DNS starting from the laminar Sexl–Womersley flow and disturbed with the optimal helical perturbation according to a non-modal stability analysis. Our results show that the optimal perturbation is unable to sustain turbulence after the first pulsation period. Spatiotemporally intermittent turbulence only survives for multiple periods if puffs are triggered. We find that puffs in pulsatile pipe flow do not only take advantage of the self-sustaining lift-up mechanism, but also of the intermittent stability of the mean velocity profile.

Keywords: unsteady shear flow; turbulence intermittency; helical instability; puff dynamics

---

1. Introduction

The dynamics and intermittency of transitional turbulence in statistically steady pipe flow have been extensively studied for over a century [1–4], and the underlying mechanisms are reasonably well understood [5]. The only control parameter is the Reynolds number ($Re = \frac{\langle u_b \rangle_t D}{\nu}$), which quantifies the relative magnitude of inertia and viscous forces in the system. Here, u_b, D, and ν denote the bulk velocity, the pipe diameter, and the fluid's kinematic viscosity, respectively. Angled brackets indicate an averaging operation with respect to time (t). Although statistically steady pipe flow is linearly stable [6,7], turbulence can be triggered with finite-amplitude perturbations [1,8]. Independently of their type [9], if successful, these perturbations result in spatially localised turbulent puffs, provided that the Reynolds number is not too high [10]. More specifically, for $Re < 2250$, puffs can remain in equilibrium for long times until they either proliferate or decay. Both processes are stochastic (memoryless) and, beyond the critical point ($Re > 2040$), ultimately lead to patterns consisting of several puffs separated by quiescent flow regions [5,11]. For $Re > 2250$, the spatiotemporal dynamics become much richer. Here, puffs may grow and split into two, as for lower Reynolds numbers, or expand continuously to become slugs (see Figure 1). In addition, laminar holes may appear inside the slugs and eventually close, leading to a merger of structures [12]. Figure 2a provides a representation of the resulting spatiotemporally intermittent behaviour of localised turbulent structures (red) and laminar islands (blue) at $Re = 2400$.

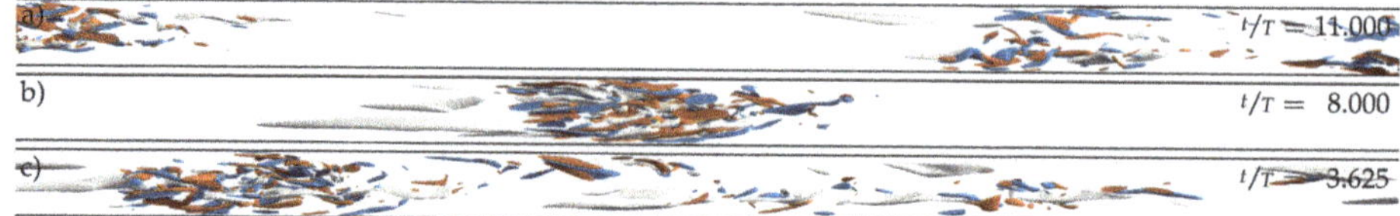

Figure 1. Instantaneous representation of localised turbulent structures in a statistically steady pipe flow ($Re = 2400$, $A = 0.0$). Grey surfaces represent low-speed streaks ($u'_z = -0.4u_b$) and blue/red surfaces represent positive/negative axial vorticity ($\omega_z = \pm 6\frac{u_b}{D}$). (**a**) Puff splitting. (**b**) Single puff. (**c**) Weak slug. The exact location and time for each snapshot are indicated in Figure 2a. The direction of the mean bulk flow (u_s) is always from left to right.

In many systems, internal fluid transport is statistically unsteady. Pumps never run perfectly uniformly, blood flow in arteries is pulsatile (due to the systolic contractions of the heart), and air oscillates in and out of the lungs while breathing. A simple mathematical model for these examples is pipe flow driven at a harmonically varying rate

$$u_b(t) = \langle u_b \rangle_t \left(1 + A \cdot \cos\left(2\pi \tfrac{t}{T}\right)\right). \tag{1}$$

In this case, two more control parameters come in to play in addition to the Reynolds number. The Womersley number ($Wo = \frac{D}{2}\sqrt{\frac{2\pi}{T\nu}}$) quantifies the relative magnitude of the viscous time scale with respect to the time scale of the imposed flow pulsation, i.e., the oscillation period T. The amplitude ($A = \frac{u_o}{u_s}$) is the relative strength of the oscillating component of the flow (u_o) with respect to the steady component of the flow ($u_s = \langle u_b \rangle_t$). For $A = 0$, the statistically steady case is recovered, whereas for large A, the purely oscillatory flow is approached (as the steady part becomes negligible). According to Sexl [13] and Womersley [14], there is an analytical solution to the Navier–Stokes equations for laminar flow through a smooth pipe and single harmonic driving. The Sexl–Womersley (SW) velocity profile ($u_{SW}(r,t)$) can be added to the (parabolic) Hagen–Poiseuille profile to obtain an analytical (laminar) solution for any combination of Wo and A. As an example, we show in Figure 3a the temporal evolution of u_{SW} for a pulsatile pipe flow at $Wo = 8$ and $A = 1$.

Understanding the transition to turbulence in statistically unsteady pipe flows remains incomplete, although progress has recently been made [15–17]. The puff dynamics for relatively small amplitudes ($A \leq 0.5$) are well understood. For $Wo \leq 5$, the flow stays for a long time in the low Reynolds number regime. A low instantaneous Re enhances the decay of puffs, and hence, puffs only survive if the mean Reynolds number is substantially increased with respect to the steady case [15,16]. For $Wo \geq 12$, the minimum Reynolds number necessary for puffs to survive tends asymptotically to the one for statistically steady pipe flow [15,16,18,19]. For intermediate Womersley numbers, the threshold decreases smoothly from the low to high Wo regime [15,16]. This can be seen, for example, in Figure 8 of Xu et al. [15].

Puffs, however, are not the only mechanism through which pulsatile pipe flow may become turbulent. A new instability was discovered recently in laboratory experiments by Xu et al. [17]. In their experiments, curvature, misalignment of pipe segments, small contractions, and, in general, finite-size geometric imperfections led to the cyclic development of sudden bursts of turbulence. At each period, helical-like structures grew and triggered turbulence during the deceleration phase of the pulsation before the flow relaminarised again during the acceleration phase. This behaviour was observed for relatively high amplitudes ($A \geq 0.5$), intermediate Womersley numbers ($5 \leq Wo \leq 8$), and mean Reynolds numbers as low as $Re = 800$. Motivated by this finding, Xu et al. [20] carried out a comprehensive non-modal stability analysis of pulsatile pipe flow. They showed that certain helical perturbations exploit an Orr-like mechanism to grow by several orders of magnitude in energy. They linked this mechanism to the inflection points of the SW velocity profile that emerge during the deceleration phase (see Figure 3a–c). Inflectional SW velocity profiles are indeed known to be linearly unstable in the quasi-steady limit [21], as long as they satisfy the Fjortoft criteria. The smaller the Womersley number, the longer the velocity profile is unstable, thus effectively providing a more fertile ground for instabilities to grow. However, as the Womersley number is reduced, the velocity profile becomes increasingly parabolic and, hence, loses its inflection points. The amplification of helical disturbances is most efficient for $Wo \approx 7$ [20], exactly in the regime where the helical instability was observed experimentally [17].

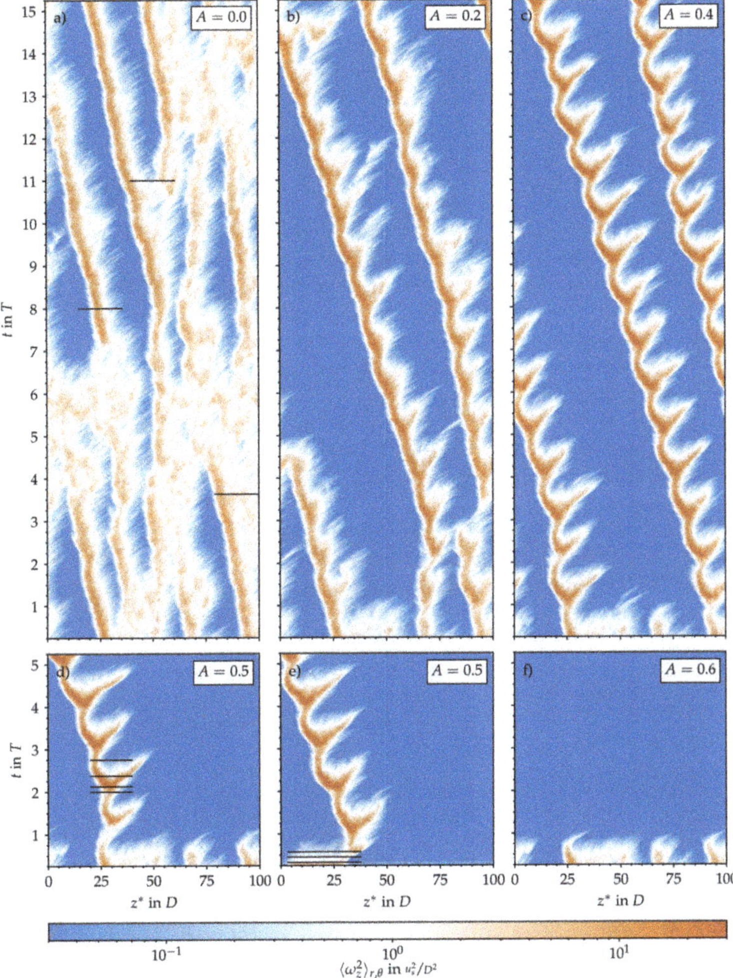

Figure 2. Spatiotemporal representation of the turbulence activity in the computational pipe domain based on the cross-sectional average of the streamwise vorticity (ω_z) plotted on a logarithmic scale and in a co-moving reference frame. Steady ($A = 0$, **a**) and pulsatile (**b**–**f**) pipe flow at $Re = 2400$, $Wo = 8$, and different amplitudes A. Initial conditions for all $A \neq 0$ were either taken from the steady case at time $\frac{t}{T} = 0.25$ (**b**–**d**,**f**) or composed of a localised helical perturbation on top of the laminar Sexl–Womersley velocity profile (**e**).

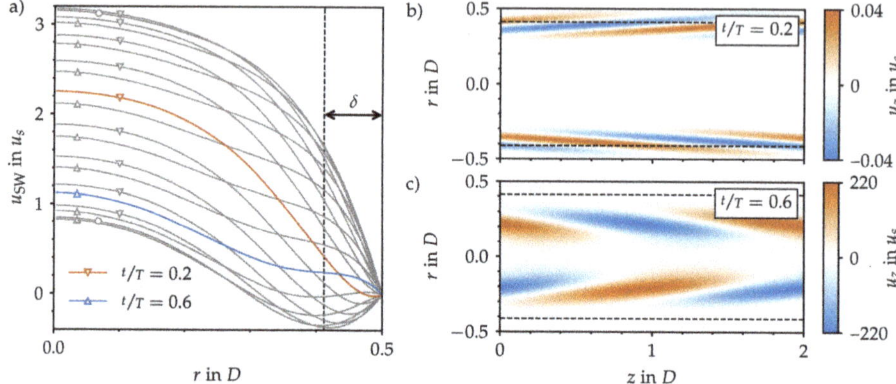

Figure 3. Sexl–Womersley (SW) flow and its optimal perturbation for ($Re = 2400$, $Wo = 8$, $A = 1.0$). (**a**) Time-dependent velocity profile (u_{SW}) for 20 equispaced points within one pulsation period (T). Circles denote the maximum and minimum peak flow (PF), whereas upward- and downward-facing triangles denote phases of acceleration (AC) and deceleration (DC), respectively. (**b**) Optimal helical perturbation during DC ($\frac{t}{T} = 0.2$) according to our transient growth analysis based on the linearised Navier–Stokes equations. To be used as initial condition in our direct numerical simulation (DNS) (Section 3.3), the helix is scaled to an amplitude of $4 \times 10^{-2}\ u_s$. (**c**) Evolution of the optimal perturbation under the constraints of the linearised Navier–Stokes equations at the later time of maximal energy amplification. Note that, in the framework of transient growth analysis, the absolute amplitude of the initial helix is not important; only the relative growth rate is of interest. The dashed lines correspond to the Stoke layer thickness (δ).

The purpose of this paper is to investigate the spatiotemporal intermittency of turbulence in pulsatile pipe flow. More specifically, we aim to characterise the intermediate regime in which helical structures and puffs are expected to compete. To that end, we perform transient growth analysis and direct numerical simulations of pulsatile pipe flow at fixed $Re = 2400$ and $Wo = 8$, as well as different pulsation amplitudes A.

2. Numerical Methodology

2.1. Governing Equations

We consider a viscous fluid with constant properties confined in a straight smooth rigid pipe of circular cross-section and diameter D. The fluid flow is driven through the pipe with a time-dependent pressure gradient, and is considered to be incompressible and governed by the Navier–Stokes equations (NSE)

$$\frac{\partial \mathbf{u}}{\partial t} + (\mathbf{u} \cdot \nabla)\mathbf{u} = -\nabla p + \frac{1}{Re}\nabla^2 \mathbf{u} + \mathbf{F}_d(t) + \mathbf{F}_p(r, \theta, z, t) \quad \text{and} \quad \nabla \cdot \mathbf{u} = 0. \tag{2}$$

Here, $\mathbf{u}$ and p denote the fluid velocity and pressure. The driving force $\mathbf{F}_d(t)$ represents a mean pressure gradient, which is adapted in a way such that the flow rate (u_b) given in Equation (1) is maintained. The additional body force term $\mathbf{F}_p(r, \theta, z, t)$ is used to model geometric imperfections in the pipe geometry, and thus to perturb the flow locally (see Section 2.4). Unless otherwise stated, all quantities are rendered dimensionless using the pipe diameter D, the statistically steady part of the bulk velocity $u_s = \langle u_b \rangle_t$ (see Equation (1)), and the fluid's density (ρ).

2.2. Direct Numerical Simulation

In order to study the departure from the laminar Sexl–Womersley flow and the dynamics of intermittent localised turbulence, we perform direct numerical simulations (DNS) using our open-source pseudo-spectral simulation code nsPipe [22]. In nsPipe, the governing Equation (2) are treated in cylindrical coordinates (r, θ, z) and discretised using a Fourier–Galerkin ansatz in θ and z and high-order finite differences in r. No-slip boundary conditions are imposed at the solid pipe wall and periodic boundary conditions in θ and z. The discretised NSEs are integrated forward in time using a second-order predictor–corrector method with variable time-step size; details are given in López et al. [22] and the references therein. We have modified nsPipe to account for a time-dependent driving force that maintains a pulsating flow rate according to Equation (1) and an additional volume force F_p to perturb the flow locally.

We use a computational domain of $100D$ in length, and the number of radial grid points and Fourier modes used in our DNS is $(N_r \times N_\theta \times N_z) = (96 \times 192 \times 2400)$. After dealiasing, this results in a spatial resolution of $\Delta\theta R^+ = 3.1$ and $\Delta z^+ = 3.8$, whereas radial grid points are clustered towards the pipe wall such that $0.06 \leq \Delta r^+ \leq 1.4$, and 14 points lie within the buffer layer based on the shear Reynolds number for the statistically steady case ($A = 0$). By comparison to the resolution used in other contemporary DNS studies in the literature and the fact that we consider only moderate A (the instantaneous Reynolds number is never $> (1 + A)Re$), we expect our choice of resolution to be sufficient for all set-ups considered here. The adaptive time-step size is roughly $\Delta t = 2 \times 10^{-3}\frac{D}{u_s}$.

2.3. Transient Growth Analysis

In order to study the linear stability of the SW profile and to determine the perturbations that grow the most on top of it within one pulsation period, we have performed transient growth analysis (TGA) for the parameter space at hand. This non-modal method returns the most dangerous perturbation in terms of energy growth out of all possible axial/azimuthal wavenumbers and pairs of initial (t_0) and final (t_f) times. To this end, the governing Equations (2) are linearised (LNSE). The LNSE and their adjoint counterpart are integrated forward and backward in time iteratively, until such an optimum is reached for each combination of Re, Wo, and A. During integration, the underlaying velocity profile develops in time (see, e.g., Figure 3a), but remains unchanged by the developing perturbations.

The LNSE and their adjoint are discretised using a Fourier–Galerkin ansatz in θ and z and a Chebyshev collocation method in r. Further details are given in Barkley et al. [23], and our TGA computations were undertaken using an in-house Matlab script.

2.4. Modelling Geometric Imperfections in Our DNS

Section 3.6 presents results from DNS in which we mimiced the geometric perturbation of the experiments of Xu et al. [17]. Inspired by the optimal baffle designed by Marensi et al. [24], we here model the effect of geometric perturbations with an additional volume force in Equation (2) of the form

$$F_p(r, \theta, z, t) = -A_p \cdot f_p(r, \theta, z) \cdot \mathbf{u}(r, \theta, z, t). \tag{3}$$

The body force F_p acts against the velocity field u and is localised in the radial, azimuthal, and axial direction by

$$f_p(r, \theta, z) = f(r) \cdot g(\theta, z) \cdot h(z) \text{ with} \tag{4}$$

$$f(r) = \frac{1}{2} + \frac{1}{\pi} \arctan(M_r(r - r_0)), \tag{5}$$

$$g(\theta, z) = \frac{1}{\pi}(\arctan(M_\theta(\theta - \pi(\theta_0(z) - L_\theta))) - \arctan(M_\theta(\theta - \pi(\theta_0(z) + L_\theta)))), \tag{6}$$

$$h(z) = \frac{1}{\pi}\left(\arctan\left(M_z\left(z - z_0 + \frac{L_z}{2}\right)\right) - \arctan\left(M_z\left(z - z_0 - \frac{L_z}{2}\right)\right)\right) \text{and} \tag{7}$$

$$\theta_0(z) = 1 + 2\Delta\theta \frac{(z - z_0)}{L_z}. \tag{8}$$

These localisation functions satisfy the constraints $\max(f) = 1$, $\min(f) = 0$, $\max(g) = 1$, $\min(g) = 0$, $\max(h) = 1$, and $\min(h) = 0$; the perturbation amplitude is given by A_p.

Due to the big parametric space in hand, we designed three simple body force set-ups and left further optimisation of parameters as future work. The first set-up is an axisymmetric force that models the effect of a small circumferential contraction similar to weak stenosis in blood vessels [25] or imperfect pipe joints in laboratory experiments [17] (see Figure 4a). The second set-up is a highly localised force that approximates the effect of a single bump or an individual roughness element (see Figure 4b). The third set-up is also a highly localised force that approximates the effect of a single bump or an individual roughness element, but this time, it is tilted with respect to the axial direction (see Figure 4c). The parameters defining the perturbations are given in Table 1. We studied the effect of the axisymmetric force on steady laminar Hagen–Poiseuille flow at $Re = 2400$ to select a suitable value of the force amplitude A_p. Our criterion was that the force must be strong enough to sufficiently disturb the flow without creating too long of a re-circulation region. For $A_p = 0.25$, we found a fair compromise between these two constraints.

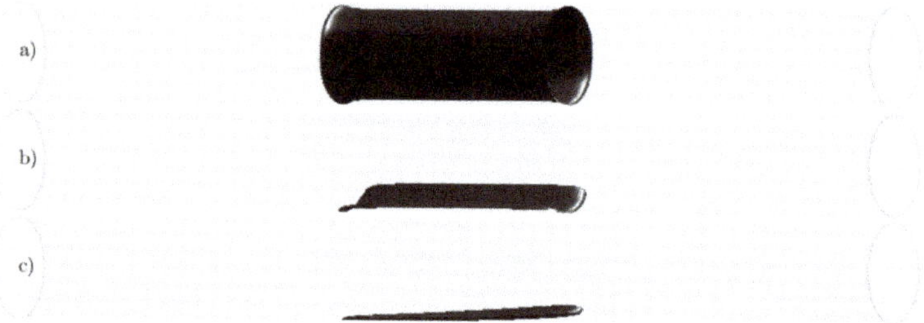

Figure 4. Geometric representation of the perturbation force (F_p) in terms of iso-surfaces (black) of the localisation function for $f_p = 0.5$. (**a**) Axisymmetric contraction. (**b**) Localised bump. (**c**) Tilted bump. See Table 1 for details. The direction of the mean bulk flow (u_s) is always from left to right.

Table 1. Parameters to control the body force term in Equation (3) to model the effect of geometric perturbations: Magnitude (A_p) and slope (M), size (L), and location in the radial (r), azimuthal (θ), and axial (z) direction. Geometric representations of the perturbations are shown in Figure 4.

	A_p	M_z in $\frac{1}{D}$	L_z in D	z_0 in D	M_r in $\frac{1}{D}$	r_0 in D	M_θ	L_θ	$\Delta\theta$
Contraction	0.25	4	2.5	10	100	0.45	20	≥ 1	0
Bump	0.25	4	2.5	10	100	0.45	20	0.25	0
Tilted Bump	0.25	4	2.5	10	100	0.45	20	0.0625	0.1

The goal of this model is to serve as a proof of concept. Our hypothesis is that geometric imperfections employed in the experiments locally modify the flow pattern causing the instability. The model satisfies this requirement, as it represents a small perturbation to the flow. It is meant for testing such a hypothesis, whereas the precise shape of its geometry plays an ancillary role. In order to faithfully reproduce the experiments of Xu et al. [17], one would need to have a boundary-fitted mesh or use immersed boundary methods. We are, however, confident that if the DNS was exactly reproducing the precise imperfections of the experiments, the exact same behaviour would be observed in the DNS.

3. Results

We first tested the effect of the pulsation on spatiotemporal intermittency by performing a DNS initialised with a snapshot of the statistically steady pipe flow (SSPF), as shown in Figure 2a. We refer to these simulations as IC SSPF. Next, we followed Xu et al. [20] and performed a linear non-modal stability analysis to identify the optimal perturbation for the parameter values of interest ($Re = 2400$, $Wo = 8$, and several A). This method produces the geometry (radial shape and axial/azimuthal wavenumbers) and the initial time (t_0) of the perturbation achieving the maximum energy amplification. We used these optimal perturbations on top of the Sexl–Womersley velocity profile as initial conditions for a second set of DNS in order to test whether puffs or helical waves were developed. We refer to these simulations as IC SWOP. In a last step, we performed a third set of DNS with the body force term in Equation (3) to mimic the experimental setup of Xu et al. [17]. All parameter combinations for which we have performed DNS are summarised in Figure 5b.

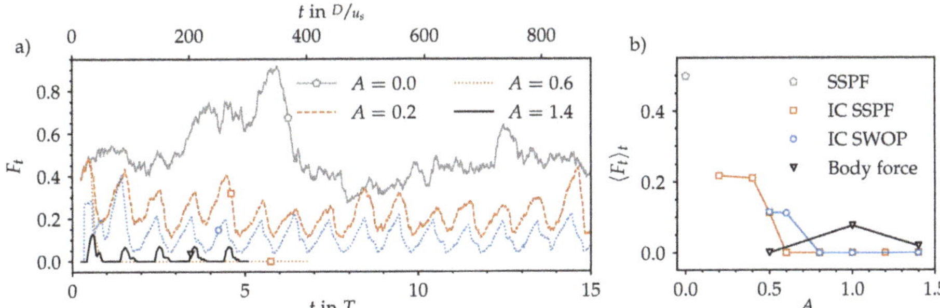

Figure 5. Turbulent fraction (F_t) in the computational pipe domain based on the axial vorticity data shown in Figure 2 and Figure 7. The threshold to distinguish turbulent from laminar regions is set to $\langle \omega_z^2 \rangle_{r,\theta} = 4 \times 10^{-2}$. (**a**) Time series of the turbulent fraction for several amplitudes A (line styles) and different numerical set-ups (symbols and colours from those in (**b**)). (**b**) Time-averaged turbulent fraction $\langle F_t \rangle_{t>2}$ for four different set-ups: The statistically steady pipe flow (SSPF) serves as reference data and as initial condition (IC) for the first set-up. The IC for the second set-up are composed out of the analytical Sexl–Womersley (SW) velocity profile superimposed with an optimal perturbation (OP). The third set-up is initialised with an unperturbed SW flow and then permanently perturbed using a localised body force (see Section 3.6).

3.1. Temporal Modulation of Statistically Steady Puff Dynamics

Figure 2a shows the typical intermittent behaviour of statistically steady pipe flow at $Re = 2400$. Here, the vorticity is viewed from a reference frame co-moving at the constant bulk speed $u_s = \langle u_b \rangle_t$. This case was run for 6000 convective time units ($\frac{D}{u_s}$) or an equivalent to more than 100 periods beforehand in order to relax from its initial conditions and to let the flow develop its typical patchy and intermittent character: Turbulence is spatially localised and surrounded by laminar regions of relative calm. The time scale of laminar–turbulent interactions is on the order of $100\frac{D}{u_s}$, as can be seen in Figure 5a, where we plot the temporal evolution of the turbulent (volume) fraction ($F_t = \frac{V_{\text{turb}}}{V_{\text{pipe}}}$) in the computational pipe domain (V_{pipe}). It changes considerably every two or three hundred time units, reflecting the interactions visible in the corresponding space–time diagram. We computed F_t based on the streamwise vorticity plotted in Figure 2a and the threshold to separate laminar regions (deep blue in Figure 2) from turbulent ones was set to $\omega_z^2 = 4 \times 10^{-2}$ in order to match the average turbulent fraction reported by Avila and Hof [12] at $Re = 2400$ (approximately 50%, as in Figure 5b).

We started all pulsatile IC SSPF runs from the same initial flow field and set the initial time to $\frac{t}{T} = 0.25$ to match the instantaneous bulk velocity of the pulsation (see Equation (1)) to the one of the steady flow. This ensured a smooth evolution from the initial condition and further allowed us to track the exact same realisations of localised flow structures in space and time as A was increased. The resulting spatiotemporal dynamics are shown in Figure 2b–d,f in a frame co-moving at the instantaneous bulk speed

$$x^*(t) = \int_{t_0}^{t} u_b dt. \tag{9}$$

Already at $A = 0.2$, the time scale of the flow modulation ($T \approx 60\frac{D}{u_s}$) dominates the dynamics (Figure 2b). In general, as the amplitude of the pulsation increases, the turbulent fraction in the flow decreases, as seen in Figure 5b. Many structures in the initial flow field decay quickly and do not survive the first acceleration (AC) phase. At $A = 0.2$, only two puffs survive after $t = 5T$, and the dynamics appear to reach an equilibrium state that repeats cyclically. The two surviving puffs grow in intensity and in length during the early deceleration (DC) phase of the flow, and they split into two in the late stages of DC before the minimum flow rate is reached. Out of these two, only the upstream puff survives the entire AC phase and reaches the peak flow rate, where this cycle starts over. Indeed, it is well known that for SSPF, only the upstream puff survives in puff interactions [26]. Overall, it appears that for $A = 0.2$, the flow is clearly self-sustained (above the critical point) and that a successful splitting event may occur at later times. However, the length of the computational domain ($100D$) may not be sufficient to accommodate three puffs without strong interactions due to the periodic boundary conditions. Similar results were obtained for $A = 0.4$ and 0.5 and are shown in Figure 2c,d; the question of whether, in these cases, the puffs will ultimately decay or successfully split would require substantially longer runs than those performed here and is not further pursued. Figure 6 shows typical localised structures at four equispaced points of the cycle for $A = 0.5$, illustrating the cyclically occurring splitting attempts. In agreement with Xu et al. [15], Xu and Avila [16], these figures show that the surviving puffs (Figure 6d,c) are very similar to the puffs in the steady case (Figure 1c) even at this relatively large pulsation amplitude. For $A = 0.6$, no turbulent structure survived the first pulsation period, and the flow fully relaminarised. We checked amplitudes up to $A = 1.4$ (see Figure 5b). In general, with increasing amplitude, the downstream puff separates farther away from the upstream puff during AC before it dies at almost the end of AC.

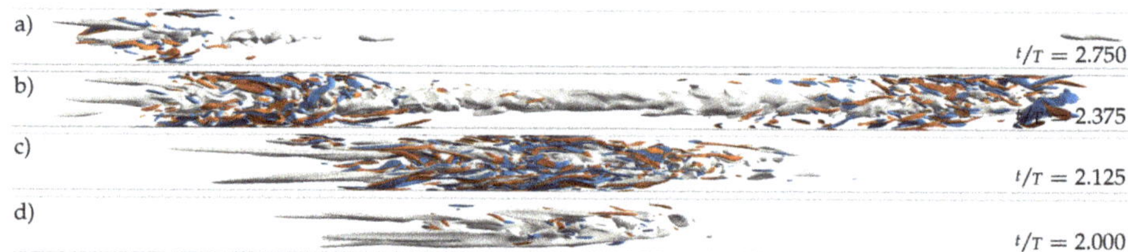

Figure 6. Instantaneous representation of localised turbulent structures in a pulsatile pipe flow ($Re = 2400$, $Wo = 8$, $A = 0.5$). Grey surfaces represent low-speed streaks ($u'_z = -0.4u_s$) and blue/red surfaces represent positive/negative axial vorticity ($\omega_z = \pm 8\frac{u_s}{D}$). (**a**) Death of downstream puff. (**b**) Splitting event. (**c**) Growing puff. (**d**) Isolated puff. The exact location and time for each snapshot are as indicated in Figure 2d. The direction of the mean bulk flow (u_s) is always from left to right.

3.2. Optimal Infinitesimal Perturbations of Pulsatile Pipe Flow

We performed a linear non-modal stability analysis of Sexl–Womersley flow at ($Re = 2400$, $Wo = 8$) and amplitudes up to $A = 1.6$, as described in Section 2.3. For $A \leq 0.4$, the optimal perturbation is the same as for statistically steady pipe flow: an axial two-roll configuration (not shown here). For $A \geq 0.5$, the optimal perturbation is a streamwise helix and the optimal initial time of perturbation is $\frac{t_0}{T} \in [0.2, 0.3]$. In Figure 3b,c, we show the optimal perturbation for $A = 1$ at the optimal time of perturbation (t_0) and at the point of maximum energy amplification (t_f), respectively. Initially, the optimal helical perturbation is localised very close to the pipe wall at the border of the Stokes layer

($\delta = \frac{1}{\sqrt{2}Wo}$), and it is tilted towards it. Within the rest of the DC phase and the first stages of AC, the perturbation rapidly grows by four orders of magnitude in energy within only 40% of the period. By the time of maximum energy amplification (early AC phase at $\frac{t}{T} = 0.6$), the optimal helix has separated from the Stokes layer and moved completely to the outer bulk region, where the stabilising effect of acceleration arrives later. This wall-normal phase lag increases with Wo [13] and can be nicely seen for the profiles close around the peak flow rate ($\frac{t}{T} = 0.5$ and 1.0) in Figure 3a. At the end of the process, the helix has been tilted opposite to its original configuration in a process reminiscent of the Orr mechanism. See Xu et al. [20] for more details and for a comprehensive parametric exploration.

3.3. Nonlinear Dynamics of Helical Perturbations

In our second set of DNS (IC SWOP), we superimposed the optimal helical perturbation scaled to a small amplitude ($4 \times 10^{-2}\ u_s$) on top of the SW profile. All simulations were started at the optimal initial time of perturbation (t_0). We used a global, as well as an axially localised, helix as initial perturbation and we varied the pulsation amplitude A whilst keeping $Re = 2400$ and $Wo = 8$ fixed. In all runs, the global helix exhibited rapid growth, followed by a breakdown into turbulence and immediate decay within the first period, in good agreement with the DNS of Xu et al. [17], for $A = 0.85$, $Wo = 5.6$, and a shorter pipe domain. Our results hence extend their findings to larger A and Wo, and are not explicitly shown here.

Using a localised helix as initial condition instead also led to a very similar fate for the helix, but only for $A \geq 0.8$ (see Figure 5b). The amplification of the local helix and its subsequent death is shown in Figures 7c–e and 8a–d. For smaller amplitudes ($A \leq 0.8$), intermittent puff turbulence emerged after the growth and decay of the initial helix and then was sustained for many periods (see Figure 5). The dynamics of the generated localised puffs are the same as described in Section 3.1 and are exemplarily shown in Figure 8e–h. The puffs that survive AC grow during early DC and attempt to split into two puffs during late DC. In the subsequent AC, the splitting downstream puff decays and leaves only the upstream puff behind to start the cycle over. Figures 2e and 7a,b compare this cycle and its initialisation phase for different amplitudes. For $A = 0.5$, a self-sustaining puff develops only from the downstream end of the amplified helix. For $A = 0.6$, puffs develop from both ends of the localised helix. Shortly thereafter, both puffs interact, which leads to the death of the downstream puff (similar to what happens, for example, in Figure 2b). For $A = 0.8$, a puff develops only from the upstream end of the amplified helix. For this case, the puff is able to survive for four periods before the flow completely relaminarises.

For both large and small amplitudes, the initial optimum perturbation energy is amplified by about two orders of magnitude (Figure 8), which is much less than in the linear case. It is worth noting that perturbations obtained with a non-linear non-modal stability analysis should yield a more effective growth [27]. These methods would help to avoid the discrepancy between linear and non-linear behaviour of perturbations at least before their complete saturation, and should be considered in future works. Our linear optimum perturbation, once introduced into the DNS, also moves towards the bulk region of the pipe; however, before it can complete the growth predicted in the linear analysis, it breaks up into turbulent spots arising at its upstream and downstream ends. For the larger amplitudes, the helix further narrows and develops a turbulent puff with a central low-speed streak before decaying. For the lower amplitudes, on the other hand, the helix opens up again and develops a turbulent spot with several low-speed streaks closer to the wall.

We used a hyperbolic tangent, as in Equation (7), to localise the radial and axial velocities of the helix perturbation in the z direction with the parameters $M_z = \frac{20}{D}$ and $L_z = 5D$. The azimuthal velocity was calculated to preserve the divergence-free condition. For a perturbation magnitude of $4 \times 10^{-2}\ u_s$, this procedure leaves a remainder of the helix in the rest of the domain, which is everywhere $< 10^{-7} u_s$. This remainder grows dramatically and results in the white bands visible in Figures 2e and 7a–e.

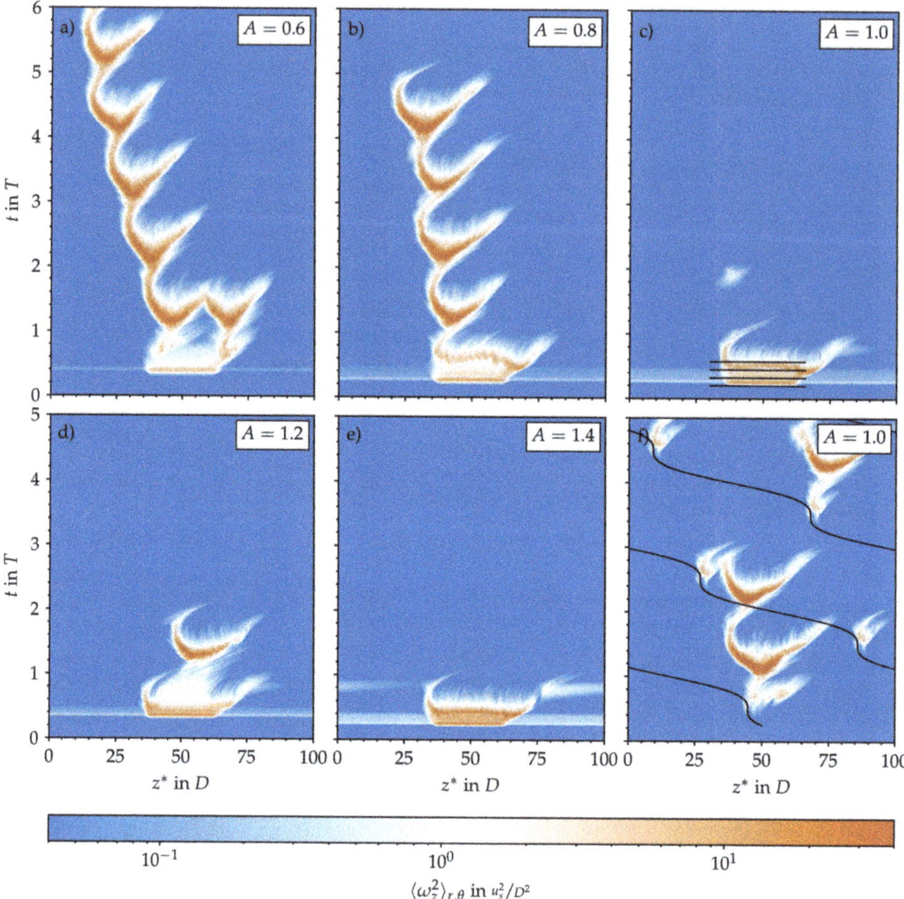

Figure 7. Spatiotemporal representation of the turbulence activity in the pipe domain based on the cross-sectional average of the streamwise vorticity (ω_z) plotted on a logarithmic scale and in a co-moving reference frame (z^*). For pulsatile pipe flow at ($Re = 2400$, $Wo = 8$). (**a–e**) For different pulsation amplitudes A, always using the SWOP initial condition. Note that the optimal time of perturbation slightly changes with A. The horizontal straight lines mark regions for which three-dimensional representations of the localised flow structures are shown in Figure 8. (**f**) For a permanent body force and the unperturbed SW velocity profile as initial condition. The curved black line represents the fixed location of the highly localised body force viewed from the co-moving reference frame. The direction of the mean bulk flow (u_s) is always from left to right.

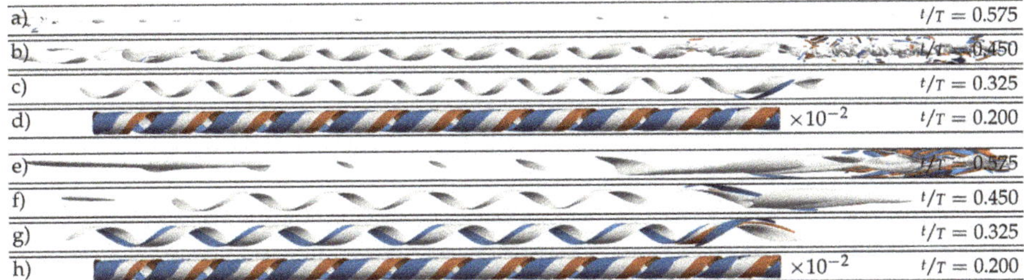

Figure 8. Instantaneous representation of localised turbulent structures in a pulsatile pipe flow DNS at $Re = 2400$, $Wo = 8$, and two different amplitudes. (**a–d**) Growth and decay of an initial helix at $A = 1.0$. (**e–h**) Development of a puff at $A = 0.5$. Both DNS were initialised at $\frac{t}{T} = 0.2$ using the SWOP initial condition. Grey surfaces represent low-speed streaks ($u'_z = -0.4u_s$) and blue/red surfaces represent positive/negative axial vorticity ($\omega_z = \pm 8\frac{u_s}{D}$). The exact location for each snapshot is as indicated in Figures 2e and 7c, respectively. (**a**) Decay. (**b**) Breakdown into turbulence. (**c**) Amplification of helix. (**d**) Localised optimal helix perturbation. (**e,f**) Birth of a downstream puff. (**g**) Amplification of helix. (**h**) Localised optimal helix perturbation. Note that the initial perturbation is two orders of magnitude smaller. The direction of the mean bulk flow (u_s) is always from left to right.

3.4. Puff Recovery Length

As has been observed in both strategies, whenever a puff tries to split, only the upstream puff survives. This is a feature common to SSPF and is related to the so-called puff recovery length, which represents the influence length the puff has downstream from its position [26,28]. Its effect can be seen in Figure 9, where the instantaneous u_z profile is presented at five axial positions and four time instants for an IC SWOP simulation at $Re = 2400$, $Wo = 8$, and $A = 0.5$. The position of the turbulent puff is presented as a shaded area in terms of axial vorticity. For all phases, the flow quickly recovers the laminar SW profile upstream of the puff location. Downstream of the puff, on the other hand, the flow needs a much longer buffer length to do so. Note the periodic boundary conditions used in our DNS.

3.5. Intermittent Production and Dissipation

In order to investigate the physical mechanisms by which puffs arise and survive in pulsatile pipe flow, we computed the production and dissipation of turbulent kinetic energy,

$$P_\alpha(r) = -\langle u'_r u'_z \rangle_\alpha \frac{\partial \langle u_z \rangle_\alpha}{\partial r} \text{ and } D_\alpha(r) = -\frac{1}{Re}\langle \nabla \mathbf{u}' : \nabla \mathbf{u}' \rangle_\alpha. \tag{10}$$

Angled brackets denote averaging with respect to α and prime denotes the fluctuation around the respective average. Here, α can be any combination of averaging in the two homogeneous directions θ and z, as well as time t, or at a fixed phase ϕ. For the cases where puffs survive, $P_{\theta,z,\phi}$ and $D_{\theta,z,\phi}$ are strongly modulated by the pulsation of the flow, as exemplified in Figure 10 for $A = 0.6$. During AC, production and dissipation are low, whereas during DC, they are high. Peak production takes place during the early DC and is very similar to steady pipe flow in terms of magnitude and wall-normal distribution. However, at the phase of maximum production, the dissipation inside the Stokes layer is much more intense than in the steady case. Right after the peak in flow rate, the mean velocity profile develops an inflection point at the wall, which satisfies the Fjortoft criterion [6]. With ongoing deceleration, the inflection point moves away from the wall and catches up with the point of peak production. Both travel together further towards the pipe centre. Near to the minimum flow rate, the unstable inflection point loses the Fjortoft condition (because of new inflection points arising in the velocity profile,) and the production collapses. Hence, it appears that the puff is taking advantage of this inflection point during DC to survive the upcoming AC.

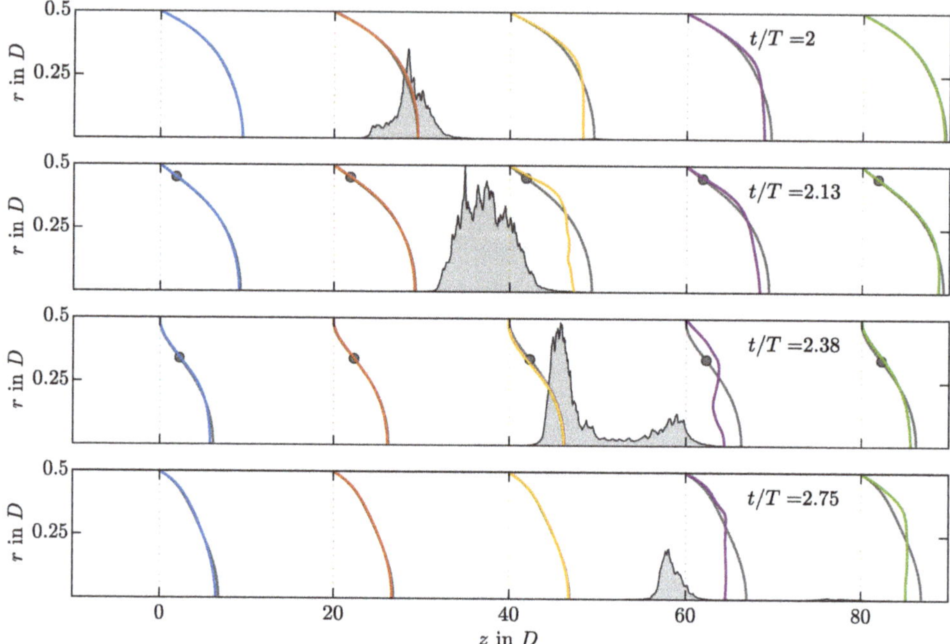

Figure 9. Instantaneous streamwise velocity profiles (u_z) at five axial locations along the pipe for an IC SWOP simulation at $Re = 2400$, $Wo = 8$, and $A = 0.5$. To not interfere with one another, they are scaled in arbitrary physical units, since, in this representation, only the development in time and deviation from the SW profile are of interest. Thus, the velocity is scaled so its all-time maximum $u_z(r, \theta = 0, z)$ is equal to $10D$. Each profile is compared with the corresponding instantaneous SW profile (grey lines, also scaled) and its inflection point (grey circles) if they fulfil the Fjortoft criterion. The shaded grey area shows the instantaneous cross-sectional average of the streamwise vorticity ($\langle \omega_z^2 \rangle_{r,\theta}$) scaled so its all-time maximum is equal to $0.5D$.

Figure 11 compares the production and dissipation profiles for the growth and decay of the localised helix during the first pulsation period for $A = 1$. Here, phase-logged time averaging is not possible, and averaging was performed only in the θ and z directions. During DC, the rate of production is negative in a small region inside the Stokes layer, meaning that turbulent kinetic energy is fed back to the mean flow and acts as an additional energy sink. This promotes relaminarisation and explains why the helix does not evolve into puff dynamics, as in the low-amplitude cases. Overall, the phenomenology is similar to that reported for oscillatory pipe flow, where negative production causes turbulence decay in cases initialised with fully developed turbulent flow fields of SSPF at high Reynolds numbers [29].

3.6. Effect of Local Geometric Imperfections

We performed a third set of DNS using the laminar SW velocity profile as the initial condition and the volume force described in Section 2.4. For this third set of simulations, we considered only the four amplitudes $A \in \{0, 0.5, 1.0, 1.4\}$. In line with the experiments of Xu et al. [17], we found no transition to turbulence at all for the axisymmetric contraction in all cases considered. By contrast, when the force is localised in all three dimensions, the response of the flow depends strongly on the amplitude of the pulsation. For $A = 0$, i.e., statistically steady pipe flow, there is no surge of turbulence or localised transition arising from the bump. This confirms that our force represents a small perturbation to the flow. As we increase the amplitude to $A = 0.5$, some vorticity is generated during peak flow rate, but no turbulent dynamics develop (see Figure 12a). The picture changes for amplitude $A = 1$. As shown in Figure 12b, during early DC, the presence of the bump is able to trigger turbulence in every period. Turbulence grows until late DC, and then laminarises. Occasionally, puffs emerge and are able to survive for more than one period if they interact (again) with the local bump due to the periodic boundary conditions used in our DNS. However, if the upstream puff interacts with a new turbulent spot arising from the bump, both die. For $A = 1.4$, on the other hand, no puffs develop, and the dynamics are solely characterised by bursts of turbulence arising at the bump, which proceed downstream as they decay (see Figure 12c). In all cases, the time at which the perturbation is triggered and grows is in agreement with our non-modal stability analysis and with the experiments of Xu et al. [17].

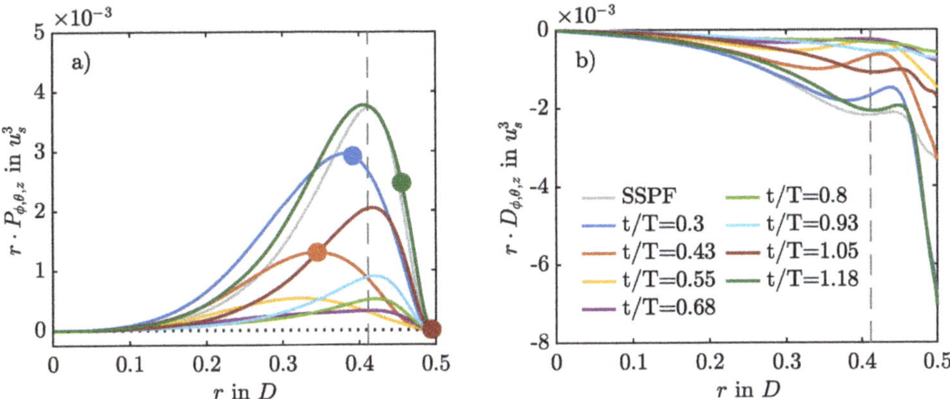

Figure 10. Production (**a**) and dissipation (**b**) of turbulent kinetic energy compared for different phases of the pulsation period for $A = 0.6$ using the SWOP initial conditions. Averages are taken over space- and phase-logged time instants ($\alpha = \theta, z, \phi$) over four periods of puff dynamics, excluding the initial period without puffs. Circles denote the existence and wall-normal location of the inflection points of the corresponding mean profile $\partial^2 \langle u_z \rangle_{\phi,\theta,z} / \partial^2 r = 0$ that satisfy the Fjortoft criterion. The vertical dashed line denotes the Stokes layer.

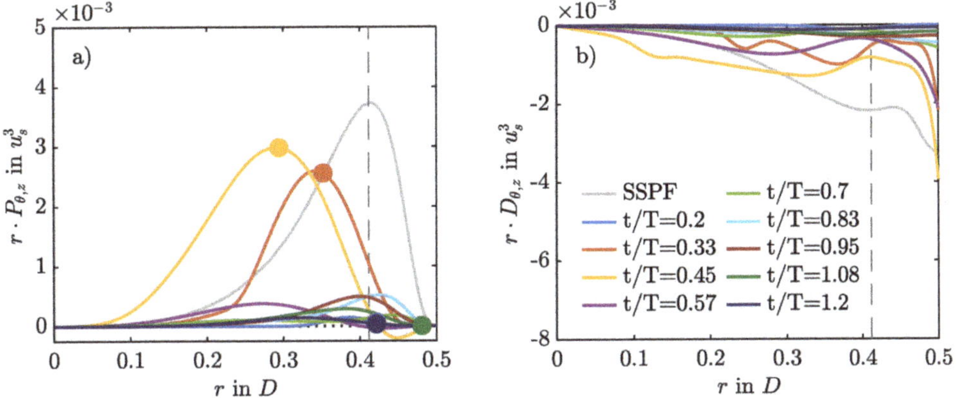

Figure 11. Production (**a**) and dissipation (**b**) of turbulent kinetic energy compared for different phases of the initial pulsation period for $A = 1$ using the SWOP initial conditions.

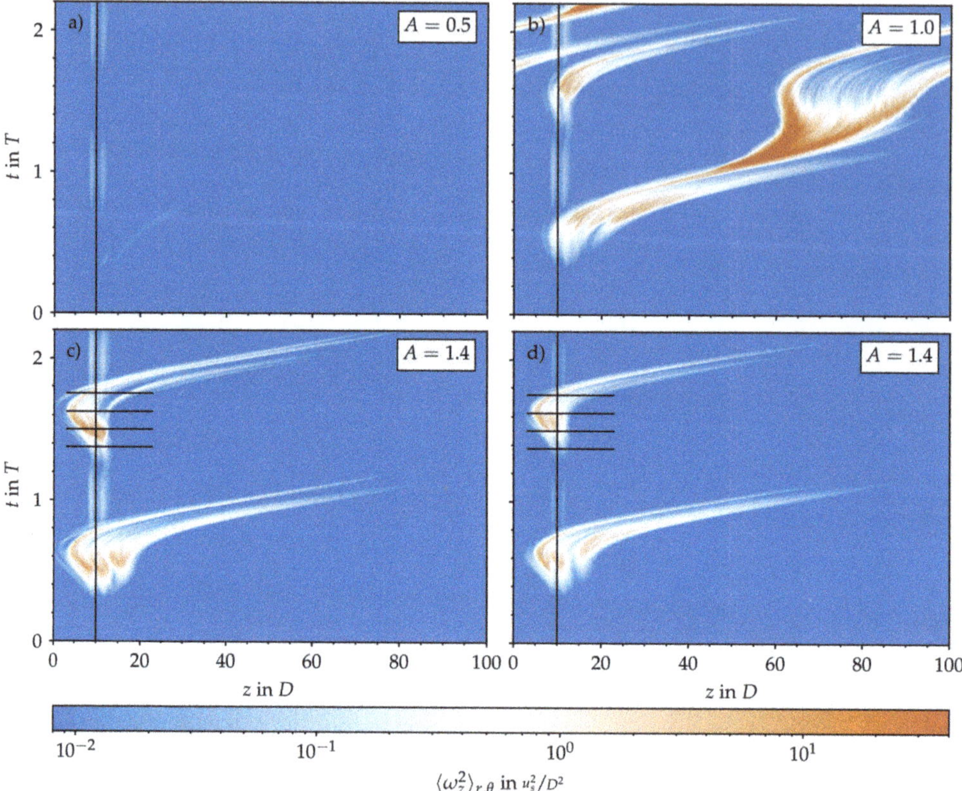

Figure 12. Spatiotemporal representation of the turbulence activity in the pipe domain based on the cross-sectional average of the streamwise vorticity (ω_z) plotted on a logarithmic scale and in a stationary reference frame. Pulsatile pipe flow at $Re = 2400$, $Wo = 8$, and different amplitudes A. Initial conditions are based on the Sexl–Womersley velocity profile, and there is a permanent body force. (**a–c**) Local bump. (**d**) Tilted bump.

Interestingly, the structures that the local bump triggers are mirror symmetric and not helical (see Figure 13a–d). They resemble structures resulting from the optimal non-modal disturbances in pulsatile pipe flow past a constriction [25]. They grow in axial length and magnitude during the late stages of DC while retaining their mirror symmetry. This is only lost in the last stages of DC, as low-velocity streaks form in the centre of the pipe. Finally, either a puff emerges from these streaks, or the flow laminarises during AC.

We also performed simulations with a tilted bump. In this case, the emerging structures exhibit not only mirror-symmetric, but also helical-like features (see Figure 13e–h). They also grow during the late stages of DC and either decay or trigger puffs depending on the pulsation amplitude. From the point of view of spatiotemporal intermittency, their evolution is quite similar to the evolution of the structures triggered by the local bump for all the amplitudes considered, as exemplarily shown in Figure 12c,d.

The fact that different geometric disturbances can trigger different structures is consistent with the non-modal stability analysis of Xu et al. [20]. The analysis showed that the instantaneous Sexl–Womersley profile is linearly unstable (in the quasi-steady limit) during most of the DC phase. Out of all the perturbations that could grow on top of this unstable profile, helical modes have the highest potential to do so. This holds for helical modes spiralling in positive and negative axial directions. That means that, if we were to disturb a flow in a way such that helical modes are excited, unless there is a preferred direction, helical modes and their swirling counterparts can grow simultaneously on top of the laminar flow profile. Our local bump represents a highly symmetric perturbation that allows this to happen, which explains why we observe mirror-symmetric structures. If we introduce some non-symmetric perturbation instead, then we see a preferred direction for the structures to swirl, as in the simulations with the tilted bump.

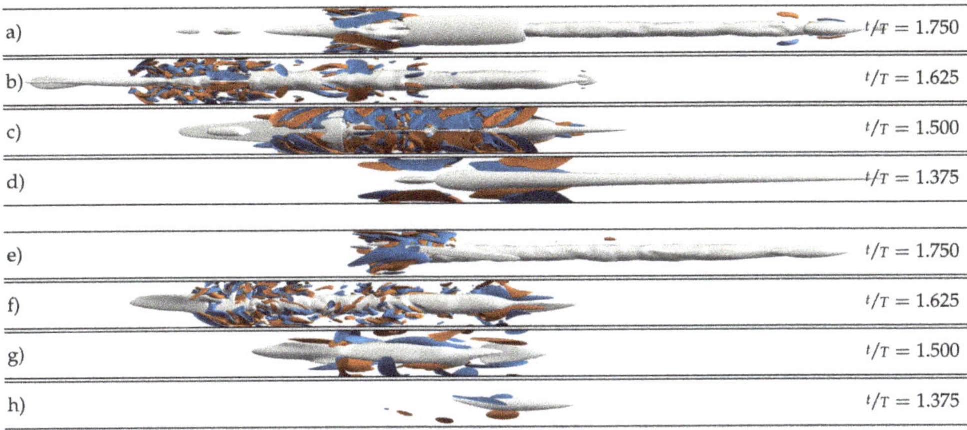

Figure 13. Instantaneous representation of localised turbulent structures in a pulsatile pipe flow DNS at ($Re = 2400$, $Wo = 8$, $A = 1.4$). The DNS was initialised at $\frac{t}{T} = 0.25$ using the corresponding SW profile and by introducing a local bump like body force, as described by Equation (3) and Table 1. Grey surfaces represent low-speed streaks ($u_z' = -0.2\,u_s$) and blue/red surfaces represent positive/negative axial vorticity ($\omega_z = \pm 2\frac{u_s}{D}$ for all panels except (**d**) and (**h**). There, it is $\pm 0.8\frac{u_s}{D}$. (**a–d**) Local bump. (**e–h**) Tilted bump. The exact instants in time are given in Figure 12c,d. The direction of the mean bulk flow (u_s) is always from left to right.

4. Discussion and Conclusions

In agreement with the experiments and simulations of Xu et al. [17], our results show that helical perturbations are able to trigger turbulence in pulsatile pipe flow, but not to maintain it. The helix perturbation grows from the instantaneous linear instability of the laminar flow profile during deceleration. However, during acceleration, the mean profile, which is close to the corresponding SW profile, is linearly stable. Without the unstable character of the profile, the perturbation no longer has its main mechanism to produce turbulent kinetic energy available, and it either completely decays or switches to puff mechanisms to survive. In either case, no helical perturbation is triggered again in the next deceleration phase.

This trend is further confirmed by the simulations that included a body force. For perturbations that seek to mimic the effect of geometric imperfections, and $A \geq 1$, turbulence is triggered intermittently every DC and dies during AC, as in the experiments. Thus, for pulsatile pipe flows that are not constantly disturbed, it is the presence of a self-generating puff mechanism (i.e., streak–vortex interaction with lift-up) that guarantees that the flow remains intermittently turbulent throughout many periods.

For puffs to survive in pulsatile pipe flow, plug-like mean profiles must be avoided, as also happens in statistically steady pipe flow [12,26,28,30]. This means that high amplitudes and/or Womersley numbers are detrimental for puffs' survival, but so are flows with a high fraction of turbulence. This includes cases initialised with a fully turbulent flow field and cases initialised with a helix perturbation in the whole domain. The former has also been shown by Feldmann [29] in purely oscillatory pipe flow at much higher Reynolds numbers. In agreement with Xu et al. [17], global helical perturbations with an initial magnitude of only $4 \times 10^{-2} u_s$ grow quickly and break up, and turbulence spreads throughout the whole pipe. The resulting highly disturbed flow, whose mean is far from the corresponding SW profile, does not allow puffs to grow. For the helical instability to be able to trigger puffs, it must be localised and surrounded by a laminar flow.

Once they have been successfully triggered, puffs take advantage of two mechanisms: the lift-up mechanism, as in SSPF, and the linear instability of the SW-like profile close to it. The former plays a leading role during late acceleration and early deceleration phases for amplitudes that result in a not-so-plug-like mean profile. The latter has a higher importance for most of the deceleration, where it compensates for the milder gradients of the instantaneous SW-like profile with production of kinetic energy due to its linear instability. The presence of puffs and their corresponding recovery length, in addition to a more intense acceleration phase, make turbulence more intermittent as the amplitude increases.

In future works, a different parametric space will be explored, and the combined effects of body force and random noises will be studied. In addition, physiological-like waveforms with longer deceleration phases will be considered, where the helical instability may have a longer time span to grow.

Author Contributions: M.A. designed the research, supervised the project, and acquired funding. D.F. performed non-linear simulations, acquired computational resources, developed the body force, analysed the data, and created the plots. D.M. performed non-linear simulations and transient growth analysis, developed and implemented the body force, analysed the data, and created the plots. All three authors discussed and interpreted the results mutually and prepared the manuscript collaboratively. All authors have read and agreed to the published version of the manuscript.

Funding: This work was funded by the German Research Foundation (DFG) through the research unit Instabilities, Bifurcations and Migration in Pulsating Flow (FOR 2688). Computational resources were provided by HLRN through the project hbi00041 and are also gratefully acknowledged.

Institutional Review Board Statement: Not applicable.

Informed Consent Statement: Not applicable.

Data Availability Statement: The data we have generated and analysed in this study will be made publicly available soon at https://pangaea.de/.

Conflicts of Interest: The authors declare no conflict of interest.

Abbreviations

The following abbreviations are used in this manuscript:

AC	Acceleration
DC	Deceleration
SW	Sexl–Womersley
NSE	Navier–Stokes equations
TGA	Transient growth analysis
DNS	Direct numerical simulation
SSPF	Statistically steady pipe flow
IC SSPF	Cases with a SSPF initial condition
IC SWOP	Cases with a SW profile and optimum perturbation initial condition

References

1. Reynolds, O. An experimental investigation of the circumstances which determine whether the motion of water shall be direct or sinuous, and of the law of resistance in parallel channels. *Proc. R. Soc. Lond.* **1883**, *35*, 84–99.

2. Rotta, J.C. Experimenteller Beitrag zur Entstehung turbulenter Strömung im Rohr. *Ing. Arch.* **1956**, *24*, 258–281. [CrossRef]

3.	Wygnanski, I.J.; Champagne, F.H. On transition in a pipe. Part 1. The origin of puffs and slugs and the flow in a turbulent slug. *J. Fluid Mech.* **1973**, *59*, 281–335. [CrossRef]

4.	Wygnanski, I.; Sokolov, M.; Friedman, D. On transition in a pipe. Part 2. The equilibrium puff. *J. Fluid Mech.* **1975**, *69*, 283–304. [CrossRef]

5.	Avila, K.; Moxey, D.; De Lozar, A.; Avila, M.; Barkley, D.; Hof, B. The onset of turbulence in pipe flow. *Science* **2011**, *333*, 192–196. [CrossRef]

6.	Schmid, P.J.; Henningson, D.S. *Stability and Transition in Shear Flows*; Applied Mathematical Sciences; Springer: New York, NY, USA, 2001; Volume 142. [CrossRef]

7.	Meseguer, A.; Trefethen, L. Linearized pipe flow to Reynolds number 107. *J. Comput. Phys.* **2003**, *186*, 178–197. [CrossRef]

8.	Darbyshire, A.G.; Mullin, T. Transition to turbulence in constant-mass-flux pipe flow. *J. Fluid Mech.* **1995**, *289*, 83–114. [CrossRef]

9.	de Lozar, A.; Hof, B. An experimental study of the decay of turbulent puffs in pipe flow. *Philos. Trans. R. Soc. A Math. Phys. Eng. Sci.* **2009**, *367*, 589–599. [CrossRef]

10.	Barkley, D.; Song, B.; Mukund, V.; Lemoult, G.; Avila, M.; Hof, B. The rise of fully turbulent flow. *Nature* **2015**, *526*, 550–553. [CrossRef]

11.	Mukund, V.; Hof, B. The critical point of the transition to turbulence in pipe flow. *J. Fluid Mech.* **2018**, *839*, 76–94. [CrossRef]

12.	Avila, M.; Hof, B. Nature of laminar-turbulence intermittency in shear flows. *Phys. Rev. E* **2013**, *87*, 063012. [CrossRef] [PubMed]

13.	Sexl, T. Über den von E. G. Richardson entdeckten "Annulareffekt". *Z. Phys.* **1930**, *61*, 349–362. [CrossRef]

14.	Womersley, J.R. Method for the calculation of velocity, rate of flow and viscous drag in arteries when the pressure gradient is known. *J. Physiol.* **1955**, *127*, 553–563. [CrossRef] [PubMed]

15.	Xu, D.; Warnecke, S.; Song, B.; Ma, X.; Hof, B. Transition to turbulence in pulsating pipe flow. *J. Fluid Mech.* **2017**, *831*, 418–432. [CrossRef]

16.	Xu, D.; Avila, M. The effect of pulsation frequency on transition in pulsatile pipe flow. *J. Fluid Mech.* **2018**, *857*, 937–951. [CrossRef]

17.	Xu, D.; Varshney, A.; Ma, X.; Song, B.; Riedl, M.; Avila, M.; Hof, B. Nonlinear hydrodynamic instability and turbulence in pulsatile flow. *Proc. Natl. Acad. Sci. USA* **2020**, *117*, 11233–11239. [CrossRef]

18.	Stettler, J.C.; Hussain, A.K.M.F. On transition of the pulsatile pipe flow. *J. Fluid Mech.* **1986**, *170*, 169–197. [CrossRef]

19.	Trip, R.; Kuik, D.J.; Westerweel, J.; Poelma, C. An experimental study of transitional pulsatile pipe flow. *Phys. Fluids* **2012**, *24*, 1–17. [CrossRef]

20.	Xu, D.; Song, B.; Avila, M. Non-modal transient growth of disturbances in pulsatile and oscillatory pipe flow. *J. Fluid Mech.* **2020**, in press.

21.	Truckenmüller, K.E. Stabilitätstheorie für die Oszillierende Rohrströmung. Ph.D. Thesis, Helmut-Schmidt-Universität, Hamburg, Germany, 2006.

22.	López, J.M.; Feldmann, D.; Rampp, M.; Vela-Martín, A.; Shi, L.; Avila, M. nsCouette—A high-performance code for direct numerical simulations of turbulent Taylor–Couette flow. *SoftwareX* **2020**, *11*, 100395. [CrossRef]

23.	Barkley, D.; Blackburn, H.M.; Sherwin, S.J. Direct optimal growth analysis for timesteppers. *Int. J. Numer. Methods Fluids* **2008**, *57*, 1435–1458. [CrossRef]

24.	Marensi, E.; Ding, Z.; Willis, A.P.; Kerswell, R.R. Designing a minimal baffle to destabilise turbulence in pipe flows. *J. Fluid Mech.* **2020**, *900*. [CrossRef]

25.	Blackburn, H.M.; Sherwin, S.J.; Barkley, D. Convective instability and transient growth in steady and pulsatile stenotic flows. *J. Fluid Mech.* **2008**, *607*, 267–277. [CrossRef]

26.	Hof, B.; de Lozar, A.; Avila, M.; Tu, X.; Schneider, T.M. Eliminating Turbulence in Spatially Intermittent Flows. *Science* **2010**, *327*, 1491–1494. [CrossRef] [PubMed]

27.	Kerswell, R. Nonlinear Nonmodal Stability Theory. *Annu. Rev. Fluid Mech.* **2018**, *50*, 319–345. [CrossRef]

28.	Barkley, D. Simplifying the complexity of pipe flow. *Phys. Rev. E* **2011**, *84*, 016309. [CrossRef]

29. Feldmann, D. Eine Numerische Studie zur Turbulenten Bewegungsform in der oszillierenden Rohrströmung. Ph.D. Thesis, Technische Universität Ilmenau, Ilmenau, Germany, 2015. urn:urn:nbn:de:gbv:ilm1-2015000634.
30. Kühnen, J.; Song, B.; Scarselli, D.; Budanur, N.B.; Riedl, M.; Willis, A.P.; Avila, M.; Hof, B. Destabilizing turbulence in pipe flow. *Nat. Phys.* **2018**, *14*, 386–390. [CrossRef]

Publisher's Note: MDPI stays neutral with regard to jurisdictional claims in published maps and institutional affiliations.

MDPI

St. Alban-Anlage 66

4052 Basel

Switzerland

Tel. +41 61 683 77 34

Fax +41 61 302 89 18

www.mdpi.com

Entropy Editorial Office

E-mail: entropy@mdpi.com

www.mdpi.com/journal/entropy